MINING IN WESTERN AUSTRALIA

Mining—then and now

The Essex Gold Mine (on left) situated on former G.M.L. 316, approximately 3.5 km south-south-east of Wiluna, was worked from 1897 to 1901 and produced 434 ozs (13.5 kg) of gold from 337 tons (342 tonnes) of ore. The leaseholders, who are possibly the men shown in the photograph, were J. K. Kimber, J. H. Lawrence and J. Cameron.

The Mt Whaleback iron-ore mine (below) is worked by the Mt Newman Mining Co. Pty Limited. In the year ending December, 1978, the mine production from this deposit amounted to 77 844 000 tonnes, of which 29 096 000 tonnes of iron-ore were crushed for railing.

MINING IN WESTERN AUSTRALIA

REX T. PRIDER
EDITOR

UNIVERSITY OF WESTERN AUSTRALIA PRESS
FOR THE EDUCATION COMMITTEE
OF THE 150TH ANNIVERSARY CELEBRATIONS 1979

First published in 1979
by the University of Western Australia Press, Nedlands W.A. 6009
for the Education Committee of the 150th Anniversary Celebrations 1979

Agents: Eastern States of Australia and New Zealand: Melbourne University Press, Carlton South, Vic. 3053; United Kingdom and Europe: International Scholarly Book Services (Europe), a division of Abacus Books Ltd, Abacus House, Tonbridge Wells, Kent TN4 PHU, England; U.S.A. and Canada: International Scholarly Book Services Inc., Box 555, Forest Grove, Oregon 97116.
Photoset by the University of Western Australia Press, printed in Western Australia by Frank Daniels Pty Ltd, and bound by Printers Trade Services, West Perth.

Mining in Western Australia.

(Sesquicentenary celebrations series)
Includes bibliographical references.
ISBN 0 85564 153 3

1. Mines and mineral resources—Western Australia—Addresses, essays, lectures. 2. Mineral industries—Western Australia—Addresses, essays, lectures.
I. Prider, Rex Tregilgas, 1910-, ed. (Series)

328.2'0994'1

Foreword

The 150th anniversary of the founding of Western Australia is a significant milestone in the history of our State.

This, the Western Australian Sesquicentenary Series, is a collection of scholarly essays on a number of important themes about Western Australia, its people and its resources, and has been published specifically to mark that occasion.

Much of the work has been researched by members of tertiary institutions and government instrumentalities and written expressly for this series.

In the list of contributors are the names of many Western Australians who are in the forefront of their particular field.

The collection is seen as a worthwhile contribution to the scholarly heritage of the State, and I should like to express my appreciation to the Series Co-ordinating Committee, under the chairmanship of Dr R. L. Vickery, for the conception and planning of the series.

13 February 1979

Sir Charles Court
Premier of Western Australia

Sesquicentenary Celebrations Series

VOLUMES IN THE SERIES

Aborigines of the West
Agriculture in Western Australia
The Beginning: European Discovery and Early Settlement of Swan River
Education in Western Australia
Environment and Science
Essays on Western Australian Politics
A History of Commerce and Industry in Western Australia
Immigrants in Western Australia
The Literature of Western Australia
Mining in Western Australia
Western Landscapes
Western Towns and Buildings
Westralian Portraits
Westralian Voices: Documents in Western Australian Social History

SERIES CO-ORDINATING COMMITTEE

Dr R. L. Vickery (Chairman)
Mr J. Skivinis (Executive Officer)
Prof. G. C. Bolton
Mr F. M. Collopy
Dr N. F. Dufty
Mr A. Gardiner
Mr E. Hinchliffe
Mr L. J. Hunt
Dr C. F. Makin
Mr K. Roberts
Dr C. T. Stannage

SERIES EDITOR: Mr E. Hinchliffe

Committee Acknowledgements

The Western Australian Sesquicentenary Series Co-ordinating Committee makes the following grateful acknowledgements for the assistance it has received in the development and preparation of this series of volumes:

To the Education Committee of WAY 79, under the chairmanship of the Director-General of Education, Dr D. Mossenson, and the Assistant Director-General of Education (Schools), Mr H. W. Louden, for its continued support and encouragement.

To the Director-General of Education for making available various resources of his department, particularly the services of two officers—the Superintendent of Publications, Mr E. Hinchliffe, to act as series editor, and the Relieving Superintendent of Curriculum, Mr J. Skivinis, to act as executive officer.

To the chairmen of the editorial subgroups and their committees for their work in the planning and preparation of the respective volumes.

To tertiary institutions for freely making available staff time and other resources.

To the manager and staff of the University of Western Australia Press for their help and cooperation.

In respect of this volume, particular acknowledgement is made to Emeritus Professor Rex T. Prider and his committee.

Contributors

J. L. BAXTER, BSc *W. Aust.,* is a Geologist (Mineral Resources Division) of the Geological Survey of Western Australia.

J. G. BLOCKLEY, BSc *W. Aust.,* is Supervising Geologist (Mineral Resources Division), Geological Survey of Western Australia.

B. E. BURSILL was a member of the Public Relations section of Mt Newman Mining Co. Pty Ltd when his contribution was being prepared.

I. R. CAMPBELL, BSc *W. Aust.,* is Chief Development Geologist, West Australian Petroleum Pty Ltd.

J. D. CARTER, BSc *Leeds,* is a Senior Geologist (Mineral Resources Division), Geological Survey of Western Australia.

K. J. FINUCANE, MSc *W. Aust.,* is a Consultant Geologist who, in earlier years, was Senior Geologist (Western Australia) of the North Australia Geological and Geophysical Survey.

I. O. JONES, BSc *Wales,* PhD *Glas.,* is Dean of Mining and Mineral Technology, Western Australian Institute of Technology.

A. M. KERR, BSc (Econ) *Lond.,* MA PhD *W. Aust.,* formerly Associate Professor of Economics at the University of Western Australia, is Professor of Economics at Murdoch University.

E. A. KIRKE, BSc *W. Aust.,* BChemEng *NSW,* is Senior Research Engineer, Alcoa of Australia Ltd.

J. H. LORD, BSc *W. Aust.,* is Director of the Geological Survey of Western Australia.

E. McGOWAN, MA *W. Aust.,* DA *CMU,* is a Senior Lecturer, Social Sciences Department, Nedlands College of Advanced Education.

R. H. MAZZUCCHELLI, BSc *W. Aust.,* PhD *Lond.,* is Chief Geochemist, Western Mining Corporation Ltd.

A. M. MURRAY, BSc MA *Edin.,* is Mining Geologist, Alcoa of Australia Ltd.

R. A. NICHOLS, ASTC DIP MMENG *Syd.,* has recently retired from the post of Chief Mining Engineer (Nickel Division), Western Mining Corporation Ltd.

E. P. O'DRISCOLL, BE *W. Aust.,* formerly of the Mines Department of South Australia, has recently retired from the post of Chief Hydrogeologist of the Geological Survey of Western Australia.

A. E. O'MEARA, BSc *NZ,* is Manager (Metallurgy), Project Group, Western Mining Corporation Ltd.

J. C. PARRY, BSc MBA *W. Aust.,* is Exploration Superintendent, West Australian Petroleum Pty Ltd.

P. E. PLAYFORD, BSc *W. Aust.,* PhD *Stanford,* is Assistant Director, Geological Survey of Western Australia.

R. T. PRIDER, BSc *W. Aust.,* PhD *Cantab.,* Emeritus Professor of The University of Western Australia, retired from the Chair of Geology some years ago, and is now Honorary Research Fellow in the Department of Geology.

M. K. QUARTERMAINE, BE *W. Aust.,* recently retired from the post of Lecturer-in-Charge of Mining at the Western Australian School of Mines, Kalgoorlie.

B. M. ROGERS is Under-Secretary for Mines, Mines Department of Western Australia.

D. L. ROWSTON, BSc *W. Aust.,* is Senior Geophysicist of the Geological Survey of Western Australia.

G. A. TRAVIS, AWASM *Kalg.,* MSc *Lond.,* is Senior Geologist, Special Projects, Western Mining Corporation Ltd.

A. F. TRENDALL, BSc ARCS *Lond.,* PhD *Liv.,* DSc *Lond.,* is Deputy Director of the Geological Survey of Western Australia.

H. W. WILLIAMS, BA *W. Aust.,* is Superintendent of Collieries, Western Collieries Ltd.

R. WOODALL, BSc *W. Aust.,* MS *Calif.,* is Director of Exploration, Western Mining Corporation Ltd.

Volume Editorial Subcommittee

Em. Prof. R. T. Prider (Chairman)

Dr E. McGowan

Mr J. E. L. Manners

Dr P. E. Playford

Contents

	Foreword	v
	Committee Acknowledgements	viii
	Contributors	ix
	List of Figures	xv
	List of Plates	xix
	List of Tables	xxi
	Preface	xxiii
1	**A historical account of the development of mining in Western Australia** M. K. Quartermaine and E. McGowan	1
2	**The geology of Western Australia in relation to its mineral resources** Rex T. Prider	24
3	**Mineral deposits of Western Australia**	
A	Copper, lead, zinc and tin J. G. Blockley	39
B	Gold R. Woodall and G. A. Travis	57
C	Iron A. F. Trendall	75
D	Nickel R. Woodall and G. A. Travis	87
E	Bauxite A. M. Murray	101
F	Mineral sands John L. Baxter	111
G	Minor mineral deposits Rex T. Prider	121

H Mineral fuels

(i) Coal — 131
J. H. Lord

(ii) Oil and gas — 145
Phillip E. Playford

(iii) Uranium — 160
J. D. Carter

I Ground-water and its importance to the mineral industry — 167
E. P. D. O'Driscoll

4 Exploration Techniques

A Geological methods — 178
K. J. Finucane

B Geophysical methods — 186
D. L. Rowston

C Geochemical methods — 191
R. H. Mazzucchelli

5 Exploitation—mining and metallurgy

A Coal-mining — 197
H. W. Williams

B Iron-ore mining — 208
B. E. Bursill

C Bauxite-mining and processing — 219
E. A. Kirke and A. M. Murray

D Underground mining — 229
R. A. Nichols

E Concentrating, smelting, refining — 246
A. E. O'Meara

F Oil-drilling onshore and offshore — 274
I. R. Campbell and J. C. Parry

6 Mining law in Western Australia — 285
B. M. Rogers

7 The economic and social significance of Western Australian mining developments — 293
Alex Kerr

8 Careers in the mining and minerals industry — 302
I. O. Jones

Figures

1.1	Production of major minerals of Western Australia 1886-1977, yearly and total	20
2.1	Geological sketch map of Western Australia	25
2.2	Main Precambrian provinces of Western Australia	26
2.3	Generalized geological map of the Kimberley Division, Western Australia	27
2.4	Generalized geological map of the North-West Division, Western Australia	28
2.5	Generalized geological map of the older Precambrian rocks of the southern part of the Yilgarn Block, Western Australia	30
3.1	The Warribanno Chimney: site of the state's first lead smelter	41
3.2	Annual productions of copper, lead, zinc and tin concentrates, 1850-1976	43
3.3	Western Australia, showing outline of geology and principal copper, lead, zinc and tin deposits	47
3.4	Contrasting methods of treating tin ore in the Pilbara	55
3.5	History of gold production in Western Australia	58
3.6	History of gold production in Western Australia, excluding the Golden Mile	60
3.7	Gold deposits in Western Australia	61
3.8	Profitability of gold-mining	70
3.9	Price of gold and Australian minimum weekly wage	71
3.10	Major iron-ore deposits of Western Australia in relation to the regional geology	79
3.11	The Turner Syncline, Hamersley Iron Province	81
3.12	Mt Tom Price, Hamersley Iron Province	83
3.13	Pannawonica mine, Hamersley Iron Province	83
3.14	Australian iron-ore production, 1960-76	85
3.15	Major nickel deposits of Western Australia	89
3.16	Kambalda-Widgiemooltha area: generalized geological map and location of nickel sulphide deposits	93
3.17	Western Australian annual nickel production	97
3.18	Producer price of nickel metal, 1969-76	98

3.19 Nickel exploration expenditure and nickel discovered in Western Australia 99
3.20 Bauxite deposits of Western Australia 102
3.21 Schematic bauxite profile, Darling Range, Western Australia 104
3.22 Heavy mineral deposits of Western Australia 112
3.23 Koombana Bay heavy mineral deposit 119
3.24 Capel South heavy mineral deposit 119
3.25 Perth Basin: geological locality map 133
3.26 Perth, Collie and Wilga Basins: stratigraphic columns 135
3.27 Collie and Wilga Basins: locality plan 137
3.28 Collie Basin: stratigraphic succession of coal seams 138
3.29 Collie Basin: typical cross-sections 139
3.30 Phanerozoic sedimentary basins of Western Australia, showing locations of principal exploratory wells and oil and gas discoveries 146
3.31 The Perth Basin: structural subdivisions and location of principal exploratory wells and discoveries 148
3.32 The Carnarvon Basin: structural subdivisions and location of principal exploratory wells and discoveries 149
3.33 The North-West Shelf: oil and gas discoveries and bathymetry 151
3.34 Northern Carnarvon Basin: oil and gas discoveries and bathymetry 151
3.35 Canning Basin: structural subdivisions and principal exploratory wells 154
3.36 Browse and Bonaparte Gulf Basins: principal exploratory wells and discoveries 154
3.37 Yeelirrie area: geological sketch map 165
3.38 Yeelirrie area: sketch section to illustrate geology and geomorphology 166
3.39 Perth metropolitan area: contour map of the water table, showing the Gnangarra and Jandakot Mounds 169
4.1 Dryblowing with dishes and shaker 181
4.2 Variation in nature of gold colours in soils 183
4.3 Aeromagnetic profile and related geology, Southern Cross area, Western Australia 189
4.4 Nickel sulphide mineralization, 80 km north-east of Kalgoorlie; laterite profile and geochemistry 193
5.1 Muja open cut, Collie Basin 204
5.2 Iron-ore-processing flowsheet 210
5.3 Pilbara: mines, railways and ports 217
5.4 Grade variations in a typical bauxite deposit, Darling Range, Western Australia 220

5.5 Total value of Western Australian mineral production by underground and surface operations 229
5.6 Juan shaft, Kambalda: headframe and surface facilities 235
5.7 Cut-and-fill stoping, Kambalda 242
5.8 Updip stoping and workings of Central Norseman Gold Corporation 243
5.9 Diesel front-end loader, Fisher mine, Kambalda 244
5.10 Treatment of Mt Charlotte gold-ore at Kalgoorlie Oroya mill 249
5.11 Gold-ore treatment, Central Norseman Corp. N.L. 251
5.12 Gold-ore treatment, Telfer 253
5.13 Kambalda nickel operations: concentrator flowsheet 257
5.14 Kalgoorlie nickel smelter: process flowchart 259
5.15 Kwinana nickel refinery: process flowsheet 264
5.16 Western Australian mineral-sand deposits: generalized treatment flowsheet 268
5.17 Western Australia: oil drill-hole localities mentioned in text 275
5.18 An early cable-tool rig in Western Australia 278
5.19 A modern rotary rig in Western Australia: *Maersk Endurer* 279
5.20 Barrow Island oilfield: production curve 283

Plates

Frontispiece: Mining—then and now

1 Bauxite deposit, Darling Range 105

2 Iron-ore deposit at Mt Whaleback and stockpile-shipping facilities, Port Hedland 212

3 Iron-ore from mine to market 214

4 Alumina works at Pinjarra and rehabilitation of mined areas, Darling Range 223

Geological map of Western Australia
(in pocket inside back cover)

Tables

2.1 Phanerozoic stratigraphic table of Western Australia 34-5
2.2 Geological events and mineral deposits in Western Australia 37
3.1 Copper, lead and zinc production from principal Western Australian deposits 44
3.2 Common ore minerals of copper, lead and zinc 45
3.3 Tin concentrates—production of principal Western Australian tin-fields 52
3.4 Major gold deposits of Western Australia: production statistics 62
3.5 Production statistics and geological characteristics of other significant Western Australian gold deposits 67-9
3.6 Nickel sulphide deposits (higher grade) in Western Australia 90
3.7 Nickel sulphide deposits (lower grade) in Western Australia 91
3.8 Nickel sulphide deposits (cumulative ore production) of Western Australia 96
3.9 Mineralogical composition of laterites 106
3.10 Chemical composition of laterites 108
3.11 Composition of heavy mineral fractions from Western Australian deposits 111
3.12 Mineral-sands production, 1975—Western Australia and Australia 113
3.13 Heavy mineral reserves in Western Australian deposits 114
3.14 Published heavy mineral reserves in Western Australia 115
3.15 Total Western Australian production of heavy mineral sands to 31 December 1977 116
3.16 Correlation of shorelines on the Swan Coastal Plain 118
3.17 Salts in sea-water 124
3.18 Analyses of Western Australian coals 140
3.19 Collie Basin: amount of total coal reserves shown to be extractable to 1978 141
3.20 Collie Basin: annual production statistics 142
5.1 Western Australian iron-ore production to 1919 209
5.2 Major Western Australian underground mining operations, past and present 230-1
5.3 Main access methods in underground mines in Western Australia 233

5.4	Vertical development by raiseboring	238
5.5	Principal stoping methods in major mines in Western Australia —past and present	240
5.6	Western Australian nickel deposits: ore types, reserves, grade, and treatment	254-5
5.7	Properties of minerals of mineral sands	266
5.8	Western Australian mineral sands operations	267

Preface

Mining has played a key part in the development of the State of Western Australia since the discovery of payable gold at Halls Creek in the Kimberley in 1885, to the spectacular exploitation of iron-ore deposits of recent years. In this volume, the history of development of the mining industry is considered, a summary of the geology of the state is given, the ore deposits and their relation to the geological framework are described, and exploration for and exploitation of these ore deposits are dealt with, as is the contribution of the mineral industry to the economic development of the state. In the space allotted in this volume, it will be impossible to deal with any of these aspects in the detail they merit, but key references are given, which will enable the reader with an interest in particular topics to pursue the subject by further reading.

The contribution of the major mineral products to the Western Australian economy during this period is illustrated by the annual and total production figures shown in Fig. 1.1. To plot the range of values of the annual mineral production from \$2294 for gold in 1886, to \$904 142 000 for iron-ore in 1977, it has been necessary to use a logarithmic scale, as indicated along the right-hand margin. Mineral production during latter years has been enormous—a reflection both of demand and of increase in Man's earthmoving ability, from the pick and shovel-wheelbarrow days of the 1890s to the highly mechanized operations with large-scale equipment at the present, as illustrated in the frontispiece of this volume.

It has been a great privilege for me to have been invited to edit this volume. My lifetime interest has been in the field of geology, a knowledge of which is indispensible for the exploration and assessment (and subsequent exploitation) of mineral deposits from the metallic minerals and non-metallic minerals and rocks, to oil, natural gas and water accumulations.

My interests have been mainly in the field of teaching, and the contributions of my graduates to mineral exploration and exploitation have been of great satisfaction to me. Their contribution to the mineral industry is reflected, to some extent, by the high proportion of contributions to this volume by those who were my former students in geology at The University of Western Australia.

I want, here, to express my thanks to the Editorial Committee and others, particularly Mr John Blockley of the Geological Survey of Western Australia, for their assistance during the preparation of this volume, and to the individual authors, who cooperated so admirably in the prompt submission of their manuscripts. My thanks are due, also, to the various mining companies and the Mines Department of Western Australia, which have been the main sources of contributors to this volume, both for making their officers available, and for the preparation of the illustrations.

In this connection, I would particularly thank the Director of the Geological Survey Branch of the Mines Department, Mr J. H. Lord, for suggesting and making available the recently published geological map of Western Australia for incorporation in this volume.

Finally, I want to acknowledge my indebtedness to The University of Western Australia which, in appointing me an Honorary Research Fellow, has made the facilities of the Department of Geology available to me, thus enabling me to complete this work.

REX T. PRIDER

CHAPTER 1

A historical account of the development of mining in Western Australia

M. K. QUARTERMAINE[1] AND E. MCGOWAN[2]
[1] *W. A. School of Mines*
[2] *Nedlands College of Advanced Education*

The history of mining in Western Australia is an eventful one and extends from the early days of settlement to the present. However, centuries before the establishment of the Swan River Colony in 1829, part of the northwest coast had long been considered to be gold-bearing. Certain maps of the sixteenth century, had labelled the northwest coast as Provincia Aurifera or Terra Aurifera, *aurifera* signifying 'gold-bearing'. It now seems that the labelling of the northwest coast as such was the result of a cartographic misinterpretation during the sixteenth century, following the translation of Marco Polo's *De Regionibus Orientalibus,* which actually referred to Lower Siam as Provincia Aurifera but which later geographers placed on the rather vague northwest coast-line of the Great Southland or Terra Australis. Not long after settlement by the British, several authenticated findings were recorded in a number of explorers' journals. Unfortunately, for reasons discussed later, the deposits of gold were not to be exploited until the 1880s.

An apparent reluctance to initiate a viable gold-mining industry becomes something of a mystery when one considers the eagerness of the colonial government to discover and make use of other mineral resources. It seems that as early as 1839 the government was offering a grant of 2500 acres to anyone discovering a 'payable' coalfield. Seven years later, a government surveyor, A. C. Gregory, became the recipient of the free land when he reported that, on the night of 9 September 1846, he had on the bank of the Irwin River, north of Perth, lit the first fire with coal from a seam in the vicinity. There is little doubt, however, that rather than the coal deposits, it was the seams of the attractive galena or lead ore on the Murchison River reported by Gregory in 1848, that attracted the most interest in the colony at the time.

The initiative in mining was taken by B. Urban Vigors, a Perth lawyer then recently returned from Adelaide, and a meeting in September 1846 resulted in the Western Australian Mining Company being formed 'for the purpose of discovering, leasing, buying, selling and working mineral lands in Western Australia'. The company was soon seeking to raise, in one-pound shares, an ambitious £10 000—an impossible figure in a colony of 4000 people. Nevertheless the first call of 4*s* raised £1383, a very creditable figure for the time. It was soon apparent, however, that this first flush of capital was to be the last. Repeated calls over the years that followed raised a few hundred pounds more, but after a history of courageous attempts and dismal failure, the last meeting of the company took place in 1885, and the first and only dividend of 8*s* 9*d* for each fully paid-up share was made in 1906.

With capital in short supply, the 1843 Land Act had raised the price of land to £1 per acre, with a minimum of 160 acres. This was still the only way to acquire land for mining, and when a further act in September 1846 placed a royalty of one-fifteenth on any proceeds from base-metal mines, the situation become worse. Gold as the 'royal' metal was entirely reserved for the Crown, a fact which the mining historian G. Blainey blames for the retardation of gold-mining in Australia before the 1850s.

One of the reasons for the eventual demise of the Western Australian Mining Company was not so much the lack of capital, nor the harsh land laws of the time, as the fact that it simply failed to produce.

In October 1846, after paying the discoverers of a lead seam at 'Kelmscott', just south of Perth, a £20 reward, the company bought 50 acres of land for £70 and began work on what was called the Mundijong mine. Soon the mine was receiving the undivided attention of the colony. The company successfully induced a German geologist, Dr Von Sommer, to come from South Australia to work the mine. Unfortunately some months later Von Sommer was lent to the government, leaving the mine in inexperienced hands, and after £526 had been spent, the lead lode was lost, a further £60 of 'dead' work being necessary before it was found again. Altogether it cost hundreds of pounds to put a few holes in the ground, and when by December 1848 calls for more capital proved fruitless, the first mine ever put down in the colony was forced to close.

One reason for the failure of the company's Mundijong venture to draw in more capital was that its calls coincided with the formation of a new company, formed to exploit the lead discoveries in the Murchison. A group in Perth took up 32 shares at £20 each, sufficient to buy the 640 acres which became the site of the Geraldine lead mine, the first practical mine in the colony. The mine was named after Governor Fitzgerald, who was also an investor in the company.

No one in the colony had had any mining experience; L. C. Burges, the first manager of the Geraldine mine, was a pastoralist from York. The lode was in the river bed and most of the diggings struck water at about 12 m. Problems such as this were serious handicaps to the new mining ventures. When the cost of a road to the coast some 130 km away was considered, it was evident that the company had insufficient capital. After producing about five tons, it closed down within a few weeks.

A second Geraldine company was formed less than two years later, but this also struggled with inexperienced labour and managers with little mining knowledge. Moreover, having commenced production at a time of rising wool prices, it was hindered by an increasing lack of capital, so that during its first three years it produced lead only to the value of £59.

By the end of 1854 a trade depression lay upon the colony. Wool, the major export, had brought in only £18 610 in the first quarter, and imports of flour over the same period had consumed about one-tenth of these earnings. Who could blame those in the colony who saw mining, or more specifically the discovery and exploitation of gold, as the panacea for all economic ills? In January 1854, following the established practice in the eastern colonies, a large public meeting was held in Fremantle, and as a result £357 was subscribed as a reward for the discovery of payable gold. It was hoped that the government would 'see fit to contribute a like amount'. However the government did not; but in June it issued an ordinance giving the colonial governor power to provide for the preservation of law and order should gold be discovered in Western Australia!

Two months later Surveyor A. C. Gregory was sent to investigate a reported gold discovery at Cardup, some 40 km south of Fremantle. He found that about half a dozen holes had been sunk, and all, the diggers claimed, had yielded gold. Gregory was adamant that there was no trace of gold, and formed the opinion that 'the specimens of rock taken from Cardup had been adulterated with the precious metal, whether accidentally or by design, to attract public attention'. But it was R. Austin, in late 1854, who stood on Mount Magnet and reported that the 'geological formation of the country east of the Murchison River has every appearance of being one of the finest goldfields in the world'. How prophetic that was. A decade later, in June 1864, Dr James Martin, traversing the north-west coast-line, also found traces of gold in the mud of the Glenelg River. He suggested that a search should be directed towards the sources of the river and 'thence southward as far as the Fitzroy River and the Provincia Aurifera of the old charts'. Notwithstanding these and other reports, however, it was to be several decades before gold was found again.

The optimism of those who believed in the future of mining in the colony was once more aroused by the discovery of rich copper veins on the Murchison in the mid-1850s. The first true copper mine in the district is thought to have been the White Peak, which was floated by John Drummond and George Shenton in 1855.* It produced only a little ore however, and closed three years later. The fact that copper was fetching nearly £100 per ton in Singapore led to several companies being formed. Often these ores were found adjacent to or in conjunction with deposits of lead ores, so that when copper prices fell, as they did after the mid-sixties, the lead ores were available to work on.

Prices for lead rose steadily, so that the base-metal industry, by now centred on Northampton, remained profitable, reaching a peak production in 1877 with an earning of £47 481. At the time this represented 13 per cent of the colony's export earnings. The mines continued to absorb both colonial and English capital right through to the mid-1880s, by which time gold had become the more attractive proposition to the mining entrepreneurs.

Notwithstanding the steady progress in base-metal mining, there was still a group who believed the discovery of a payable goldfield in the settled districts would be the culmination of the colony's mining hopes. News of gold discoveries in Queensland and South Australia provided a new spark, and in January 1862, following reports of gold finds in the Northam district, another 'gold reward' meeting was held. It was

* The Wanerenooka copper mine at Northampton was discovered about 1856, not 1842 as has been widely published. (Ed.)

agreed that the public would offer a reward of £2500 for the discovery of gold. This time the government took the hint and, in the following month, the *Government Gazette* confirmed a reward of £5000 for the discovery of a 'workable goldfield within 150 miles of Perth before the 31st December 1862', on condition, however, that 5000 ounces were cleared at the Fremantle Customs House and actually shipped to Great Britain before 1 July 1863. Not everyone agreed with the government's action however. As several pointed out, prospecting was an expensive business which few could afford and it would have been far more appropriate if the government had financed parties of experienced prospectors from amongst the colonists. It was not surprising, therefore, that the first real nibble at the bait came from outside the colony. In May 1862 the colonial secretary in Perth received a letter from Edward Hargraves, who had achieved fame as the discoverer of the first goldfield in Australia (Ophir in New South Wales). Always restless, but shrewd enough to realize that he had little hope of duplicating his earlier success in what was to him a practically unknown corner of the continent, Hargraves asked for a commission of £500 and travelling expenses to carry out over a period of six months a survey of the colony's gold-bearing potential.

The Legislative Council unanimously accepted the offer, and in October 1862 Hargraves landed at Albany. He received a rather divided reception from the colonists. Quite unperturbed, Hargraves showed his confidence by saying: 'I do not think I will pass within five miles of a goldfield without discovering it.' His survey covered more than 3200 km and took him north-northeast from Albany through to the Murchison. He completed his report on the 'non-auriferous rocks of Western Australia' in February 1863, and his final comment—that he did not think that a goldfield existed on or anywhere near his track—was the death knell for many hopes in the colony.

There was further criticism directed at specific sections of the Hargraves' Report. In one part it was stated that the prospector had spent 86 days travelling in a 'spring cart'. This was equated with a rate of '25 or 30 miles per day', and practical miners ridiculed this as being a 'mere cursory survey'.

The one pearl of wisdom which Hargraves did leave was his comment on the colony's mineral land laws. He urged the governor to prevent valuable mineral lands being locked up and rendered useless, as was the case under the existing regulations. However, not only did the Legislative Council see fit to ignore this advice, but introduced even more stringent laws, which required mineral land to sell at £5 per acre (deposit £1), with a selection not less than 80 acres. It was a law which caused the *Perth Gazette* to comment that it was likely to check enterprise altogether and leave mineral land unexplored. 'After all', said the editor, 'who would sink a shaft on spec and pay the Government £400 for the right to do so'. He proved right and for nearly a decade the only capital entering the base-metal industries in the colony came from Britain, for many capitalists within the colony believed that only a reef of gold could repay such expenditure on land. By early 1870 reports of finds of gold began to accumulate in Perth, several of which claimed the reward offered by the government, the distance of a discovery from Perth having been extended to 480 km. However, the condition under which the discoverer was to be paid had become more difficult, because in 1872 the Act was further modified so that the reward would not be paid until 10 000 ounces of gold had been cleared and shipped through Fremantle, and the onus was on the claimant to prove that this had been done.

In 1870 Mr H. Y. L. Brown was appointed the first Government Geologist* at a salary of £600, and one of his jobs was to indicate likely areas to find gold. His choice was at a place called Peterwangy on the upper Irwin River, where a minor rush had occurred in mid-1870 following a find by a Mr Beresford (or Brelsford), a Victorian digger who had hopes of claiming the reward. However, Brown's report, following government-sponsored digging in the area, was not favourable, and the rush petered out. Nevertheless the gold was there, even if in limited amount, so it was a legitimate rush, the first in Western Australia and the first real gold discovery. Gold fever soon appeared elsewhere, however, when traces of gold were found at Kelmscott, Newcastle (Toodyay) and Boyup, and the Rockingham Mining Company claimed 'encouraging results from their Two-to-One mine at Serpentine'. Unfortunately it seems that this last mine struck trouble, and nothing more was heard of the mine nor the company.

During his commission, Brown traversed most of the southwest of the colony, finding traces of gold in several districts, notably Tallering, Weld Range, Stirling Range, Mt Burra, Eyre River and Phillips River, and finally on the great bend of the Murchison River (which Austin had found and Hargraves had missed). In his final report, however, Brown pointed out that to prospect all these areas would cost time and money.

In the southwest of the colony, another Victorian, W. G. Simpson, was sufficiently impressed with an ore discovery to consider forming a company.

Although this particular company did not eventuate, other discoveries led to the formation of the first two gold-mining companies in the colony. The first, at Cardup in early 1874, petered out before a hole was dug; the other, the Standard Gold-Mining Company, was floated in Albany on a capital of £1500 and became the first company in the colony to attempt to exploit a gold discovery. Expensive machinery was brought from Victoria and a part of a then recently-purchased government battery was acquired to treat the gold ore expected from a reef at Kendenup, some 65 kilometres north of Albany. But the mine proved to be a duffer—the first but certainly not the last in the colony—and it closed a few years later.

Overall the decade from 1870 to 1880 was one of steady progress for the pastoral, timber, pearling and lead-mining industries, with the first proving the main area of expansion. While the area of alienated land did not even double during the period, the area under lease trebled to 45 million acres. It was in the process of gaining this extra land that the first major indications of an extensive goldfield in Western Australia were uncovered in the last great unexplored area of the colony, the Kimberley region in the north. Although intensely rugged, with precipitous valleys, it nevertheless had sufficient promise from a pastoral point of view to satisfy the increasing clamour in the south for more land.

* Dr Ferdinand von Sommer had been engaged by the colony as a geologist from 1847 to 1851.

In February 1879 Alexander Forrest, a young government surveyor, was asked to lead a party in a survey of the Kimberleys. Forrest named but could not penetrate the King Leopold Ranges and was forced to make his way east along the Fitzroy River, where, some 500 km from the coast, he came upon and traversed the Ord River, the creeks of which were to become the centres for a number of gold-mining camps a few years later. The party continued east to meet the Adelaide-Port Darwin telegraph line, then turned north and rested at the Union gold-mine at the Pine Creek gold-mining centre, south of Darwin. Here Forrest discussed with the manager, Adam Johns, the similarities between the gold-bearing rocks at Pine Creek and some he and his party had seen during their recent survey of the Kimberleys. It is therefore no coincidence that the first party to find gold in the Kimberleys two years later contained a number of miners and staff from the Union gold-mine.

However, the discovery of gold had not been the motive for Forrest's expedition, and his report of 20 million acres of some of the finest country in Australia 'with open grazing plains and running springs at every mile', was widely acclaimed by both government and people. The discovery came at a time when there was a pastoral boom, a serious drought in inland Queensland, and a scarcity of land for expansion in the southern sections of the continent, so that the Kimberleys held great hope for graziers.

Once more as pastoral parties moved through the Eastern Kimberleys, attention was drawn to reports of gold. These had little effect on the southwestern areas of the colony, for they came in the middle of a pastoral and property boom in Perth. As the consumer market expanded, there were far better avenues nearer home for the investment of capital and energy by the non-pastoralists. Ironically, it was the other depressed colonies which were to provide most of the capital and labour in the Kimberley rush.

Following a declining production on the Pine Creek goldfields, Johns, Saunders and Quinn set out in late 1881 to explore the Kimberleys for gold. After investigating the west coast, they followed Forrest's tracks east along the Fitzroy. Knowing what they were looking for, they found gold in most of the creeks of the Upper Ord. But a lack of water (it was now August/September) precluded washing for gold, and this, together with the fact that Johns was ill, forced the party to leave the area. Back in Yam Creek in the Northern Territory, Phil Saunders wired Perth in November 1882 that he had found gold on 'the head watershed of the Ord River'.

The Western Australian Government wanted proof, but negotiations with Saunders to lead a government party failed, owing to some misunderstanding on both sides. The government eventually secured the services of Mr E. T. Hardman, an experienced geologist from the Geological Survey of Ireland. During May 1883, Hardman joined John Forrest's surveying party in the Kimberleys, specifically to make a survey of the areas of auriferous country. In a preliminary report in 1884, he recorded having found gold on his second visit to the Kimberleys. His final report on the Kimberleys in early 1885 was all that any miner could wish for. It was the 'most detailed and exhaustive report on any goldfield'. Its maps were elaborate and colourful, with the patches of auriferous country coloured, appropriately enough, in gold. Hardman left the colony in August 1885, after surveying parts of the southwest. On his way home to Ireland, he toured goldfields in Victoria and sent back to Perth two further reports, which, he alleged, strengthened his conviction about the gold potential of the Kimberley. The comparison, he noted, is highly favourable as to the future of the Kimberley as a gold- producing region. His activities may have also led to the discovery of tin near Greenbushes.

But interest was not centred exclusively on the Kimberleys. In June 1882 a J. J. Harwood had written to the Colonial Secretary claiming the £500 reward for discovering a 'goldfield' at Wongoong (Wungong), south of Fremantle, and requesting a lease on the ground. Although Harwood was granted a claim of a piece of ground 600 feet by 40 feet, his claim for the reward was disallowed until he had proved the existence of gold. His company spent £500 on the Three Wonders mine before it closed in late 1883. Subsequent examination showed the presence of pyrite-bearing quartz veins which assayed only a trace of gold.

On 17 April 1884 the *Government Gazette* carried details of the colony's first goldfield regulations, but they produced little more than uncommitted interest. This atmosphere was soon to change however. In late September 1885 the government received a telegram from Hall, Slattery and party, reporting the discovery of numerous deposits of gold. On 18 May 1886 the government officially recognized the colony's first goldfield. As a producer of gold and a lurer of capital and labour, the Kimberleys must compare very poorly with the earlier discoveries in the eastern colonies and with the other fields that followed. However, the discoveries came at a time in Australian mining history when the established fields in the east were being discredited, when capital was anxious and labour disillusioned with the diminishing yields. The Kimberleys held promise, and it was Western Australia's turn to become the safety valve for the enterprising and the irresponsible, looking for new frontiers to conquer. The radicalism and energy as well as the capital and labour of the 't'othersiders' were the ingredients necessary to create the new industry of gold-mining in Western Australia.

One figure of 10 000 quoted as taking part in the Kimberley gold-rush appears to be a gross exaggeration, and from the evidence available it seems that no more than 1500 to 2000 people were on the field. Even this number was present only for a period covering the middle and latter part of 1886. By late 1886 most could detect that the field would sustain no more than half of that number.

By mid-1888 most prospectors had left the field, and mines continued to close until finally the major mine, the Ruby Queen, was abandoned in 1889, leaving nearly £5000 worth of machinery to the bush and the Aborigines. By the time of the Kalgoorlie boom in the mid-1890s there were only seven mining leases operative in the Kimberley goldfields and the revenue from mining leases in the district amounted to only £143. Of the three mines left in operation—Mt Bradley, Ruby Queen (taken up by a Perth syndicate) and the St Lawrence—only the first two mines had taken advantage of the government's offer of a bonus of £4 per foot after 100 feet to encourage mining activity and continuity. In the following year seven mines received the bonus, and Warden Cummings noted that 'many mines simply worked with a view for the bonus only'; and he later informed the minister that government aid was of 'no real benefit to the field, because it leads men to look for Government aid too readily'.

It is of interest at this point to take a look at the base metal industry of the Northampton Mineral District, which stretched from the Geraldine mine on the Murchison, a distance of more than 160 km south to the Irwin River. In the period prior to the Kimberley gold discoveries, lead worth £337 916 and copper £123 487 were produced in what was the colony's most significant metal industry before the exploitation of gold. Certainly in the period prior to the late sixties, copper proved to be the most valuable of the base metals. After this date the lead mines became the most prolific and lucrative producers, reaching their peak during the late seventies, so that in 1877 the earnings from the export of lead represented 13 per cent of the colony's export trade.

Coal was another mineral the colony was seeking, not that it sought to emulate the coal-based industrial progress of England, but because some in the government believed that future industries and the developing rail transport would use this in preference to local hardwood. Also some were conscious that more steam-ships were using the shipping routes and saw a potential industry in providing these ships with coal. In 1887, aware that the Irwin coal deposit was not economic, the government announced a reward of £1000 for the discovery of a coalfield. This reward was divided between Arthur B. Perrin and the estate of D. A. Hay in 1889. Perrin is credited with the actual find in the bed of the Collie River in 1883, but at the time he did not report it. Between 1892 and 1895 a number of bores, and a 300 m tunnel, proved to the government that the seam was workable. The 200 km^2 Collie Coal Mining District was constituted on 20 February 1896 and thrown open for selection on 23 March. The rail reached Collie in 1898 and production commenced from a number of small collieries near the end of that year.

Despite all the deficiencies, the discovery of auriferous deposits in the Kimberleys was an important event in Western Australian history. It forced Hargraves's assertions into the background and reawakened interest in the gold-bearing potential of the colony. Further south, on what was to be the Pilbara Goldfield, new discoveries were announced. Stemming from the Christmas day, 1887, discovery of gold by young Jimmy Withnell on his father's property at Mallina,* 130 km east of Roebourne, three centres, Egina and Pilbara and Mallina were soon hives of activity. After rain fell during July and August 1889, extending over the Pilbara and filling the creeks inland as far as Nullagine, the move into the new distant field began in earnest, emptying the coastal zone and shifting the scene of activity to the headwaters of the Nullagine and Oakover Rivers, over 500 km in from the coast.

* Jimmy Withnell was not looking for gold. He was working on a sheep-station and had picked up a stone from a low quartz hill to throw at a crow. This stone, and others nearby, were flecked with gold. The find was reported to the Roebourne government resident who in his excitement telegraphed the news to the colonial secretary in Perth. 'James Withnell picked up a stone to throw at a crow', was all he had written on the form by the time Withnell's bank manager came in and urged him to come and see the specimens. Off he went, and off went the truncated message also, to a puzzled minister who promptly wired back: 'What happened to the crow?'

The march of industry continued and companies invariably followed the paths of the pioneering prospectors, so that by mid-1888 there were three companies exploring the reefs in the Pilbara; but inexperienced management and selection of the wrong treatment process led to their early demise.

Conditions in the Pilbara and out to the Nullagine, however, were hot, dry and dusty, with the centre at Marble Bar achieving an unenviable reputation for its long succession of days (and nights) over the Fahrenheit century mark. It is hardly surprising that, as reports of rich finds further south reached them, prospectors once more began to seek new trails. Though the ranges and creeks of the Ashburton and Gascoyne held promise, it was the Murchison which became the main attraction. It was here that the existence of gold had been postulated by Austin as early as 1854. In January 1888 more positive proof of gold came from Mick Bourke (or Burke), T. Sampey, G. Woodley and G. E. J. Inkpen who, with J. Watts, not only found gold on Annean Station but also sank a shaft to follow a reef. The mounting evidence of widespread gold less than 800 km northeast of the capital induced a Perth syndicate led by F. C. Monger to bring over J. F. Connelly, a New South Wales surveyor and mining engineer. He left Geraldton in May 1890, and within a month had found a rich source of the metal. Although he reported his claim at the Mount Gould police station in July 1890, he was beaten to the registration of the first claim on the Murchison by the partnership of Gilles McPherson and J. Peterkin, who had followed Connelly to Nannine after reports of his success in the area.

In May 1887, some 320 km east of Perth, a Charles Glass had reported finding traces of gold in pieces of quartz when putting down a well on his property. On the basis of this information, two parties had been formed, one sponsored by the Northam Pastoral Association and the other by a group of Perth businessmen headed by the mayor of Perth, Dr Scott. The Perth-backed group was the first off the mark, with its members, Richard Greaves, Edward Payne and H. Anstey, agreeing to prospect to the east of Glass's discovery, if necessary, for 30*s* per week and one-eighth of any gold found. They were out during most of the middle part of 1887 and found encouraging colours, and even a suspected reef at Cordelia, near Lake Deborah. It was not considered good enough, however. Then Greaves fell ill, left the party and returned to Melbourne for treatment. Thus Anstey had to report the finds, which were to gain for him £500 reward for the discovery of payable gold on the Yilgarn field.

The Northam-backed party returned after a fruitless journey. When they heard of Anstey's find, they again went inland to locate and extend the area of the discoveries. Golden Valley was found by B. Colreavy and H. Huggins in December, and in January 1888 Toomey and Riseley came across the reefs at Southern Cross, which they named after the constellation which had led them there. About the same time W. M. P. Parker found gold scattered throughout a series of low hills later named after him.

There was no great rush of labour to the new field, even though four new gold-bearing districts were opened by the end of the year. These centres were at Eenuin, Golden Valley, Southern Cross and Parker's Range. Little work had been done because, it was argued, 'lack of water prevented work.' The recognition of this factor by the government and its allocation of £635 for the immediate construction of dams and tanks en route to and on the Yilgarn field were only the beginning of such aid. As it became obvious that the gold from the Yilgarn mines was not to be easily won and that better machinery and higher wages were necessary to lure miners to the fields, demands for more government support and a liberalization of the mining laws became more vocal. The government was not long in indicating that it would give its support by providing water and telegraph lines, by liberalizing the mining and company laws, and ultimately by building a rail link to the coast.

Faced with rising costs, Southern Cross mine owners announced in mid-1892 a wage reduction among their miners. This produced the first miners' strike in the colony's history and forced many away from the mining centres. Some moved back to the coast, but others, more optimistic, or more reckless, moved into the semi-desert areas to look for gold.

Typical of such were the partners A. Bayley and W. Ford, who headed east along Hunt's track. Surveyor Hunt had reconnoitred a route in 1864 some 225 km east of Southern Cross to the Hampton Plains, noting the position of a number of native wells and improving them, thereby providing an invaluable service for those who came after. Bayley and Ford deviated from the track somewhere near Mount Burges and within days had found gold. (Even then they were not the first, for a claim was found and pegged and dated 1888; however, the identity of this prospector remained unknown.) One other group, consisting of T. Talbot, D. Fosser and H. Baker, was camped near Bayley and actually gathered some £1700-worth of gold nuggets on the site of Bayley's claim before he managed to peg it. Nevertheless when Bayley eventually settled his reward claim, the Talbot party on Bayley's advice pegged out next to him on Bayley's Reward South.

The arrival of Bayley back at Southern Cross on 17 September 1892 with more than 500 oz of gold, separated only by patches of quartz, was responsible for the most impulsive and intensive rush to date. It left Southern Cross a deserted shanty town, the strike-bound mines forgotten, and a week or so later the rush was reinforced from the coast and from other fields.

In November, the Dundas Goldfield was discovered by Mawson and Kirkpatrick, although the Norseman reefs were not discovered until August 1894. The discovery of gold at Coolgardie was the beginning of a new phase in the colony's mining history. The extensive alluvial deposits at Fly, Pig and Billy Can Flats were an overwhelming attraction for the gold seekers, yet at the same time the extraordinary richness of the reefs attracted new risk capital, this time from outside the colony. But it was Bayley's reward claim that dominated the scene. In a matter of weeks it yielded more than 1000 oz, much of it from the reef and most of it crushed by hand.

Many prospectors went no farther than Coolgardie, but for most it was only the beginning of the adventure. Soon after the first winter rains came in June 1893, there was a spontaneous movement out from the new centre as some two thousand gold-seekers edged into the wilderness. Among those who joined a rush, following a report of gold at Mount Youle some 80 km north-east of Coolgardie, were three Irishmen: Hannan, Flanagan and Shea (or O'Shea). They camped at Reedy's Dry

Camp and, during a search of the locality, found alluvial and reef gold at the base of a low hill that was soon to take the name Mount Charlotte. Returning to Coolgardie, Hannan claimed title to a reward claim of twenty acres—the maximum, since his find was more than 20 miles from the nearest workings (Blainey 1963, p. 180).*

Even the discovery of the nearby Golden Mile could not satisfy all those who poured into Coolgardie and beyond. Migrants and capital arrived to overwhelm the resources of the colony and even the dangers and discomfort of inland travel did little to dispel the enthusiasm. Unfortunately the failure within months of three of Coolgardie's leading mines, the Londonderry, the Wealth of Nations and later Bayley's Reward, removed much of the glamour and optimism from the 'Old Camp' and attention became focused upon the sister field, Hannans, or Kalgoorlie, as it came to be called.

The failure of the mines in the Coolgardie district had virtually halted the flow of capital to them, but at the same time it served to stimulate investment in the Kalgoorlie mines, particularly after April 1895, when the Great Boulder sank into a previously undetected lode that averaged 10 oz to the ton—the first indication of the great prize the Golden Mile was to reveal in the years to come.

Two prospectors from South Australia were the finders of the Boulder lodes. Sponsored by the Adelaide Prospecting Syndicate, the discoverers, W. G. Brookman and S. W. Pearce, arrived in Albany with £80, all that was left of the £150 originally gathered to prospect the Western Australian goldfields. On 4 July 1894, acceding to the advice of the more experienced Pearce, the syndicate pegged reef claims covering some 500 acres. Difficulties in complying with the statutory labour conditions, which if obeyed would have required the employment of more than 300 men, eventually forced the syndicate (now known as the Coolgardie Gold Mining & Prospecting Syndicate) to seek more finance. The original 15 shares in the syndicate were increased to 100 and then to 3000. A visit to the leases by George Brookman, an Adelaide financier and a brother of one of the prospectors, confirmed the perilous position of the group.

In Melbourne, later in the same year, the Ivanhoe Company was floated for £150 000 in 10*s* shares. This was followed by a further division of the leases and the Lake View and Boulder groups were floated as the Lake View Consols (two 24-acre leases) for £40 000 in 10*s* shares, of which 20 000 were taken up in Melbourne. The Lake View is credited with having the first battery on the Golden Mile.

Towards the end of 1894 a number of other mines—Hannans Brown Hill, Golden Horseshoe, and those in the Brookman group—struck the rich lodes underlying the area. All were situated within a mile radius on rich ground embracing a complex series of gold-bearing lodes and veins—the so-called Golden Mile.

* There seems to be doubt concerning the size of Hannan's Reward Claim, as Blainey (1963, p. 180) says it was 20 acres because 'Under Western Australian mining law, those who found gold more than twenty miles from the nearest workings, could take a reward claim of twenty acres. The Irishmen were twenty-four miles from Coolgardie and, being anxious to get their twenty acres . . .'. Casey and Mayman (1964, p. 56) say it was 'about ¾-acre'. According to Surveyor Robert Gledden (8 Sept. 1893, Fieldbook 2, p. 9), who surveyed the claim, it was 500 feet long by 50 feet wide, making an area of 0.574 acre. (Ed.)

After 1894 the industry, supported by overseas capital, expanded rapidly. Altogether, it is estimated that 538 companies were floated in London to exploit the gold-mines of Western Australia. Of these, 35 were wound up soon after flotation. The remaining 500 or so collected an aggregate capital of £61 million, of which £48 million went to mining companies and the rest to land, finance and exploration companies. Of the 500 companies only 33 paid dividends to 1897. These totalled £748 000, and of this amount £127 000 was issued by one company (Bayley's Reward) during the first eighteen months or so of its life. More than half the balance came from two companies on the Golden Mile, Great Boulder £320 000 and Lake View Consols £125 000.

For many of the smaller mines, amalgamation promised a profitable future. But this practice allowed more and more companies to pass into the control of a handful of groups, and by 1900 it left British shareholders in possession of 22 of Western Australia's bigger and better mines, and in control of possibly another hundred of the lesser ones. During the twenty-year period from 1895 to 1915, these British-owned mines produced £66 million in gold (56 per cent of Western Australia's output) and paid out dividends equalling a third of the value of Kalgoorlie's total output. Naturally the dividends of the companies differed. For instance, Great Boulder paid out 60 per cent of the value of gold won, whereas Associated Gold Mines paid only 21 per cent of the value of its output.

Meanwhile prospectors were still exploring the outback, and the centres of Gwalia and Wiluna were discovered in 1896, although the lodes of the latter centre were not recognized until 1904.

The continuous boom of the eighteen-nineties did not progress too far into the new century. Production reached its peak in 1903 with gold worth £8 770 719. The most important factor controlling the subsequent decline was the fall in values. For instance in 1903-4 the average value of the ore being mined was 78*s* per ton. By 1914 this figure had been halved to 37*s*. There were, of course, other factors behind the fall-off in production. After 1896 had come the challenge of the sulphide and telluride ores. Faced with these metallurgical problems, companies on the Golden Mile spent an estimated one million pounds in seeking a solution. This money was not entirely wasted, for Kalgoorlie became a centre of research in mining and metallurgy and was responsible for many innovations.

The Bullfinch deposits some 3 km east of the Golden Valley road and 35 km north of Southern Cross were the only major gold deposits found during the first two decades of the twentieth century. The early reports of values up to 100 oz to the ton by prospectors C. E. Jones, D. L. Doolette and V. F. Shallcross in December 1909 provided the tonic the flagging gold-mining industry needed. During the first months after the registration of the finds, the number of Miners Rights issued at Southern Cross expanded from the usual handful to more than 900 and the number of leases taken up quadrupled from 101 (1526 acres) to 472 (9118 acres).

The area in which the gold had been found turned out to be reefing country and, urged on by the government's promise of an early connection with the Goldfields Water Scheme, a number of companies were formed to exploit the reefs. The leading company, Bullfinch (W.A.) Proprietary, began with a flotation of £954 000, but like most of the other newly-formed companies either forgot or ignored many of the lessons of the nineties. In May 1910 the Bullfinch, after spending a mere £800 on putting down a shaft, experienced labour problems. The ore, being pyritic, had to be sent to Kalgoorlie, and when falling values were observed, the company's shares collapsed from a peak of 22*s* to 6*s*.

About this time, a number of major gold-mining companies, especially a dozen or so left on the Golden Mile, began a new phase of technical improvements. In some cases, these took the form of improvements in machinery, with more and heavier stamps being added to existing batteries. A number of mines, however, adopted some form of dry processing—for example, new ball and tube mills and roasting furnaces.

During 1903 the government completed the Mundaring-Kalgoorlie water pipeline and began to extend branch supply lines north and south of the main.

With the outbreak of war in 1914 several companies met the challenge of falling values and rising costs by letting parts of their mines to tributers—parties of miners who paid a royalty on gold won. A company could gain a number of advantages from this practice: the mines were kept in production (which, with the use of labour, satisfied the law); machinery earned its keep by being let out to the miners; there was little or no outlay by the company; and, what was more important, dividends continued from the acceptance of royalties. For the tributers, the advantages were more obvious: being practical miners, they were skilful at placing holes for firing; they wasted little timber; they used smaller amounts of explosives; and they worked alongside their men and hence got more out of them without incurring the expense of management or offices. On the other hand, the use of tributers contributed to selective mining, since they broke only rock of value. There was little heed paid to the future, with most parties functioning on weekly or monthly contracts. Developmental or 'dead' work was reduced to a minimum, so that in time knowledge of the reserves of the mines became very scanty.

Partly owing to the exigencies of the 1914-18 war (rising costs of mining requisites, and the higher wages and improved conditions secured for the workers by their unions), more mines, especially those in the outback, were forced to close, so that by 1925 the Sons of Gwalia mine at Leonora was the only big mine left operating outside the Golden Mile. The companies alleged that the industry as a whole was being held back by accelerating costs, which in the long term would mean selecting only ore of higher value and also putting many men off. The truth of this prophecy was to be revealed in the years that followed. The year after the Burnside Award of 1921, the number of men employed in the industry declined by 15 per cent, tonnage by 30 per cent, yield by 16 per cent and dividends by 28 per cent. At the same time, the value of the average ore mined increased from 85*s* 10*d* to 107*s* 5*d* per ton in 1923, and coupled with this, working costs increased from 64*s* 4*d* to more than 80*s* per ton over the same period.

The yearly increase in values was impossible to maintain. By 1925, when the value had fallen to 85*s* 8*d* per ton, many mines closed. Records show that by that year the number of big mines, that is, those producing more than 12 000 oz per year, had been reduced to eight, seven of which were on the Kalgoorlie-Boulder Belt, the exception being the Sons of Gwalia at Leonora. H. W. Gartell, a mining engineer on the goldfields at the time, reported that because of increasing costs, too much gold was being left in the ground because it had become uneconomical to mine, and he was left in no doubt that the 'industry was dying'.

The great companies, it seemed, had reaped the harvest of easily won gold and were now prepared to relinquish the mines and cut future losses. If the industry was to remain a viable proposition, it was up to the government to assist to the full. The government in the past had shown a marked reluctance to help with direct financial aid those mines which had issued millions in dividends. After the war all governments had showed concern over the lack of development by making development costs a taxable deduction. This assistance had been offered under the Mining Development Act until by 1923 the government was owed £65 728. No doubt the most important single item of government aid was that granted to the Sons of Gwalia mine after a disastrous fire in January 1921, which destroyed most of its surface plant and halted its mining operations for nearly three years.

During the years immediately after the 1914-18 war, government-sponsored parties of prospectors traversed the fields, seeking new deposits of gold. Nothing of major importance eventuated, although a number of independent prospectors uncovered deposits that were rewarding enough to permit the formation of companies and the commencement of mining operations. These, such as the Celebration Mine on the Hampton Plains Estate, about 30 km south of Kalgoorlie, and the St Ives Reward on Lake Lefroy, some 70 km southwest of the same centre, persisted for a few years but added little to the state's mining wealth.

The election of a Labor Government in April 1924 raised hopes for long-awaited changes in the government's attitude towards the gold-mining industry, since Philip Collier, the premier, represented Boulder and the minister for mines, Michael Troy, represented Mount Magnet. Two other members of cabinet, S. W. Munsie and J. Cunningham, represented Hannans and Kalgoorlie respectively. However, the most important measure undertaken was the appointment of a Royal Commission to 'consider the conditions affecting the mining industry and to make recommendations with a view to improvements'.

The Kingsley Thomas Commission (Thomas, a mining engineer from South Africa being the sole commissioner) was critical of mine management for a lack of planning, misdirection of capital, the continuance of dividends despite falling profits, lack of attention to development of reserves, and failure to upgrade worn-out and outmoded machinery and plant. Needless to say the companies retaliated with claims of high wages and accelerating costs (despite intensive programmes of mechanization), higher tariffs on much-needed imported machinery and the need for more government assistance in the form of lower charges for water, power and transport charges as well as reductions in the overall level of taxation.

Amalgamation was one recommendation accepted by the Lake View and Star (LVS), which took over most of the leases on the west side of the Boulder Dyke to become the owner of the biggest property in Kalgoorlie. The report had also advocated a change from steam generators to oil-generated electricity. This the LVS company did in 1933, when it set up a Crossley Premier plant with an aggregate horsepower of nearly 6000, which resulted in a saving of some £23 000 per year. The company was also the first to use the new flotation method of treating gold ore. Further north, a new mine at Wiluna set out to emulate the Kalgoorlie company by utilizing the latest method of mining and treating gold. This infant company thus contributed to a revival of the industry.

'It is clear', said G. B. Snooks, a Western Australian research economist, writing on the period, 'that the poor performance of the agricultural and pastoral industries in the 1930s would have been a greater disaster if it had not been for the existence of large deposits of gold-bearing ore in Western Australia.' There is little doubt that the better price for gold was a major stimulating factor. Over the decade this increase was 121 per cent, which meant a price increase from the traditional 84*s* 11½*d* to 188*s* 6*d* per oz. The Commonwealth Government also assisted in this price movement with the Bonus Scheme it announced in October 1936, which scheme offered gold-mining companies a bonus of 20*s* per oz (soon reduced to 10*s*) on increased production.

By the end of 1933, about seventy companies had been floated to secure capital for the purchase and development of Western Australian gold-mines. The majority were part of a number of interlocking companies, which became known as the de Bernales Group, and which were largely financed with British capital. But finance also came from South Africa, via the New Consolidated Goldfields, to the Lake View and the Wiluna mines, and American money provided the basis for the low-grade Big Bell mine near Cue, south of Meekatharra.

The Americans hoped to use the new flotation process to treat the low-grade ore and make 10*s* per ton profit. The government agreed to construct a rail-link to the mine from the main Perth-Wiluna line. Although the mine did not begin production until October 1937, in its first full year it treated 85 958 tons of ore worth 2.36 dwt per ton to secure 10 140 oz of gold. This was to give shareholders their first dividend of £34 000.

A number of companies came from the eastern states to exploit Western Australia's gold deposits. One of the biggest was the Western Mining Corporation, which still remains in operation. Registered in Melbourne in March 1933, with a capital of £1 million, mainly from American and British sources, the company, once it had settled in Western Australia, took up exploration concessions on more than 88 000 square miles of country.

The most powerful and certainly the most interesting flotations were those undertaken by Claude de Bernales in London. 'He was', wrote Professor Blainey, 'the prince of the gold promoters in the thirties.' But even prior to this date the financier had been a familiar figure on the Western Australian mining scene. He arrived from London in the middle of the last gold boom in 1897, and in 1903 had married a widow, the owner of the Kalgoorlie Foundry. Soon he began to specialize in the business of buying and selling mining machinery. Throughout his travels he acquired the leases of closed-down mines as well as the discarded machinery. In 1924 he formed in Melbourne a family concern, the Australian Machinery and Investment Company, to hold the accumulation of forfeited leases. With capital gained from the sale of the Wiluna leases and the promise of a government gold bonus, de Bernales moved to London where he formed a number of companies, which bought the family-held gold-mine leases from him.

Sporadic reports of gold finally culminated in the finding of the largest nugget in the state. The Golden Eagle (so named because of its shape) was uncovered on 15 January 1931 by Jimmy Larcombe, a sixteen-year-old boy working a claim with his father at Larkinville near Coolgardie. The lump of gold weighed 1134 oz and was later sold to the government for £10 876. To the prospectors and goldfielders its discovery vindicated their long belief that there remained plenty of gold waiting to be unearthed, even close to the big centres. At a time of increasing unemployment, the news acted like a magnet, attracting more than 1000 men to the vicinity before the end of the year.

The new Labor Government, which took office in May 1933, inaugurated the State Prospecting Scheme. The Scheme also had the support at a time of rising unemployment, of the Commonwealth Government, and by the end of the thirties some £255 860 had been spent by both governments. However, prospectors had returned gold worth only £164 168 and had repaid the state no more than £26 635.

The better prospects for employment had also induced larger numbers of men to move to the goldfields. Among them were many migrants from southern Europe. Competition for work increased the long-standing antagonism and clashes between the different ethnic groups prevailed. In early 1934, a series of riots occurred in the Kalgoorlie-Boulder district, and resulted in damage and loss in gold production to the value of £200 000.

Industrial relations throughout the ensuing decade were relatively peaceful. Doubtless this was due to the continuing prosperity of the industry, and the fact that the Prosperity Bonus prevailed after December 1934. However, in early 1935 a serious dispute eventuated over the question of the hours of work per week. The companies wanted alternating 40-hour and 48-hour weeks, thus avoiding the uneconomical Saturday-morning shift, but the miners' unions wanted the straight 44-hour week. The resulting five-week stoppage was the worst ever experienced and cost possibly half a million pounds in lost profit and wages. But on the whole the decade was a profitable one for all concerned.

For the gold-mining industry, the Second World War was the beginning of the end. Unlike the previous war, gold was not considered to be a strategic mineral. Certainly the most important element contributing to the decline of the industry and the closure of so many mines was the fall in the quality and quantity of the industry's work-force. Over the years 1939-45 manpower decreased from 14 595 to 4818. The outback mines, the newest of the mines in the industry, suffered more than most. By the end of 1942, 18 of the larger outback mines (such as Triton with 310 and Youanmi with 223) had been forced to close. As well, more than 40 other mines, including those on the Kalgoorlie-Boulder Belt, which in 1939 had employed 13 226 men, were operating on a reduced scale with 5366. The following year the situation worsened. The number of producing mines dropped from 731 in 1940 to 324 in 1943. Those producing more than 5000 oz were reduced from 36 in 1940 to 21, and the only mine producing more than 100 000 oz in 1940, the Lake View and Star, produced just 87 196 oz throughout 1943.

The gold-mining industry was also affected when an important allied industry, the firewood industry, suffered similar reductions in manpower. In early 1942 the Kalgoorlie Electric Power and Lighting Corporation, with a daily consumption of over 1700 tons of wood, found its wood supply reduced to some 1400 tons, and cut out nearly half its boiler strength. Since most mines in Kalgoorlie were dependent upon the corporation's power, this cut into the production of gold.

A further blow came in mid-1948, when the leading supplier of firewood and mining timber on the Boulder Belt, worried by labour problems and the need for capital to replace worn-out equipment, and by an increase from 1*s* 6*d* to 2*s* 10*d* per ton in firewood freight on the Eastern Goldfields, declared that it was going into voluntary liquidation. A stunned industry soon called a meeting of firewood and timber consumers and appointed a committee to decide whether to lease the company assets or make a straight-out purchase. But despite the acquisition of these power sources by the mining companies, the continued dependency on wood fuel only served to delay the inevitable modernization to oil-fuelled equipment.

As a result of this trouble over firewood, coal-fired boilers were installed and coal was supplied from Collie by Western Collieries Ltd, a company formed in 1944 from the Goldfields Coal Syndicate. But neither wood nor coal could compete with diesel power, and soon each of the Kalgoorlie mines installed its own diesel power-plant and the Kalgoorlie Electric Power and Lighting corporation closed. Western Collieries, however, survived with the help of government contracts. In 1920 Amalgamated Collieries had been formed in order to acquire all the small companies in Collie, but in 1930 its monopoly ceased with the establishment of the Griffin Coal Mining Co. After the formation of Western Collieries, the influence of Amalgamated Collieries began to decline and the company eventually ceased business in 1960.

The role of *coal* in the economy of the state has varied. Originally mined for ship bunkering, railway locomotives, heating and power generation, Collie coal was found to be inferior to imported coal because of its bad storage characteristics. It was therefore not in demand for export. Its use in railways also declined in favour of diesel fuel. Even in power generation it has had severe competition from fuel oil and natural gas, but, because of the growing world shortage of these fuels as well as rising prices, the present trend is for a return to coal. From a mere 3508 tonnes in 1898, annual production rose to 2.3 million tonnes in 1976-77.

Throughout most of the period of the 1939-45 war, Minister for Mines Panton continually asserted that Canberra's treatment of the gold-mining industry would mean not only the collapse of an industry but also the closing down of 20 fair-sized towns with a population of nearly 60 000, living in over 15 000 dwellings scattered over 1300 km of auriferous country in the outback. Throughout this period the industry laboured under a tremendous strain. There was an overall increase in wages and materials of 43 per cent, whereas over the period of the war, the price of gold increased by a mere 12 per cent. Under a number of acts, both the state and commonwealth governments were forced to come to the aid of the mining industry. Rents and payments on mining tenements were waived and the costs of maintaining a number of mines were accepted by the state. However, the trends introduced by the war—a diminishing labour force, a fixed price for gold and accelerating costs—continued to operate after the ending of hostilities in 1945. Notwithstanding this, the general air of confidence that permeated the industry overshadowed the doubts. As in the years after the First World War, it was believed that the overseas demand for gold, held back by the war, would soon outstrip present production and inflate the price of gold.

To cater for the anticipated developments in the industry, many Western Australian companies regrouped. Others, closed because of the war, reopened. In Coolgardie and on the Hampton Plains Estate new shafts were put down to prove surface indications. The Lake View and Star was still considered to be the premier mine in the state. The doyen of the Kalgoorlie mines, the Great Boulder Proprietary on 121 acres, had shortly before increased its ore reserves. The Perseverance, though small, was still considered to hold the best 24-acre lease on the belt, because few of its lodes had been worked below the 1000-ft level and estimates of its reserves were thought to be conservative.

However, the Mt Charlotte mine on the northern end of the Boulder Belt (including the original reward claim pegged by Paddy Hannan) was destined in later years to become the most unusual mine. There had been little development on the mine, but some 25 bore holes had secured evidence of about 1.2 million tons of 5.3 dwt per ton free-milling ore, a potential which provoked widespread interest among the older companies in Kalgoorlie.

Outside Kalgoorlie, four other areas held mines with prospects: Central Norseman, the Sons of Gwalia near Leonora, Wiluna, and Big Bell near Cue. Sons of Gwalia was the oldest, having been registered in January 1898. It held 511 acres, but falling yields through declining values and labour difficulties had halted dividends and the company was waiting for a solution to its labour problems and for an increase in the price for gold.

For many of the companies, the phase of post-war optimism was short-lived. The hoped-for intake of men to the pre-war figure of 14 000 did not eventuate and the maximum labour-force of the post-war years was only half this number. This was a frustrating period for the companies, for the repeal of the Commonwealth Gold Tax in 1947, followed by the increase in gold prices engendered by Britain's devaluation of sterling in September 1949 (which led an immediate increase from 215*s* 3*d* to 309*s* 10*d* for the Australian gold producers), had stimulated plans for an expansion.

For the decade 1953-63, despite a reduction in labour in the mines of more than a third, production of gold continued at above 800 000 oz per year, indicating a vastly increased productivity within the industry. This may have been the result of the improving skill of miners and the spur provided by the contract system, but it was mostly due to the use of improved rock-breaking tools and facilities and possibly the extraction of rich ore pillars. Lighter and more portable one-man drills equipped with tungsten carbide-tipped bits and the adoption of the new explosive, ANFO, had increased the footage rate. There was also an increase in the rate of ore removal from the levels with the use of bigger electrically driven locomotives and diesel-powered load-haul-dump vehicles underground and also the change from steam-driven to electric-winding gear.

It was during this period—and this probably contributed to some of the labour shortage on the goldfields—that minerals other than gold began to be exploited. Today, iron ore, bauxite, nickel, mineral sands and petroleum in particular have supplanted gold as the leading mineral in the mining economy of the state (Fig. 1.1).

It also was proving difficult for the gold-mining companies to keep their costs down, especially those connected with wages. Over the period 1958-65, the wages for the general labourer in the gold-mining industry increased by 37 per cent.

Since 1945, many big outback mines have been forced to close. Several, such as Big Bell and Wiluna, closed before the first post-war decade was out, despite Government assistance. In fact by 1976 only two gold mines, Central Norseman and Hill 50 at Mt Magnet, were still operating as outback mines, and of these Hill 50 was to close in February 1976.

With the closure of so many small or outback mines, the importance of the state battery system to the industry had diminished. Although in 1970 there were still seventeen public batteries (including a base metal plant at Northampton) operating, they treated only 22 810 tons (a far cry from the 100 000 tons or more crushed annually during the 1930s). However they did produce gold worth $274 019, which gave to the government $108 770 in revenue. Since the inception of the system in 1898, the state batteries have treated 3 656 573 tons of auriferous rock to produce gold worth $39 280 114. Capital expenditure has been $1 878 931 and total expenditure has exceeded revenue by $7 034 396.

However, with accelerating costs forcing the closure of many mines, the industry wanted something more tangible and called for a guaranteed policy of assistance to make the production of gold a more viable, long-term proposition.

In many ways, the product of gold-mines is unique. It is, for instance, one commodity for which the price has remained fixed for a very long time. The price was fixed in 1717 by Sir Isaac Newton, Master of the English Royal Mint, who agreed to buy and sell gold at 84*s* 11½*d* and it remained stable until the 1930s, when amplified demand and currency changes forced an upward trend. In Australia this movement was reinforced by a devaluation of the Australian pound. A further devaluation in 1949 again improved the price. However, mounting costs soon eliminated any benefit to the mines. The problem was also compounded by Australia's adherence to the price fixed by the International Monetary Fund. Governments of other commonwealth countries, Canada, South Africa and later Rhodesia, permitted their gold producers to sell gold on the open market at enhanced prices. Militant lobbying eventually secured this concession for Australian gold producers, and the Gold Producers' Association was formed. During the first six months of the association's

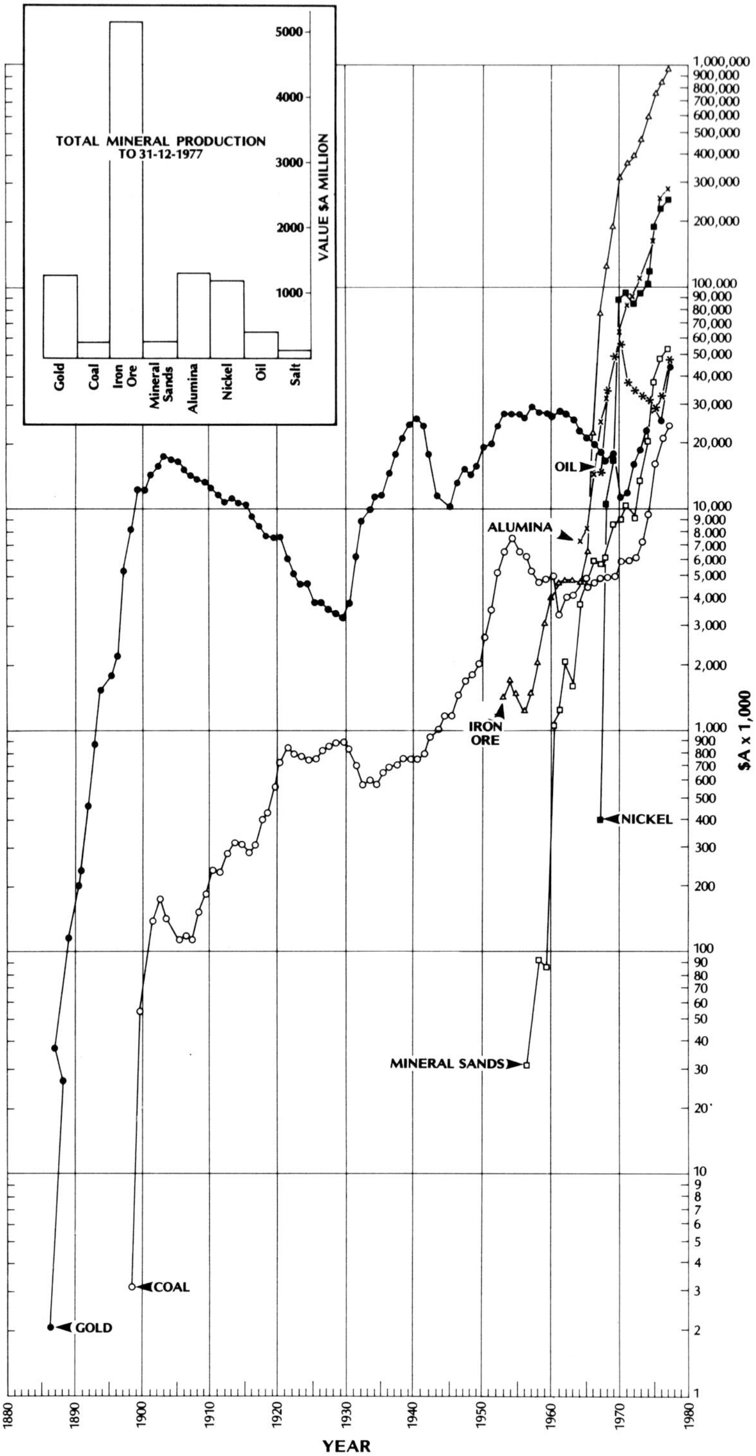

Fig. 1.1 Production of major minerals of Western Australia 1886-1977, yearly and total

operation, to July 1952, Western Australian mine owners sold their gold for a premium of £A539 358, or nearly 21*s* per oz.

The ultimate solution was a guaranteed subsidy sufficient to outweigh increasing costs. Since 1954, following the introduction of the Gold-Mining Industry Assistance Scheme, the producers in Western Australia have secured some $29.5 million in subsidies from the Commonwealth Government. At the same time the sale of gold on the open market has persisted. Unfortunately for producers the premiums gained are considered to be profit and the subsidy is reduced accordingly. Since April 1973, owing largely to the rising prices for gold (to more than $200 in January 1979), no producer has been in receipt of a subsidy.

In June 1973 Gold Mines of Kalgoorlie and Lake View and Star joined to form a new company: Kalgoorlie Lake View. Central Norseman continues to operate profitably owing to improving economics. Two major new investments have been the Telfer project, a new mine some 200 km east of Nullagine, and Kalgoorlie Mining Associates, a combination of Kalgoorlie Lake View and an American company, Homestake Gold Corporation. The main object of the latter merger is to revitalize the biggest gold-mining venture now in the state, Mt Charlotte at Kalgoorlie. This mine was scheduled to close in December 1976, but a 17½ per cent depreciation of the Australian dollar in November of that year gave it a new lease of life. Unfortunately these are the only current signs of optimism. Great Boulder closed in March 1976, Hill 50 at Mt Magnet in February 1976 and Marvel Loch in September of the same year.*

Though the suspension of Part IV of the Bank Act now permits anyone in Australia to own, buy and sell gold, thus serving to stimulate sales, other factors have had the opposite effect. It has been argued that, despite an apparent generous price for gold, insufficient capital has been accumulated or generated to promote new exploration or developments in the mines. There is an old adage which still appears to hold true, that *gold-mining is profitable when the price of gold per ounce exceeds the weekly wage of labour.* Enormous sums are currently needed to replace and extend existing treatment plants to cope with reserves of average low-grade ore (less than 4.0 dwt per ton in 1976). Furthermore, with no guarantee of gold price stability, it would be foolish and futile to expect investment in gold-mining; this, in the long run must mean the demise of the century-old industry.

One of the minerals recognized early in the history of the state was *iron-ore* which was mined in a small way from minor deposits in the Darling Range. The large deposits in the northwest went largely unrecognized, although in 1890 iron-ore from Koolan Island was mentioned as having been used as ballast in pearling luggers.

The first lease on Koolan Island was granted in 1907, and subsequent lessees on this and nearby Cockatoo Island did little, until in 1936 the Yampi Sound Mining Company indicated its intention of selling ore to Japan. The result was that the Commonwealth Government placed an embargo on the export of iron-ore, giving as its reason the acute shortage of such ore in Australia.

* At the time of going to press (Sept. 1979), the price of gold had risen to more than $300 per oz, and consideration is again being given to re-opening the mines of the Golden Mile at Kalgoorlie. (Ed.)

A deposit at Koolyanobbing, about 56 km northeast of Southern Cross, and first recognized in about 1916, was opened up in a small way in 1950 to supply ore to the charcoal-iron industry at Wundowie. As part of the agreement for the establishment of an integrated iron-and-steel industry at Kwinana, the Koolyanobbing leases were granted to Australian Iron and Steel Ltd, and production commenced in 1967.

In 1927, Hoskins Bros had taken up leases on Cockatoo Island, and these were taken over from Hoskins Bros by Australian Iron and Steel when this company was formed the following year. Unfortunately the company's first expedition to the island in 1929 was shipwrecked, and the company thereafter concentrated its mining activities in South Australia. However, in 1944 it returned and began feasibility studies and construction that ultimately led to the first shipment of ore from Cockatoo Island to Newcastle in July 1951, and from Koolan Island, for export, in 1967.

In the 1950s it began to be recognized that Western Australia had a vast iron-ore potential, especially in the Pilbara, and that a market existed in Japan. Exploration began to reveal the true picture, with the result that the embargo was lifted in December 1960.

The first company to benefit was Western Mining Corporation, which exported ore from a small deposit at Koolanooka (near Morawa) through the port of Geraldton in March 1966. The exporting of ore from deposits at Goldsworthy (100 km east of Port Hedland) and Tom Price (270 km south-southwest of Port Hedland, but shipped through Dampier 250 km to the northwest) began in 1966. Newman (350 km south of Port Hedland) was producing by 1969, and Pannawonica (90 km south of Cape Lambert) by 1972.

Ore deposits are also being worked in conjunction with the original mines—for example Shay Gap, Mt Enid and Paraburdoo—but other large orebodies are awaiting a suitable economic climate and government sanction before producing. These include Marandoo, McCamey's Monster, Wittenoom, Area C, Rhodes' Ridge, Marillana, Koodaiderrie, Deepdale, Weld Range, Tallering, Mt Jackson and Mt Gibson.

The presence of large tonnages of *bauxite* in the Darling Range was recognized early in this century, but it was not until the late 1950s, when the Western Mining Corporation investigated the potential of the area and was able to interest outside capital, that it became a viable proposition. Trading under the name of Alcoa of Australia Limited, the company's first production from its Jarrahdale/Kwinana complex came in 1963. A second refinery has been established near Pinjarra to treat ore from the nearby deposits of Del Park and a third is proposed for Wagerup.

On the Mitchell Plateau in the Kimberley Region, the Alumax Bauxite Corporation has outlined a large bauxite deposit, but its exploitation awaits the necessary markets and agreements.

In the late 1950s and early 1960s, the profitability of the gold-mines in Kalgoorlie was declining, but suddenly *nickel* became the glamour metal, following a find at Kambalda only 50 km away. This find started the 1966-to-1970 mineral boom, which did so much good for and some harm to the Western Australian mining industry. The initial effect on Kalgoorlie was that, at a time when experienced manpower was limited, men were enticed away to mine nickel. Kalgoorlie's only hope of salvation in the face of rapidly escalating costs and an artificially low price for gold, was to increase productivity, and this it could not do without assistance from man-power. In the long term, however, the discovery of nickel meant a new lease of life for Kalgoorlie. Nickel discoveries, not necessarily all economic, were announced from Norseman to Wiluna and east to Laverton and the border, and Kalgoorlie became the distribution centre for a vast nickel province, and appears set for a long life independent of gold.

As in the old gold days, overseas capital and expertise poured in to augment the meagre local supply, and further nickel finds were developed near Windarra, Scotia, Widgiemooltha, Spargoville, Nepean and Agnew. Other deposits have been discovered but are awaiting a more favourable economic climate before mining begins.

The mining of *heavy mineral sands* in Western Australia began in 1949 at Cheyne Bay on the south coast, but it was 1956 before this industry became significant. In 1956 mines were established at Koombana Bay and later at Capel, Yoganup and Wonnerup. Following the discovery in 1970 of heavy-mineral-sand deposits on a pastoral property at Eneabba, production began in 1971 and the minerals were shipped through Geraldton. A further deposit at Jurien was worked for a time in 1975-76, but this operation has now been abandoned.

Early popular opinion was that *petroleum* would never be found commercially in Australia—the country was 'too old'. Nevertheless, attempts were made without lasting success at Roma (Queensland), Lakes Entrance (Victoria), in the Kimberleys and at Byro (Western Australia). In 1952 a more concerted effort was begun by WAPET (West Australian Petroleum Pty Ltd), a consortium of local and overseas companies, and fairly soon established the existence of oil in a wildcat bore at Rough Range, Exmouth Gulf. This spurred oil exploration by locally based and overseas companies in all sedimentary basins in Australia and led to a producing oil-field on Barrow Island (1967) and a producing gas-field near Dongara (1971) as well as other major finds in the eastern states. In turn these have led to the deep-water exploration of the North-West Shelf and the subsequent discovery of its vast gas-field, and has stimulated further interest in exploring the Exmouth Plateau in still deeper water. From an industry which could 'never be', oil/natural gas now ranks fifth in Western Australia in terms of value produced.

A summary history of mining of different mineral deposits in Western Australia can be seen from Fig. 1.1, which shows the value of the annual and total production of the state's main mineral products since mining began.

FURTHER READINGS

Battye, J. A. 1937. *Cyclopedia of Western Australia.* Adelaide: Hussey & Gillingham.
Blainey, G. 1963. *The rush that never ended.* Melbourne: Melbourne Univ. Press.
Casey, G. & Mayman, T. 1964. *The Mile that Midas touched.* Adelaide: Rigby.
Raggatt, H. G. 1968. *Mountains of Ore.* Melbourne: Lansdowne Press.
Uren, M. J. 1948. *Glint of Gold.* Melbourne: Robertson & Mullens.

CHAPTER 2

The geology of Western Australia in relation to its mineral resources

REX T. PRIDER
Department of Geology, University of Western Australia

The mineral resources of a region are directly related to the geology (including geomorphology) of that region, and to some extent to its present and recent past climates, which may have resulted in secondary enrichment or impoverishment of existing ore deposits. Although geology may be described as the science of the earth in all its aspects, we shall confine our attention here to the distribution and broad structural relations of rocks of different ages which make up Western Australia—factors which, since they control mineral deposits of all kinds, are all-important to the mining industry.

About two-thirds of the land area of Western Australia is occupied by the ancient Precambrian basement, which is composed of a complex of igneous, metamorphic and sedimentary rocks formed more than 600 million years ago. Most of our mineral deposits of economic importance, except coal, oil, natural gas and water, and superficial deposits such as lateritic iron-ore, manganese and bauxite deposits, black sand, and various alluvial accumulations, occur in these Precambrian rocks. The remainder of the state is occupied by sedimentary basins (see Fig. 3.30, page 146), in which Phanerozoic (Palaeozoic and later) sediments are developed. It is in these younger sedimentary basins that artesian water, coal, oil and natural gas are likely to occur. Finally, there are the still younger superficial deposits—laterites, salt-lake deposits, and soils on which much of the economy of this country depends. It will be convenient therefore, in outlining the geology of the state in relation to its mineral resources, to consider it under the three main headings:

(a) Precambrian basement,
(b) Phanerozoic basins,
(c) superficial deposits.

The distribution of the solid rocks (omitting superficial deposits) is shown in the accompanying generalized map (Fig. 2.1) and in the geological map of Western Australia accompanying this volume.

PRECAMBRIAN BASEMENT

The Precambrian basement includes the Archaean and Proterozoic rocks all of which originated more than 600 million years ago. The Archaean is a complex of crystalline igneous and metamorphic rocks, dominantly granites and gneisses, with

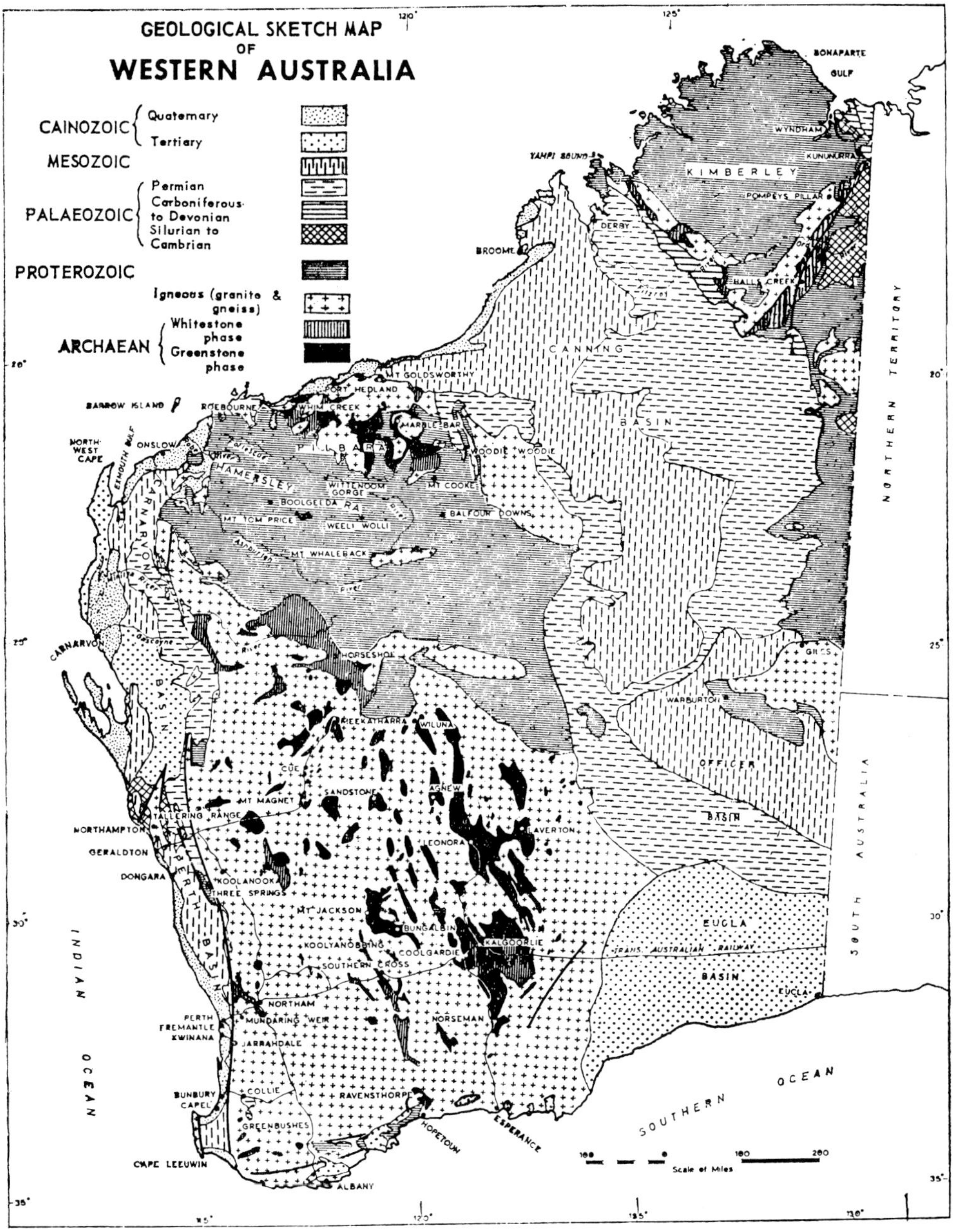

Fig. 2.1 Geological sketch map of Western Australia (from Clarke *et al.* 1971). The Archaean is subdivided into (*a*) Greenstone Phase (= Volcanic Assemblage)—which includes the greenstones of the 'Kalgoorlie-Yilgarn' System and the Warrawoona Group; (*b*) Whitestone Phase (= Sedimentary Assemblage)—which includes the whitestones of the 'Kalgoorlie-Yilgarn' System and the Gorge Creek Group; (*c*) Granite and gneiss—which includes the metasedimentary and meta-igneous gneisses (Older Granites) and later intrusive granites (Younger Granites).

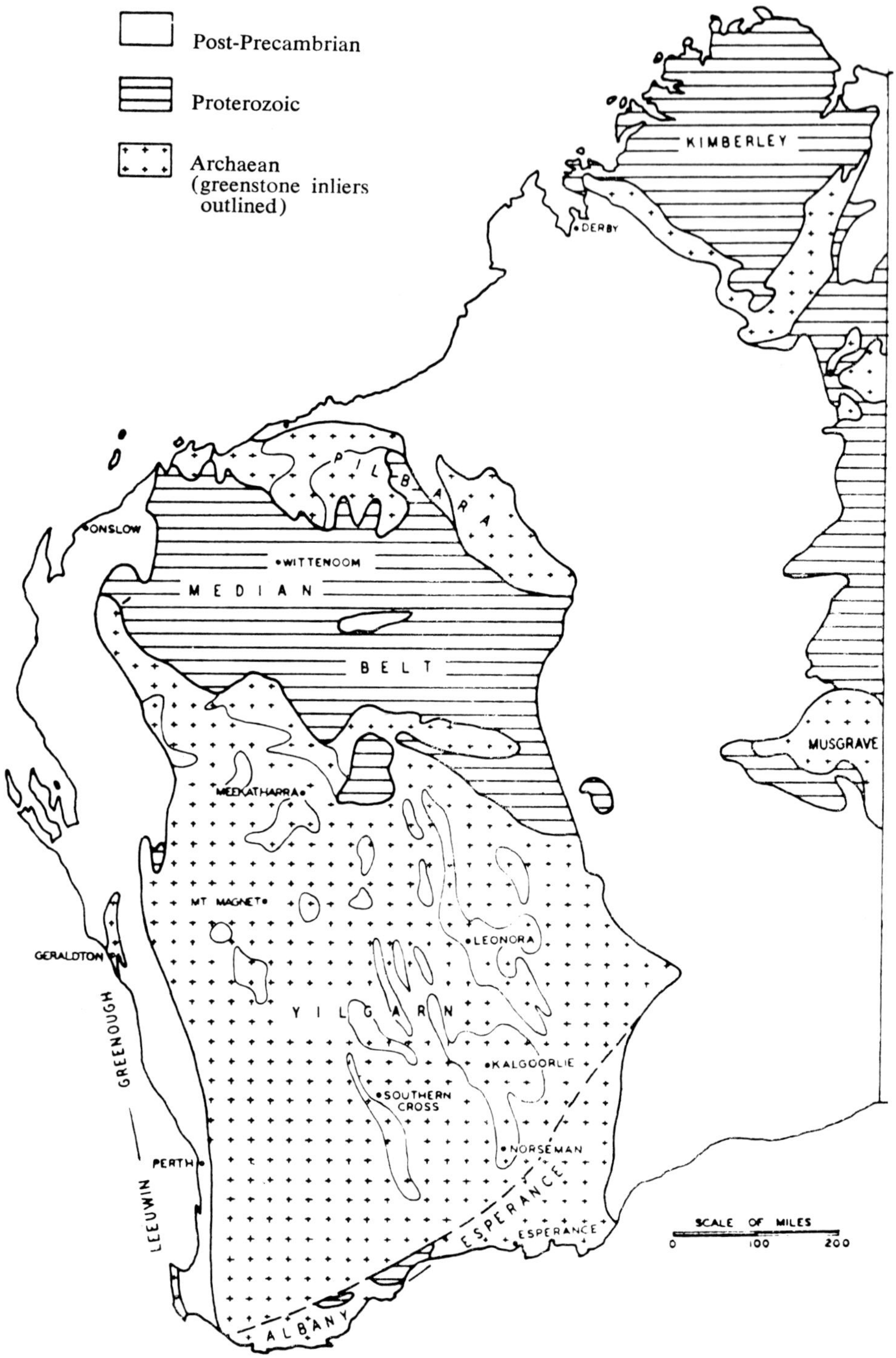

Fig. 2.2 Main Precambrian provinces of Western Australia (after Prider 1965, p. 56).

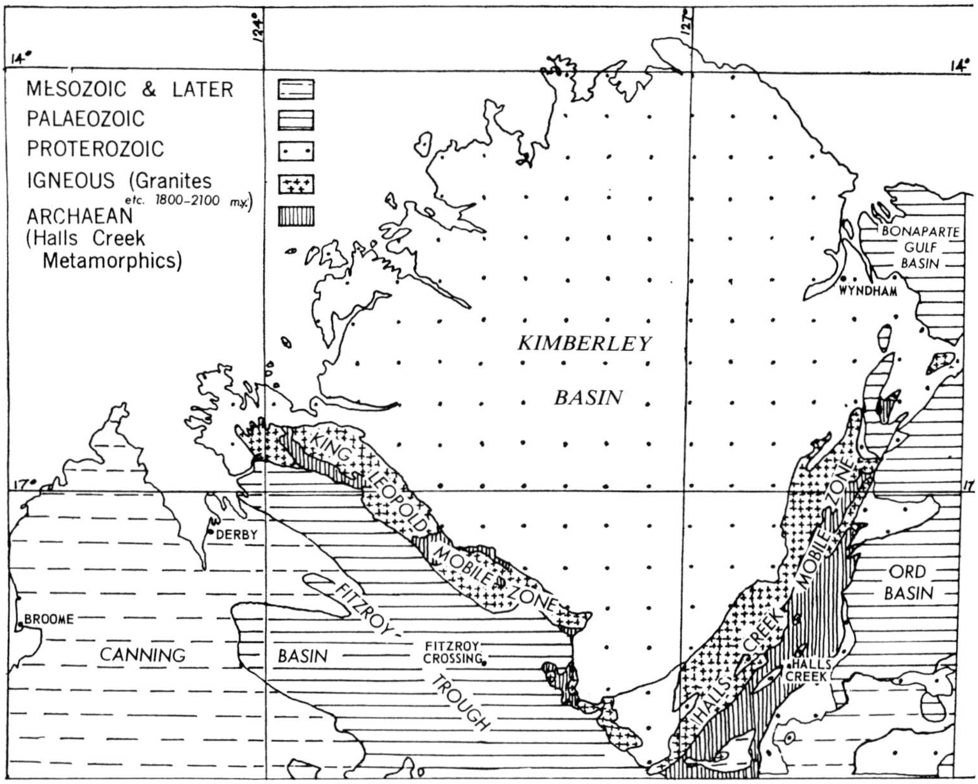

Fig. 2.3 Generalized geological map of the Kimberley Division, Western Australia (from Clarke *et al.* 1971).

minor amounts of schistose metamorphosed acid and basic volcanics and sedimentary units. In places, particularly in the Pilbara, this Archaean complex is overlain unconformably by sedimentary and volcanic rocks of Proterozoic age, which do not exhibit the extensive metamorphism so characteristic of the older Archaean complex. The time-boundary between the younger Precambrian (Proterozoic) and older Precambrian (Archaean) is approximately 2400 million years ago. Within the different areas occupied by the Precambrian rocks (Fig. 2.2), the same generalized sequence can be distinguished.

In the Kimberley (Fig. 2.3) the oldest rocks, the *Halls Creek Group* and *Lamboo Complex*, are metamorphosed igneous and sedimentary rocks of the King Leopold and Halls Creek Mobile Belts flanking the Proterozoic Kimberley Basin. These are intruded by granite and carry auriferous and base-metal ore deposits. They are either Lower Proterozoic or Archaean and are overlain by unmetamorphosed sediments and volcanics with basic igneous intrusives. The Precambrian age of all these rocks is evidenced by the fact that in the East Kimberley they are overlain by

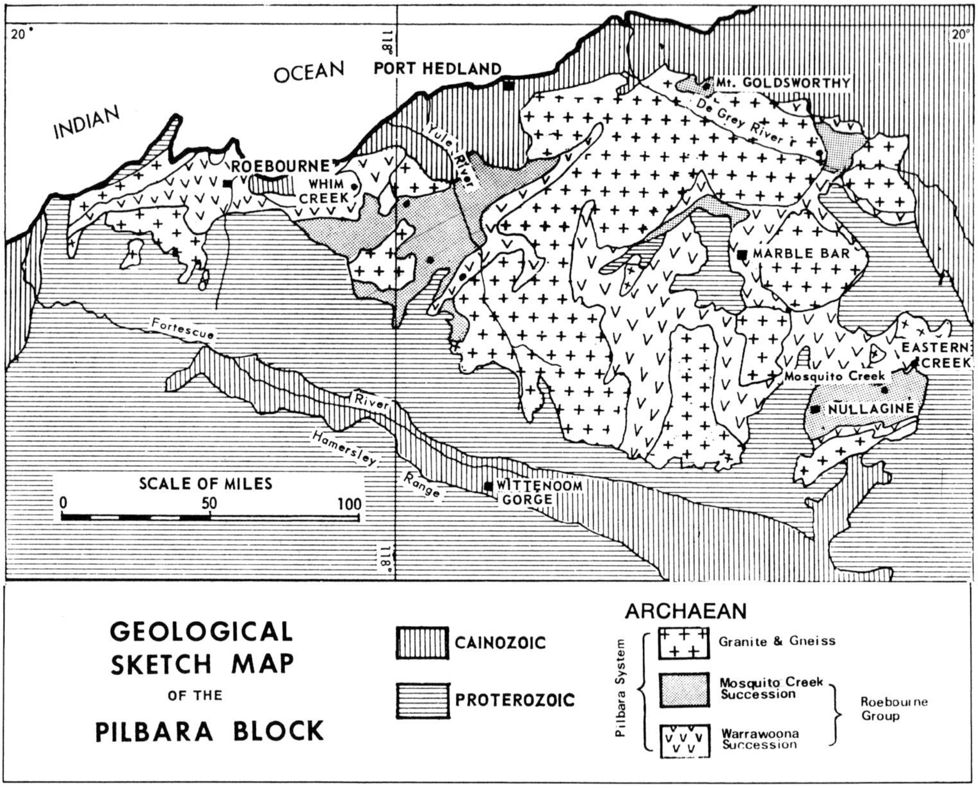

Fig. 2.4 Generalized geological map of the North-West Division, Western Australia (from Clarke *et al.* 1971).

sedimentary rocks containing fossils of Middle Cambrian age. This is the only area in Western Australia where the Precambrian age of the rocks of this crystalline complex can definitely be proved on stratigraphical evidence alone. In the southern part of the state we find a similar sequence of crystalline schists with intrusive granites, and by lithological correlation (which is not a very sound method), we assume that they are Precambrian, although they cannot actually be traced through from the Kimberley. During recent years the Precambrian age of these rocks has been confirmed by actual age determinations based on the decay of radioactive elements which occur in them. This work indicates that the bulk of the granitic intrusions of the southern part of the state and in the Pilbara crystallized from a molten state some 2600 to 3100 million years ago. Some granites, however, such as those in the vicinity of Albany and along the south coast, are much younger, being emplaced approximately 1100 million years ago.

The Precambrian sequence in the North-West Division (Fig. 2.4) appears to be the most complete that is present in Western Australia, and is as follows:

The oldest is a volcanic assemblage (*Warrawoona Group*) which consists mainly of metamorphosed and deformed submarine basaltic lavas, acid volcanics and tuffs interbedded with thin units of chemically-deposited sedimentary rocks (chert, jaspilite and banded iron-formations). Overlying the volcanics is a sedimentary succession (*Gorge Creek Group*) of banded iron-formation, sandstone, shale and conglomerate with subordinate volcanics. The banded iron-formation constitutes the protore of important iron-ore deposits such as those at Mount Goldsworthy. The sedimentary assemblage is unconformably overlain by acid volcanic rocks (part of the *Whim Creek Group*), which at Whim Creek form the host to disseminated copper-lead-zinc deposits. All parts of the above succession have been intruded during several intrusive episodes by ultrabasic, basic and acid magmas yielding a variety of igneous rocks. Intense folding and metamorphism have transformed many of the rocks into various types of schists, slates and quartzites. The succession is invaded by granitic igneous rocks—older granite gneisses (age about 3100 million years) and younger massive granites (age 2600 million years)—and most parts carry auriferous ore bodies, possibly genetically related to the intrusive granites. Other end-stage products of the younger granites are the coarse-grained pegmatites, which are important carriers of tantalum (in tantalite), beryllium (in beryl), lithium (in spodumene and lepidolite), tin (in cassiterite) and various radioactive elements. Still younger (Proterozoic) sedimentary rocks, such as conglomerates, sandstones, shales and banded iron-formations with interbedded basic igneous rocks, were deposited unconformably on the highly-folded, granite-intruded volcanic and sedimentary assemblages.

In the Median Belt between the Pilbara and Yilgarn Blocks (Fig. 2.2), this thick Proterozoic succession consists of a number of distinct groups. The three lower groups (the *Fortescue, Hamersley* and *Wyloo Groups*) are of Lower Proterozoic age, the youngest of these (the Wyloo Group) being intruded by granite aged approximately 1700 million years. The two upper groups (the *Yeeneena* and *Bangemall Groups)* are of Middle Proterozoic age, the youngest (the Bangemall Group) being about 1100 million years old. The Yeeneena Group rocks are intruded by granite 600 million years old near the Telfer gold deposit at the Paterson Range. Of these Proterozoic rocks the Hamersley Group is most important economically, since most of the iron-ore deposits of the Hamersley Iron Province occur within, or have been derived from, the thick jaspilites (banded iron-formations) within this group. Except in occasional narrow belts (such as the Ophthalmia Range) marginal to the Archaean blocks, these Proterozoic rocks have not suffered the intense folding that affected the older rocks, and consequently they are generally flat-dipping to horizontally-bedded unmetamorphosed sediments. Such sediments cover very extensive areas in the North-West (Fig. 2.4) and they are similar in many respects to the flat-dipping Proterozoic sediments of the Kimberley Basin. The final episode in the Precambrian history of the North-West Division was the intrusion of dolerite dykes and sills into all of the earlier rocks.

Coming to the southern half of the state we find a similar sequence to that in the North-West. In the part of the Precambrian Shield extending south of lat. 26°S. the

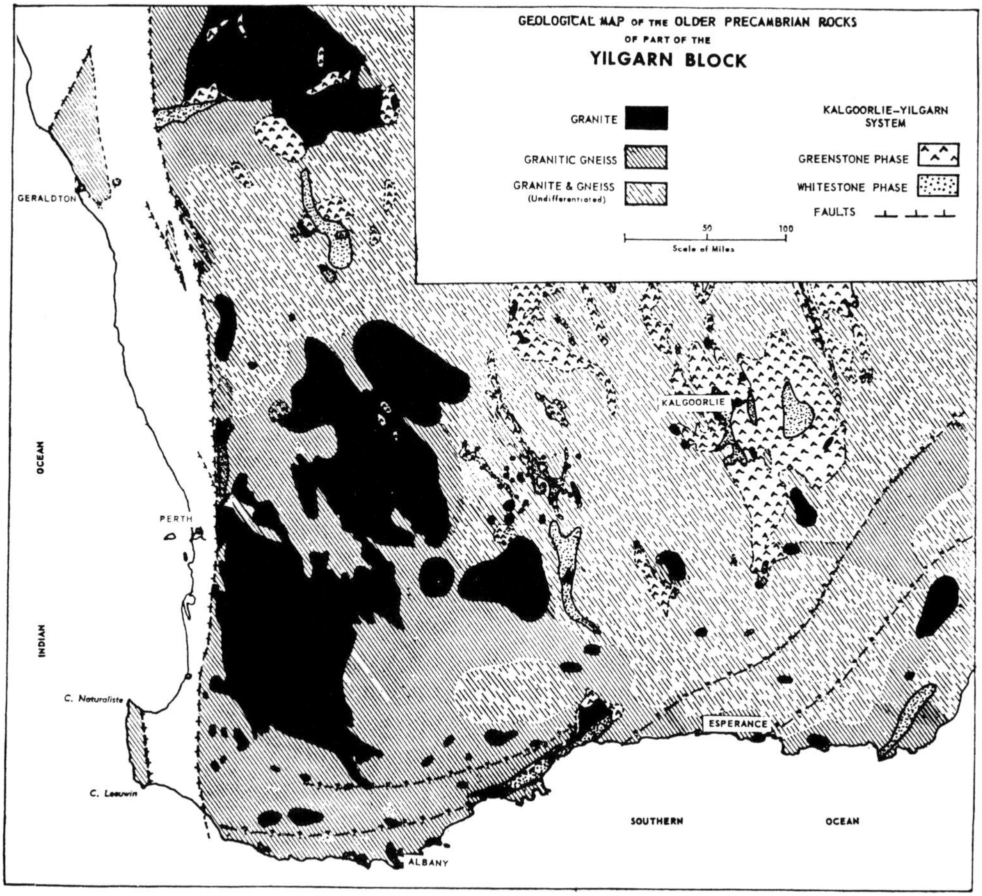

Fig. 2.5 Generalized geological map of the older Precambrian rocks of the southern part of the Yilgarn Block, Western Australia (from Clarke *et al.* 1971)

oldest rocks that are recognized are the greenstones of the various gold-mining fields, which occur in comparatively narrow belts elongated in a general north-northwest direction (Fig. 2.5). These greenstones, which are for the most part metamorphosed submarine basaltic lavas, contain interbedded ultramafic lavas and jaspilites and are overlain by metamorphosed sedimentary rocks. This system of rocks has been called the 'Kalgoorlie-Yilgarn System'. From the mining point of view, it is the most important system, for the auriferous ore deposits of the main mining fields are confined to it, as are the known nickel deposits. It also contains, in the jaspilites, important iron-ore deposits such as those of Koolyanobbing in the Yilgarn. After the Kalgoorlie-Yilgarn rocks had been formed, they were intensely compressed into tightly-closed folds with north-northwest-trending axes and steep westerly dips. During this period of intense earth-movement approximately 2900 million years ago, granite magma concordantly intruded these older rocks, and alkaline solutions permeated them, producing the granitic gneisses which occupy the bulk of the southern half of Western Australia. Subsequently, about 2700 million years ago, granite magma was again intruded as in the Pilbara, and associated with these granites are economically important pegmatites.

The Kalgoorlie-Yilgarn System in the southern half (the Yilgarn Block) of the state appears to be the equivalent of the Archaean of the Pilbara Block. The Proterozoic in the Yilgarn Block is represented by a narrow strip of slightly-altered, steeply-dipping sedimentary rocks along the Darling Scarp, and flat-dipping sediments on top of the older crystalline rocks of the western part of the Yilgarn Block as at Watheroo, Yandanooka and the Billeranga Hills near Morawa. Proterozoic igneous activity is represented by Early Proterozoic layered complexes and Late Proterozoic basic dykes. The former, aged 2420 million years, take the form of basic-ultrabasic layered complexes, such as the large east-west dykes of the Norseman-Laverton greenstone belt, which have some prospects for the occurrence of nickel and chrome deposits. The latter are dolerite dykes which, like most of those in the Pilbara, intruded all of the Precambrian rocks about 550 million years ago. They occur throughout the Precambrian Shield, but are most abundant near its western margin, where some are quarried and crushed for road metal.

The Yilgarn Block is flanked to the south and east by the Albany-Esperance Province extending north-easterly to the Musgrave Block in Central Australia (Fig. 2.2). In this province the rocks are granitic gneisses and metasediments (schists and quartzites) similar to those of the Yilgarn Block, but they have a markedly different structure, striking east-northeast almost at right angles to the regional strike of the Yilgarn Block. These rocks are intruded at Albany, Porongurups, Esperance and Doubtful Island Bay by granites and pegmatites, with a radiogenic age of 1100 million years. In this province there is an east-west belt of isoclinally-folded metasediments, the *Mt Barren Beds,* on the westerly prolongation of which lies the Stirling Range consisting of almost horizontally-bedded low-grade metasediments—phyllites and quartzites (the *Stirling Range Beds*)—in which the original sedimentation structures are preserved. These rocks are probably of Middle Proterozoic age. All the south coast gneisses and metasediments (including the Stirling Range Beds) are intruded by dolerite dykes as in the Yilgarn Block.

There is a comparatively narrow strip of crystalline rocks along the western margin of the Perth Basin. The southern section (the Leeuwin Block) extends from Cape Leeuwin to Cape Naturaliste and the northern section (Northampton Block) extends from the vicinity of Geraldton to the southern margin of the Carnarvon Basin. The former is a belt about 20 km wide of isoclinally-folded gneisses with a regional northwest trend and a radiogenic age (probably the age of metamorphism) of 650 million years; the latter, which includes the Northampton Mineral Field, consists of northwest-striking, high-grade metasedimentary granulites containing segregation pegmatites aged 1000 million years. In the Northampton Block basic dykes occurring in a northeast-trending swarm represent the Late Proterozoic period of igneous activity of other parts of the shield.

Table 2.1

Phanerozoic stratigraphic table of Western Australia

		CANNING BASIN	BONAPARTE GULF BASIN	CARNARVON BASIN		PERTH BASIN		SOUTH COAST, EUCLA BASIN, & OFFICER BASIN
				Northern Section	Southern Section (near mouth of Murchison)	Northern Part	Southern part and Collie Basin	
QUAT-ERNARY	Holocene	Sands and alluvium	Sands and alluvium	Sands and alluvium	Sands and alluvium	Sands and alluvium	Sands and alluvium	Sands and alluvium Roe Car
	Pleistocene			Exmouth Ss, Bundera Car	Tamala Ls	Tamala Ls Pelsart Ls	Tamala Ls Rottnest Ls Peppermint Grove Fm Guildford Clay	
TERTIARY	Pliocene	Borda Ss Pender Bay Cg						
	Miocene			Cape Range Gp { Vlaming Ss Pilgramunna Fm Trealla Ls Tulki Ls Mandu Car Lamont Ss				Nullarbor Ls Abrakurrie Ls
	Oligocene	Oakover Beds Lawford Beds	White Mountain Fm					
	Eocene — Paleocene	Lampe Beds Lake George Beds		**Cardabia Gp** { Merlinleigh Ss (in eastern part of basin) Giralia Car Jubilee Car Cashin Car Pirie Car Wadera Car Boongerooda Grnsd.	Pindilya Fm		Kings Park Fm	Wilson Bluff Ls Hampton Ss Plantagenet Gp
CRETACEOUS	Upper			Miria Marl Korojon Car Toolonga Clut	Toolonga Clut		Poison Hill Grnsd Gingin Chalk Molecap Grnsd Dandaragan Ss	Bejah Clst Samuel Fm Madura Fm
	Lower	Anketell Ss Frezier Ss Parda Fm Melligo Ss Broome Ss Mowla Ss		Winning Gp { Gearle Sltst Windalia Radl Muderong Sh Birdrong Ss Barrow Fm Wogatti Ss	Winning Gp { Alinga Fm Windalia Radl Birdrong Ss		Osborne Fm Leederville Fm South Perth Sh Donnybrook Ss	Loongana Ss
JURASSIC	Upper	Jarlemai Sltst Alexander Fm Wallal Fm		Learmonth Fm and		Yarragadee Fm	Yarragadee Fm	
	Middle			Dingo Claystone		Champion Bay Gp { Kojarena Ss Newmarra-carra Ls Bringo Sh Colalura Ss	= Cadda Fm	
	Lower	Millyit Ss				Chapman Gp { Moonyoon-ooka Ss Greenough	= Cockleshell Gully Fm	

	Upper			Mungaroo Fm		Lesueur Ss		
TRIASSIC	Middle	Erskine Ss		Locker Sh		Woodada Fm		
	Lower	Blina Sh Culvida Ss			Kockatea Sh Wittecarra Ss	Kockatea Sh		
PERMIAN	Upper	Liveringa Gp				Wagina Ss	Cardiff Mb Collieburn Mb	
	Lower	Noonkanbah Fm Poole Ss Nura Nura Mb Grant Gp	Keep Inlet Beds	Kennedy Gp { Binthalya S.Gp, Mungadan Ss, Coolkilya Grwk } Byro Gp { Baker Fm, Nalbia Grwk, Wandagee Fm, Quinnanie Sh, Cundlego Fm, Bulgadoo Sh, Mallens Grwk, Coyrie Fm } Wooramel Gp { Billidee Fm, Moogooloo Ss, Cordalia Grwk } Callytharra Fm Lyons Fm	Lyons Fm	Carynginia Fm Irwin River Coal M High Cliff Ss Fossil Cliff Mb Holmwood Sh Nangetty Fm	Collie Coal M = Sue Coal M Ewington Mb Stockton Fm	Paterson Fm
CARBON-IFEROUS	Upper	Anderson Fm	Border Creek Fm Point Spring Ss	Harris Ss Yindagindy Fm				
	Lower	Fairfield Gp	Milligans Beds Septimus Ls Enga Ss	Williambury Fm Moogooree Ls				
DEVONIAN	Upper	Virgin Hills Fm Sadler & Pillara Ls Mellinjerie Ls	Burt Range Fm Ninbing Ls Cockatoo Fm	Willaraddie Fm Munabia Ss				
	Middle	Poulton Fm		Gneudna Fm Nannyarra Grwk				
	Lower	Tandalgoo Red Beds						Wanna Beds Lennis Ss
SILURIAN	Upper			Dirk Hartog Fm				
	Lower	Carribuddy Fm		Tumblagooda Ss	Tumblagooda Ss	Tumblagooda Ss		
ORDOVICIAN	Middle	Goldwyer Fm Willara Fm						Table Hill Volc
	Lower	Gap Creek Fm Emanuel Fm	Pander Grnsd					
CAMBRIAN	Upper		Carlton Gp					
	Middle		Tarrara Fm					
	Lower		Antrim Plateau Volc					

ABBREVIATIONS
Fm: Formation; Gp: Group; Mb: Member; S.Gp: Sub-Group; Car: Calcarenite; Clst: Claystone; Clut: Calcilutite; Cg: Conglomerate; Coal M: Coal Measures; Dolo: Dolomite; Grnsd: Greensand; Grwk: Greywacke; Radl: Radiolarite; Ls: Limestone; Ss: Sandstone; Sh: Shale; Sltst: Siltstone; Volc: Volcanics

Soils and drift sands. Western Australia, an area of 2 525 500 km^2 extending from lat. 13° 44′S to lat. 35° 08′S, although having little variety in its broad physical features, has a very considerable variation in climates, from the tropical areas of summer rainfall in the north through a central and inland province of low rainfall to the temperate areas of winter rainfall in the south. Moreover, throughout this enormous area there is very considerable variation in the nature of the country rocks. As the nature of the soils that are developed is dependent on these two factors, climate and parent rock, it is apparent that there will be very considerable variation in the soils over this extensive area.

Soils resulting from the weathering of rocks *in situ* will contain a concentration of the weather-resistant heavy minerals of the parent rocks. Such minerals include economically important minerals such as gold and cassiterite (tin ore). Such residual accumulations are called eluvial deposits. If the material is transported by running water, there may be sorting of heavy from light minerals, and such heavy mineral accumulations are called alluvial deposits. Most of the so-called 'alluvial' gold deposits of Western Australia are not true alluvials, but are residual eluvial deposits.

There are considerable areas of Western Australia covered by drift sand, which may be in the form of parallel red sand dunes or, in the southern part of the state, extensive sandy plains, probably residual from the weathering of granite. These sandplains are often underlain by lateritic material and they probably represent the leached A-horizon of the laterite profile. These extensive sand-covered and highly-weathered areas are of significance to the mineral industry insofar as they obscure the underlying rocks and make exploration more difficult.

Coastal sand deposits are of considerable economic importance. Along the south and west coasts there are beach-sand deposits, in which there is a natural concentration of heavy minerals such as zircon, monazite, xenotime, rutile and ilmenite. Such deposits are at present being exploited at Capel, Bunbury, and Eneabba for their ilmenite content, which is valuable because of its low chrome content. Other heavy minerals such as zircon, monazite and xenotime are also being exported overseas.

Salt-lake deposits. These, together with the coastal sand deposits, represent the youngest of the geological formations developed; indeed, they are in course of formation at the present time. They are evaporite deposits resulting from the evaporation of lake waters in the areas of internal drainage. Soluble salts produced by rock-weathering are leached out by rain and running water and transported by streams to these lakes. During the long dry summers most of these lakes dry up and the soluble salts are deposited, yielding accumulations of gypsum and common salt. In a few of these inland lakes (e.g. Lake Campion) hydrated potassium aluminium sulphate (alunite), which is a valuable source of potash for fertilizers, has been formed. Valuable potash reserves occur at Lake MacLeod, north of Carnarvon. This 2700-km^2 coastal lake contains marine brines rich in potash salts (the last salts to crystallize on the evaporation of sea-water) saturating the mud of the lake floor.

More complete descriptions of the geology of Western Australia are contained in 'The Geology of Western Australia' (*Geol. Surv. West. Aust. Mem. No. 2,* 1975) and Clarke *et al.,* 1971.

Table 2.2

Geological events and mineral deposits in Western Australia

Era	*Main geological events*	*Economic mineral deposits*
CAINOZOIC	15. Weathering and erosion (Present day)	Beach sand deposits, salt, gypsum, sands and clays, peat, alluvial deposits (gold, tin, etc.), unconfined water
	14. Sedimentation (Pleistocene and Holocene)	Ilmenite and other black sand minerals, limestone
	13. Peneplanation (mid-Tertiary) and laterite formation	Bauxite Iron and manganese ores Alluvial tin and gold Clays
	12. Sedimentation (older-Tertiary)	Confined water
MESOZOIC	11. Sedimentation (with basalt outflows in Cretaceous)	Confined water Oil and gas Coal Basalt for aggregate stone
PALAEOZOIC	10. Sedimentation, earth movements, periods of erosion	Coal Oil and gas
PROTEROZOIC	9. Basic igneous intrusions	Road metal (blue metal) Lead, zinc and copper
	8. Sedimentation and minor granitic magma intrusions	Iron-ore (of Yampi Sound) Alluvial gold (of Nullagine and Telfer)
	7. Chemical sediments (banded iron-formations)	Iron-ore and blue asbestos (in Hamersley Basin)
ARCHAEAN	6. Pegmatite and quartz vein formation from end-stages of granitic intrusions	Tin, tantalum, tungsten, beryllium, lithium minerals Gold and silver in early Archaean country rocks (1 and 2)
	5. Intrusion of granitic magma (2700 million years ago)	Aggregate and building stone
	4. Granitization—conversion of pre-existing rocks into granitic rocks Contemporaneous with intense folding, fracturing and metamorphism of pre-existing rocks	Aggregate and building stone
	3. Intrusion of basic magma forming stratiform layered basic/ultrabasic igneous complexes	Nickel, copper and chromium in ultrabasic rocks, gold
	2. Sedimentation including banded iron-formation	Iron-ore
	1. Eruption of submarine basaltic lavas, acid volcanics and chemical deposition of banded ferruginous cherts	Copper in acid volcanics Iron-ore

THE MINERAL DEPOSITS

Mineral deposits are directly related to the geology—the nature of the rocks and their structural relationships—of the areas in which they occur. Table 2.2 summarizes the sequence of events represented in the geological history of Western Australia and the mineral deposits associated with each. This table, read in conjunction with a geological map of any area, will indicate the economically important deposits which could possibly occur in that particular geological environment.

The ore deposits of Western Australia are dealt with in the next chapter. Throughout this book many unfamiliar localities may be mentioned, and the reader is advised to consult the *Map and Gazetteer of Western Australia,* a publication of the Mines Department of Western Australia. The map contained in this publication records all economically important mineral occurrences reported till 1970. For more comprehensive information on individual ore deposits in Western Australia the reader should consult Knight (1975, 1976), Leslie *et al.* (1976) and Traves and King (1975).

REFERENCES

Clarke, E. de C., Prider, R. T. and Teichert, C. 1971. *Elements of Geology for Australian Students,* 4th rev. ed. 215-321. Nedlands: Univ. W. Aust. Press.

Geological Survey of Western Australia, 1975. *Geology of Western Australia, W. Aust. geol. Surv. Mem.* **2,** 541 pp.

Knight, C. L. 1975. *Economic Geology of Australia and Papua New Guinea—1. Metals.* Australas. Inst. Min. Metall. Monogr. Ser. 5, 1126 pp.

——— 1976. *Economic Geology of Australia and Papua New Guinea—4. Industrial Minerals and Rocks.* Australas. Inst. Min. Metall. Monogr. Ser. 8, 423 pp.

Leslie, R. B., Evans, H. J. & Knight, C. L. 1976. *Economic Geology of Australia and Papua New Guinea—3. Petroleum.* Australas. Inst. Min. Metall. Monogr. Ser. 7, 541 pp.

Mines Department of Western Australia, 1970. *Map and Gazetteer of Western Australia,* 36 pp., 1 map.

Prider, R. T. 1965. Geology and mineralization of the Western Australian Shield. In *Geology of Australian Ore Deposits.* 8th Comm. Min. Metall. Congress (ed. J. McAndrew): 56-65.

Traves, D. M. & King, D. 1975. *Economic Geology of Australia and Papua New Guinea—2. Coal.* Australas. Inst. Min. Metall. Monogr. Ser. 6, 398 pp.

CHAPTER 3

MINERAL DEPOSITS OF WESTERN AUSTRALIA

In this chapter the main minerals produced in Western Australia are considered from the aspects of historical development, geological occurrence and contribution to the State's mineral production. It will not be possible in the space available here to deal with mineral occurrences in the detail warranted but works referred to in the text will lead readers to more extensive accounts of ore deposits in which they may be particularly interested.

3 A

Copper, lead, zinc and tin

J. G. Blockley
Geological Survey of Western Australia

HISTORICAL BACKGROUND

The metals copper, lead, zinc and tin have played an important part in helping man attain his present civilized state. Their use in tools and weapons before the discovery of efficient methods of smelting iron-ore enabled man to cultivate crops, hunt game and resist enemies more readily than had his earlier implements of stone and wood.

Copper was probably the first metal to be used in tools, for it was available to early man in its native form and could be readily reduced from its ores. There are records of copper being used in Egypt as early as 12 000 BC, and by 4000 BC its use was widespread in Europe. The obvious disadvantage of copper, its softness, was overcome by the discovery of bronze, an alloy of copper and tin. This alloy was probably made in the first place by the accidental smelting of tin-bearing copper ore, but there is evidence that even in ancient times bronzes with different properties were deliberately made by varying the proportion of tin in the metal.

Lead pans were used to hold plants in the hanging gardens of Babylon (600 BC) and lead water-pipes were widely used by the ancient Romans.

Zinc, which is more difficult to extract from its ores than either of the other metals, was probably unknown as a metal in ancient times, although the Romans made brass, an alloy of zinc and copper, by adding calamine (zinc carbonate) to copper ore during smelting. Commercial production of zinc metal did not commence in Europe until about 1740.

Tin was probably smelted from its ores before 3500 BC, but supplies seem to have been small until 1500 BC. Phoenician trading-ships carried tin ingots from Britain to the Middle East in 500 BC. Tin metal was used in ancient Egypt for making ornaments, and on at least one occasion for wrapping a mummy. The use of tin for coating steel seems to have been discovered in Bohemia in the thirteenth century, where tin plate was used for making parts of armour.

Following the collapse of the Roman empire, there was a general decline in the mining of all metals—a decline which persisted, except for the exploitation of precious metals during the settlement of the Americas, until the beginning of the Industrial Revolution. The increased demand for metals needed by growing industries coincided with the early years of the Western Australian Colony, and no doubt prompted the settlers to bear in mind the possibility of establishing a profitable mining industry.

Copper, because of its ductility and high electrical conductivity, finds its greatest use in wire and cables in the electrical and communications industries. Other important uses are in car radiators, alloys, chemicals and fertilizers.

The greatest use of lead is in storage batteries, but the metal is also employed in pipes, sheets, ammunition, solder and type-metal. Lead-based compounds are used as pigments and as additives to motor spirit.

Zinc is used chiefly to make galvanized iron and alloys such as brass, die metal and casting metals. Its compounds are used in rubber, paints and ceramics.

About half of the world's tin output is made into tinplate, which is used mainly for manufacturing 'tin' cans. Most of the remainder is used in solder, bearing-metal, bronze and in various compounds employed in the ceramic and textile industries.

COPPER, LEAD AND ZINC

HISTORY

The earliest mining venture in Western Australia was an unsuccessful attempt in 1846 to mine small copper and lead-zinc deposits in the Darling Range near Armadale. The first commercial mineral deposit in the state was a vein of galena found on 16 October 1848 in the bed of the Murchison River by J. Walcott, a member of an expedition led by the explorer A. C. Gregory. The find was quickly developed as the Geraldine lead mine and by 1850 the colony was exporting lead concentrates. By 1853 a lead smelter was erected at a site now known as the Warribanno Chimney (Fig. 3.1), 5 km south of the mine, and pig lead was added to the export list. Further deposits of lead and some of copper near the Geraldine mine and at White Peak were also developed in the early 1850s.

Meanwhile, Gregory's favourable report on the mineral and agricultural potential of the country attracted a number of settlers to the area north of Geraldton, then called Champion Bay. Among these was James Drummond, son of the colony's first botanist, who took up a squatting lease on the Bowes River. In 1855 a shepherd employed by Drummond found copper ore near Wanerenooka Spring, and in the

Fig. 3.1 The Warribanno Chimney, 8 km north of Ajana, site of the state's first lead smelter. The chimney is built over a shaft which connects to a tunnel driven in from the side of the hill.

following year the Wanerenooka mine was established. Other discoveries of copper and lead followed quickly, and the centre of mining activity shifted from Geraldine to this new locality, which in 1864 was named Northampton. Skilled miners were brought in from Cornwall and Wales, and by the late 1860s, Northampton was a flourishing town producing a considerable part of the colony's export earnings. When in the early 1880s most of the mines closed because of falling metal prices, many of the miners became farmers and helped make Northampton the centre of an important agricultural district.

The discovery of payable gold in the Kimberley region in 1885 attracted new interest to Western Australia's mineral potential, and resulted in a wave of prospecting which extended to almost all parts of the state. Although mainly interested in gold, these prospectors found a large number of deposits of copper and lead-zinc ores. The more important discoveries of copper were made at Whim Creek (1888), Murrin Murrin (1898), and Ravensthorpe (1899); and those of lead-zinc at Uaroo, Braeside and Narlarla, all in 1901.

As a result of these discoveries, the 1890s and 1900s saw a period of active copper-mining, but many of the lead deposits, because of their remoteness, could not be mined economically at that time. Their development had to await the higher metal prices and the cheaper mining and transport methods which became available after the Second World War, especially during the 1950s. These factors also allowed some of the old Northampton mines to reopen after the War, as they had previously done during and after the First World War, when demand for lead was high.

The expansion of agriculture into sandplain areas in the 1950s and 1960s created a need for fertilizers carrying trace element additives, particularly copper. This provided a local market for copper ores which had previously been shipped overseas or to the eastern states, and allowed many smaller deposits to be developed for the first time, and some older mining areas to be reopened. The more important new mines were Thaduna and Copper Hills, and of the reopened mines those at Ravensthorpe and Kumarina were most productive. The Ravensthorpe mines, which also produced copper concentrates for smelting, remained open until about 1971, but most of the other mines closed in the late 1960s.

Following the decline in nickel exploration after 1971, many companies began searching for other metals. In particular, it was recognized that the state's Archaean greenstone belts might contain stratabound copper-zinc deposits similar to those in the Canadian Shield. The subsequent search disclosed several such deposits, of which the more important were Golden Grove and Teutonic Bore. Some older deposits, such as Mons Cupri and Murrin Murrin, were also recognized as belonging to this important class of mineralization, and further attention was focused upon them.

At present the only copper produced in Western Australia comes as a by-product of nickel mining. There is some intermittent production of lead and zinc from the Northampton Field, but the only major development based on these metals is taking place at Sorby Hills in the Kimberley Division.

Annual amounts of copper, lead and zinc mined from Western Australia are

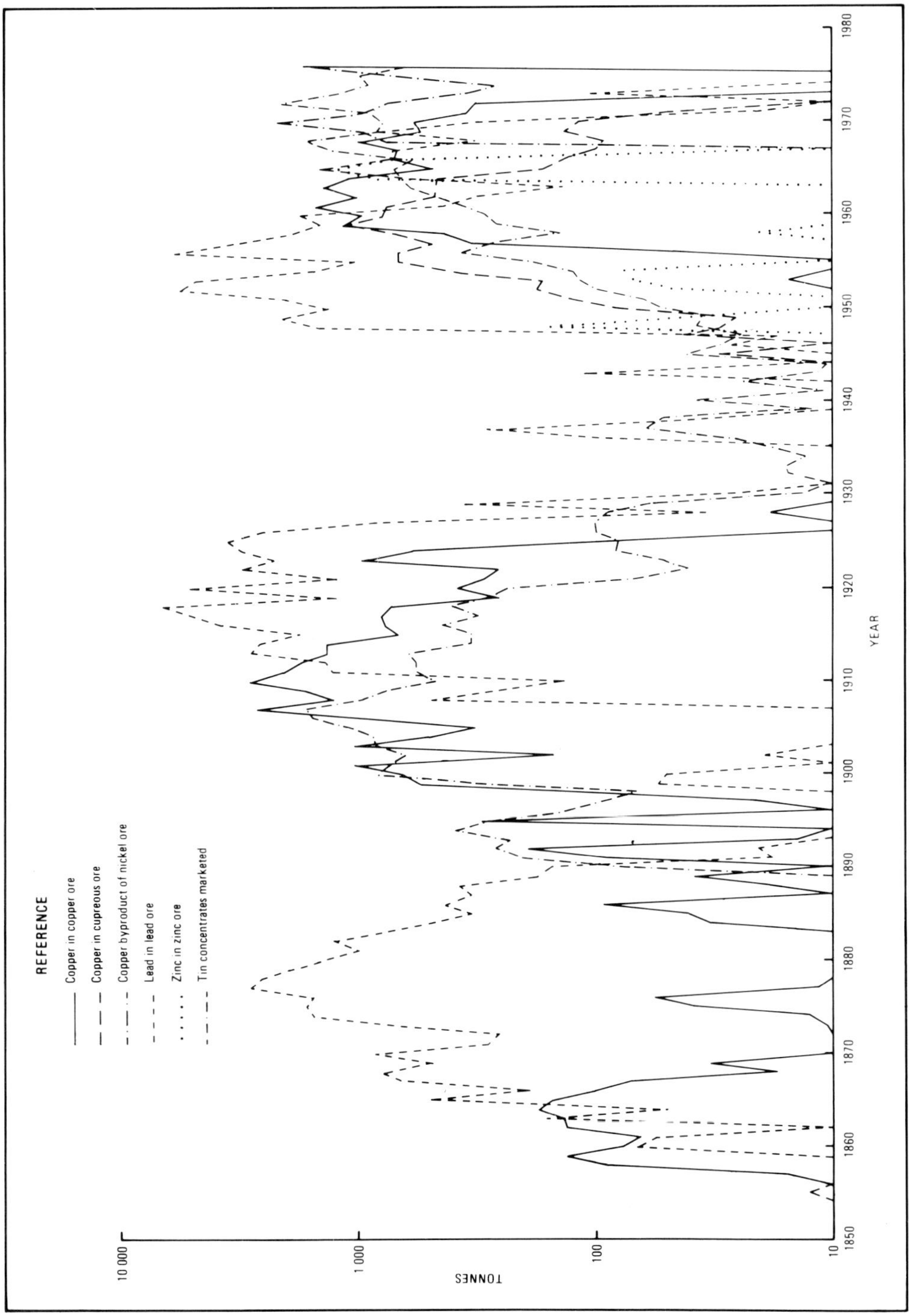

Fig. 3.2 Annual productions of copper, lead, zinc and tin concentrates, 1850-1976

shown graphically in Fig. 3.2. Productions from each of the more important deposits or centres are shown in Table 3.1 which also gives state and Australian totals.

It can be seen from Fig. 3.2 that production of these metals has been intermittent, rising to sharp peaks at times of high demand and falling to almost nothing when metal prices are low. As a result the mining of copper, lead and zinc has had little permanent effect on the distribution of population within the state. The only long-standing settlements resulting from the discovery of these metals have been Northampton and Ravensthorpe, and in each case, these have had alternative industries such as farming to sustain them between periods of mineral production.

When the Western Australian production of copper, lead, and zinc is compared to the Australian totals, it can be appreciated that this state has played only a minor role in the extraction of these metals.

Table 3.1

Production of copper, lead and zinc from principal deposits in Western Australia, with state and Australian totals

Deposit	*Copper (tonnes)*	*Lead (tonnes)*	*Zinc (tonnes)*	*Associated metals*
Braeside		3 200	25	920 kg Ag
Copper Hills	1 800			
Gabanintha	350			790 kg Au
Ilgararie	790			
Kooline		2 700		825 kg Ag
Kumarina	580			
Mons Cupri	200			
Murrin Murrin	4 600			3 385 kg Au
Narlarla		2 110	2 870	1 160 kg Ag
Northampton	3 700(1)	100 000(1)	40	165 kg Ag
Ravensthorpe	20 100			3 960 kg Au
Thaduna	2 350			2 520 kg Ag
Uaroo		2 170		1 210 kg Ag
Whim Creek	11 350			
Whundo-Yanrey	2 500			
Total W. Aust.	59 000(1, 2)	111 000(1)	2 950	
Total Australia (3)	4 290 000	19 760 000	16 050 000	

(1) Includes estimates of early Northampton production.
(2) Includes about 8000 tonnes of copper derived as a by-product of nickel mining.
(3) To end of 1974.

Table 3.2

Common ore minerals of copper, lead and zinc in Western Australia

	COPPER			LEAD			ZINC		
	Mineral	*Composition*	*% Cu**	*Mineral*	*Composition*	*% Pb**	*Mineral*	*Composition*	*% Zn**
Oxidized ores	Malachite	$Cu_2CO_3(OH)_2$	57	Cerussite	$PbCO_3$	77	Smithsonite	$ZnCO_3$	52
	Azurite	$Cu(CO_3)_2(OH)_2$	55	Anglesite	$PbSO_4$	68	Hydrozincite	$Zn_5(CO_3)_2(OH)_6$	59
	Cuprite	Cu_2O	89	Pyromorphite	$Pb_5(PO_4)_3Cl$	76			
	Chrysocolla	$Cu_2H_2(Si_2O_5)(OH)_4$	36						
	Atacamite	$Cu_2Cl(OH)_3$	59						
	Native Copper	Cu	100						
Supergene ores	Chalcocite	Cu_2S	80						
	Covellite	CuS	66						
Primary ores	Chalcopyrite	$CuFeS_2$	34	Galena	PbS	87	Sphalerite	ZnS	67
	Bornite	Cu_5FeS_4	63						

* Maximum content of pure mineral

NATURE OF THE ORES

The common ore minerals of copper, lead and zinc found in Western Australia are listed in Table 3.2. Primary ores, unaffected by weathering processes, consist of sulphides of the commercial metal variously mixed with iron sulphides, quartz, altered rock material and, in places, barite or calcite. Oxidized ores, mainly carbonates and oxides, are found in the upper parts of orebodies, where oxygen-charged ground-water has oxidized the primary sulphide minerals. Supergene ores are found in many copper deposits in Western Australia. They form where copper leached by acidic ground-water from the near-surface part of a lode is redeposited lower in the ore channel to enrich the copper present in the primary ore.

Lead and zinc ores commonly occur together and in most places contain payable amounts of silver, although the Northampton lead-zinc ores are unusually poor in this metal. Copper occurs in several associations. The more important in this state are with lead-zinc ores as at Northampton, with gold ores as at Murrin Murrin and Ravensthorpe, or with zinc and minor lead as at Golden Grove.

TYPES OF DEPOSITS

Ores of copper, lead and zinc are commonly found in a number of different types of deposits. In Western Australia their most common occurrence is as vein deposits, in which the ore minerals, together with iron sulphides and gangue, are deposited in open fissures, or replace crushed and brecciated rock material along faults. Such deposits are younger than their host rock.

Important quantities of copper ores have been won from stratabound deposits in Archaean rocks. In these deposits the ore is generally parallel to the bedding or volcanic layering of the host rock, and was probably introduced at the same time as the host rock was formed. A special category of stratabound deposits, generally known as the Mississippi Valley type from the area in which it is best described, is represented by a few deposits in the Kimberley region. These are lead-zinc-silver orebodies formed as cavity fillings or replacements in limestone or dolomite.

Disseminated deposits in which the ore minerals appear as isolated grains or in numerous small veinlets throughout the host rock have been relatively unimportant in Western Australia, but may assume a greater role in the future.

DISTRIBUTION AND GEOLOGY

Fig. 3.3 shows the principal deposits of copper, lead and zinc in Western Australia superimposed on an outline map of the state's geology. With two exceptions, the deposits lie within Precambrian rocks, with copper being confined mainly to rocks of Archaean age, and lead-zinc to those belonging to the Proterozoic era. The exceptions are deposits of lead-zinc minerals found in Palaeozoic limestone in the Canning and Bonaparte Gulf Basins.

The following brief descriptions, grouped by deposit types, summarize the principal geological features of the more important centres of past production and of

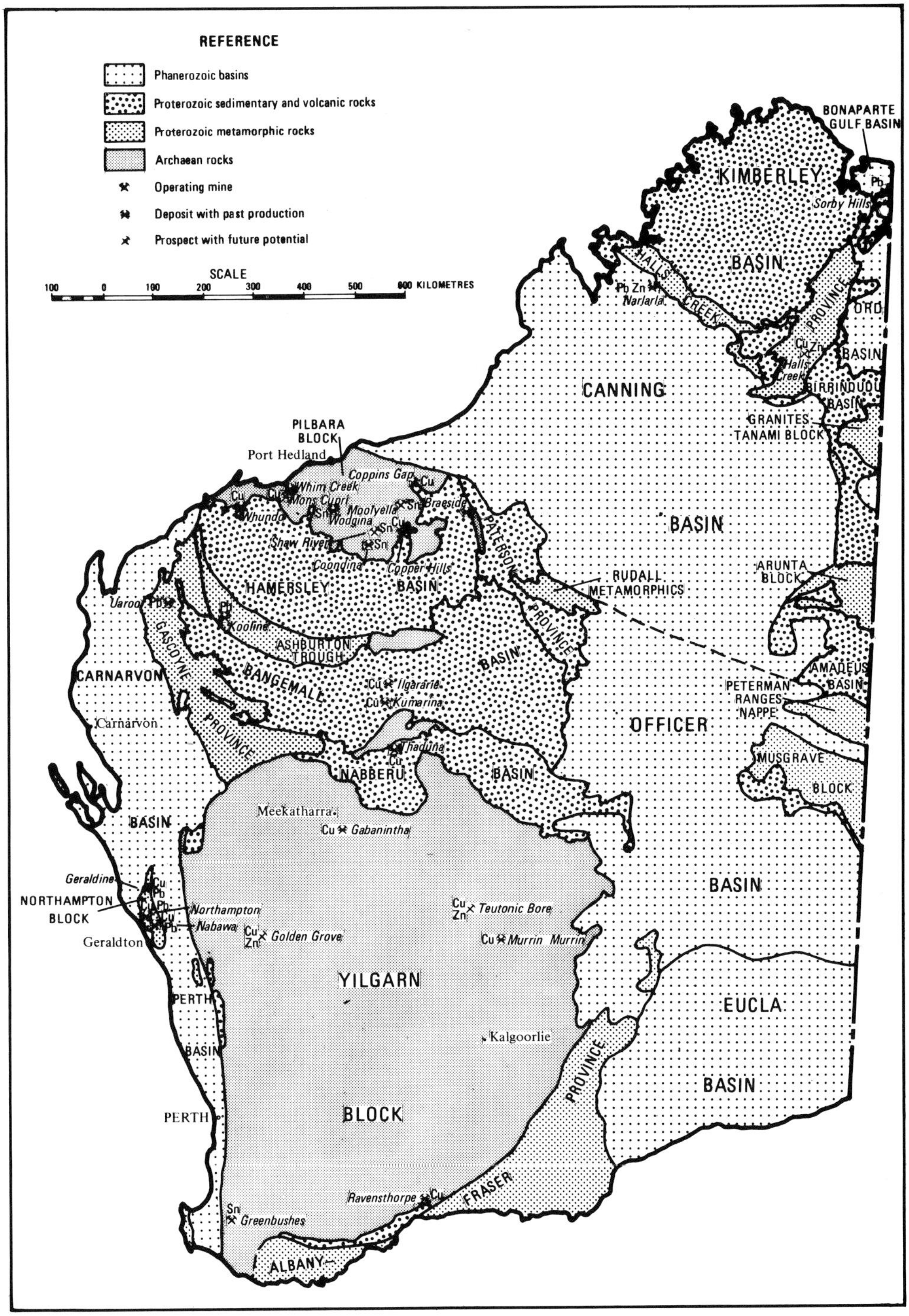

Fig. 3.3 Western Australia, showing outline of geology and principal copper, lead, zinc and tin deposits

some newer prospects. Deposits of these metals are widespread in Western Australia and more detailed information can be found in Low (1963), Blockley (1971) and Marston (in press).

(i) *Vein deposits*

The Northampton Mineral Field includes the mining centres of Northampton, Galena (formerly Geraldine) and Nabawa. Early production records are incomplete, but the field is estimated to have yielded about 100 000 tonnes of lead and 3700 tonnes of copper. It lies in the fault-bounded Northampton Block of Middle Proterozoic granulite, gneiss and intrusive dolerite. The lead and copper ores occupy fissures and breccias formed on a series of steeply-dipping, mainly north-northeasterly-striking faults. Many of the orebodies are alongside or within dolerite dykes which follow the same fracture system as the ore deposits. This had led several writers to postulate a genetic relationship between ore and dolerite.

The lead lodes consist of galena and sphalerite, with minor marcasite and chalcopyrite in a quartz gangue which may also contain barite and calcite. The copper lodes contain chalcopyrite in the primary zone, but most of the ore mined has come from the supergene or oxidized parts of the lodes.

The larger mines of the field are the Baddera at Northampton, the Surprise and Mary Springs at Galena, and the Narra Tarra and Protheroe at Nabawa. Each of these mines has produced, or has the potential to produce, 12 000 to 15 000 tonnes of lead. The largest copper mine was the Narra Tarra, which yielded about 1800 tonnes of copper during the 1920s.

The copper-gold deposits of Ravensthorpe in the Yilgarn Block have yielded about 19 000 tonnes of copper from quartz veins cutting metamorphosed Archaean lavas and sediments near the eastern edge of a granodiorite batholith. The primary copper ore comprises quartz, gold and chalcopyrite, with minor pyrite and pyrrhotite. In the oxidized zone the ores mined were mainly ferruginous copper carbonates. Supergene ores were not important at Ravensthorpe.

At Braeside in the northeastern part of the Hamersley Basin, orebodies containing galena, sphalerite and minor pyrite and chalcopyrite within quartz veins occur in basalt and tuff of the Proterozoic Fortescue Group. Most of the ore won from the field came from the Ragged Hills mine.

During the 1950s about 30 small lead-mines worked at Kooline in the Ashburton Trough yielded 2700 tonnes of lead and 825 kg of silver. The orebodies are galena-bearing quartz veins, with calcite, barite and traces of copper minerals, occurring mainly along the cleavage planes of metamorphosed greywacke and shale.

Most of the lead produced from the Uaroo centre in the northwestern Gascoyne Province came from the Silver King mine. This deposit was worked in the early part of the present century, its high silver content making it economic despite its remote location. The ore follows a quartz vein cutting metasedimentary rocks.

At Copper Hills about 15 000 tonnes of cupreous ore containing 1800 tonnes of copper were mined from faults cutting acid tuffs and porphyries in the southern part of the Pilbara Block. Another important source of cupreous ore was the Thaduna mine, which between about 1958 and 1971 yielded 33 550 tonnes of ore containing 2350 tonnes of copper. The lode is a fault zone cutting siltstone and greywacke in the Nabberu Basin. It is unusual in having a high graphite content, which impedes normal methods of concentrating the ore by flotation. The chief ore minerals were malachite, azurite, cuprite, and in some of the deeper openings chalcocite. The primary ore proved to be uneconomic when tested by drilling.

The Kumarina and Ilgararie deposits follow fault zones cutting sandstone and shale of the Bangemall Group. Some lodes follow the contacts of dolerite dykes. Most of the ore mined comprised malachite, azurite, cuprite and chalcocite.

The copper-gold deposits of Gabanintha are typical of many in the northern part of the Yilgarn Block. They are quartz veins cutting metabasalt and amphibolite close to a granite/greenstone contact. Most of the copper ore came from the oxidized and supergene zones, the primary ore generally being too low grade to be mined.

(ii) *Stratabound deposits*

The best documented stratabound deposit in Western Australia is the Whim Creek copper mine in the western Pilbara Block. This mine yielded about 11 350 tonnes of copper from a conformable mineralized zone 10 to 14 m thick in Archaean slate and siltstone deposited between two episodes of volcanism (Reynolds *et al.* 1975). The primary ore minerals are pyrite and/or pyrrhotite, chalcopyrite, sphalerite and minor galena. About two-thirds of the deposit has been altered by weathering to malachite, azurite, cuprite and chrysocolla. The orebody probably originated as a subaqueous hot-spring deposit related to the last stage of a volcanic cycle.

After Whim Creek and Ravensthorpe, the copper deposits at Murrin Murrin have been the most important in the state. About 4050 tonnes of copper were extracted from conformable lenses of gold, chalcopyrite and sphalerite-bearing massive pyrite in a sequence of basic-to-acid lavas and associated sediments. Most of the ore mined came from the oxidized and supergene zones of the orebodies. The grade became uneconomic in the primary zone, although some of the deposits were subsequently worked for pyrite as a source of sulphur for manufacturing sulphuric acid used in making superphosphate fertilizer.

Small stratabound deposits of copper-zinc ore in the Whundo area of the West Pilbara have yielded about 2500 tonnes of copper. The orebodies are lenses of chalcopyrite and sphalerite-bearing pyrrhotite formed at the contact of a predominantly basic sequence of volcanic rocks below and a mainly acid sequence above. The greater part of the copper production came from a shoot of oxidized and supergene ore mined during 1976.

The Golden Grove deposit located south of Yalgoo is the most encouraging prospect found during the widespread search for copper-zinc deposits in the early 1970s. Following its discovery in 1971, it was investigated by a consortium of Australian and overseas companies which reported reserves of about 13.5 million tonnes averaging 3.6 per cent copper. The mineralization lies near the top of a volcanic pile within a 100 m-thick lode horizon occurring below a chert bed in a thick sequence of acid tuffs. In the primary zone, the mineralization comprises pyrite, magnetite, chalcopyrite, pyrrhotite and sphalerite, with minor galena. Malachite and azurite are present in the oxidized zone and chalcocite and digenite are found in some zinc-rich shoots. The mineralized shoots are broken up by faults and intrusions of dolerite and dacite. Although the prospective orebody is conformable to the enclosing rocks in the broad sense, there are local departures from conformability, owing to movement of the ore minerals during metamorphism of the deposit. The sulphide minerals were almost certainly deposited by late-stage metal-rich volcanic fluids at the time that the host rocks were formed.

The Teutonic Bore prospect was discovered during 1976 in a greenstone belt which had previously been the subject of intensive nickel exploration. Its only surface expression was an insignificant gossan cropping out in a largely soil-covered area. Its discovery highlights the principle that geologically prospective ground can rarely be said to be completely explored for minerals; new ideas or techniques may always result in new discoveries.

Stratabound copper-lead-zinc deposits are currently being investigated near Halls Creek. In the Little Mount Isa prospect, located about 34 km northeast of Halls Creek, the primary mineralization comprises pyrite and pyrrhotite, with sphalerite, galena and chalcopyrite. Geologically the deposit resembles the famous Mount Isa mine in Queensland.

The Narlarla lead-zinc deposits are the state's best known examples of the Mississippi Valley type of mineralization. They occur in an Upper Devonian limestone reef complex forming the Napier Range at the northern margin of the Canning Basin. The orebodies fill çavities of uncertain origin in the limestone. In the primary zone they consist of lenses and irregular masses of galena and sphalerite, with minor marcasite and copper minerals in a gangue of dolomite. Minerals in the oxidized zone included cerussite, smithsonite and hydrozincite.

The most significant prospect found during a widespread search for other Mississippi Valley type deposits in the Kimberley region is near Sorby Hills, in the Devonian of the Bonaparte Basin northeast of Kununurra. Evaluation of the lead mineralization in limestone is still in progress.

(iii) *Disseminated deposits*

The Mons Cupri deposit near Whim Creek yielded about 200 tonnes of copper from a network of copper-bearing chalcedony veins within a chloritized rhyolite breccia. The copper is considered to have been introduced by late-stage volcanic fluids trapped within the volcanic pile by an overlying bed of slate. A stratiform, lead-zinc-silver deposit which has not yet been mined lies stratigraphically above the copper.

At Coppins Gap, in the northern part of the Pilbara Block, drilling has outlined a deposit of disseminated copper-molybdenum mineralization within a stock of altered granodioritic porphyry which intrudes iron-formation, chert and metamorphosed volcanic rocks near the edge of a large granitic batholith. The bulk of the primary copper mineralization takes the form of chalcopyrite disseminated throughout the silicified granodiorite, but some is present along with molybdenite in quartz veins and as disseminations in metabasalt close to the porphyry. This prospect has similarities to the 'porphyry' copper-molybdenum deposits which yield a large portion of the world's supply of these metals.

TIN

History and Production

The first commercial discovery of tin ore in Western Australia was made in 1888 by Mr D. Stinton, who observed the mineral cassiterite near Greenbushes Well while hunting kangaroos. It is recorded that he had earlier been advised to prospect for tin in this general area by Government Geologist, E. Hardman. Stinton formed a syndicate to exploit his discovery and commenced production in 1889. Other prospectors quickly followed, but mining during the early years of the field was hampered by falling tin prices and unhelpful mining laws. However, at the turn of the century a rise in the price of tin and a change in mining legislation brought a sharp increase in activity, and production rose to 800 tonnes in 1906. As the shallower alluvial deposits were worked out, production fell steadily, until in 1921 it was only about 50 tonnes per year.

In the same year that Stinton made his Greenbushes discovery, prospectors were combing the newly found Pilbara fields in search of alluvial gold. At several places their panning dishes disclosed tin ore, but because of the overriding interest in gold and the legal difficulties in getting mining titles to minerals other than gold, no tin was mined until 1893. This first production came from Eley's find in the southern part of the Shaw River tin field.

Subsequent discoveries were made at Moolyella (1898), Cooglegong (1900) and Wodgina (1902). Production followed a similar pattern to that at Greenbushes, reaching a peak of over 800 tonnes in 1907 and then declining to less than 50 tonnes per year by 1920. Because of the shortage of water for mining and concentrating the ore, the early prospectors left much lower-grade ground untreated.

The slump in tin-mining which followed the First World War continued from about 1921 to the mid-1950s, most production during these three and a half decades coming from mines worked by individual prospectors and small syndicates.

In the years immediately after the Second World War there was a sharp increase in the price of tin and also a more general availability of earth-moving machinery. These factors combined to make the relatively shallow Pilbara deposits once more economical for mining, and during the mid-1950s resulted in a minor tin boom centred on Marble Bar. This renewed production reached its peak in 1967, when 1250 tonnes of tin concentrates were sold, but has since fallen to around 100 tonnes per year. One significant development was the discovery in 1966 of a new tin field at Coondina in an area assumed to have been reasonably well prospected in the past.

Although the 1950s saw some renewed interest in the Greenbushes field, the real revival there did not begin until a large bucket dredge was installed in about 1966.

hydraulic sluicing, dredging and earth-moving machinery to a maximum depth of about 15 m.

Most tin production from the Pilbara Block has come from the geologically similar Moolyella and Shaw River fields. Primary tin mineralization occurs in layered, albite-rich aplite/pegmatite dykes intruded as swarms around the margins of 'younger granites'. Only a few tonnes of tin concentrate have been mined from pegmatite, the greater part of the production coming from shallow alluvial deposits formed in creeks draining the tin-bearing pegmatites. These creek-beds and adjacent eluvial deposits are mined by means of earth-moving machinery and the ore is trucked to central treatment plants for processing. Mining methods such as hydraulic sluicing and dredging are not used in the Pilbara because of the general shortage of water.

The Coondina deposits were first reported in 1966, although they had apparently been worked on a small scale by Aborigines before that time (Fig. 3.4b). Earlier production came from several small creek-beds draining an area of layered aplite/pegmatites related to a 'younger granite', and by about 1968 the field was considered to have been worked out. However, the discovery of tin-bearing cemented alluvium in a fossil stream channel in 1969 gave new life to the field.

The Wodgina field is a roof pendant of greenstones within a large granitic batholith. Most of the tin produced came from albite pegmatites cutting metamorphosed basalt, ultramafic rocks and siliceous sediments. The most important producer was the Mount Cassiterite mine, where the richest ore was found in tourmaline-mica veins alongside pegmatite and quartz. Alluvial deposits were comparatively unimportant at Wodgina. In addition to tin Wodgina has been an important producer of tantalite and beryl; also some uranium-bearing minerals occur there.

Significant quantities of tin concentrate have beem won from pegmatites and associated alluvial deposits at Tabba Tabba, Strelley and Pilgangoora. Alluvial cassiterite has been mined at Abydos and Friendly Creek, the latter deposits comprising in part an older Tertiary stream channel cut by the present drainage.

Fig. 3.4 Contrasting methods of treating tin ore in the Pilbara Block.

Top: Cooglegong Tin Pty Ltd plant at Spear Hill. The ore is screened to remove larger stones then pumped as a slurry to the two revolving flat cones (left of photo), where concentration of the cassiterite is effected by the action of water washing the lighter material over the edge of the cone.

Bottom: Aboriginal prospector, Mollie Dann, uses the yandy dish to concentrate cassiterite from crushed ore. In earlier times Aboriginal women separated grass seeds from sand using the same method, which involves a dexterous movement of the wrists to 'throw' the lighter material away from the heavier.

REFERENCES

Blockley, J. G. 1971. The lead, zinc and silver deposits of Western Australia. *W. Aust. geol. Surv. Miner. Resour. Bull.* **9.**

——— in press, The tin deposits of Western Australia with special reference to the associated granites. *W. Aust. geol. Surv. Miner. Resour. Bull.* **12.**

Low, G. H. 1963. Copper deposits of Western Australia. *W. Aust. geol. Surv. Miner. Resour. Bull.* **8.**

Marston, R. J. (in press). Copper mineralization in Western Australia. *W. Aust. geol. Surv. Miner. Resour. Bull.* **13.**

Reynolds, D. G., Brook, W. A., Marshall, A. E., & Allchurch, P. D. 1975. Volcanogenic copper-zinc deposits in the Pilbara and Yilgarn Archaean Blocks. In *Economic Geology of Australia and Papua New Guinea—I. Metals* (ed. C. L. Knight). Australas. Inst. Min. Metall. Melbourne, Monogr. 5(1), 185-95.

3 B

Gold

R. Woodall and G. A. Travis
Western Mining Corporation Ltd

HISTORY

The discovery and development of the gold-mining industry was the basis for much of the early economic growth of Western Australia. It brought people, capital, water, railways and communications to the vast interior and created employment opportunities in both coastal cities and inland towns.

The discovery of gold in the Kimberley district near Halls Creek in 1885 encouraged an influx of prospectors from the goldfields of the eastern states. Widespread prospecting resulted in many new discoveries, and by 1890 gold was being mined in tbe Kimberley, Pilbara, Murchison, Ashburton and Yilgarn districts from small but rich finds.

The discovery of the bonanza of Bayley's Find at Coolgardie in 1892 started the great gold-rush that enticed prospectors in large numbers to the arid interior east of Southern Cross. New finds were made almost daily as prospectors fanned out from Coolgardie. In 1893 gold was found at Kalgoorlie, the field which was to yield subsequently more gold than the combined output of all other deposits in the state. By 1904 the prospectors had been so effective that almost all of the state's significant outcropping gold deposits had been found.

The succession of spectacular gold finds and rapidly increasing production fuelled an investment and stock-market boom. Most of the finance came from England, and between 1886 and 1907 British investors are estimated to have poured between \$80 and \$100 million into Australian gold-mines (Blainey, 1963). A great deal of this money came to Western Australia, and in the three years 1894-96 a total of 960 new Western Australian mining and prospecting companies were floated on the London Stock Exchange.

Although only a very small percentage of the investors gained the gold riches they sought, the economic benefits to the state were enormous. In 1881, 53 years after the establishment of the colony, the then agricultural and pastoral-based population had grown very slowly to a meagre 29 700 souls; but under the stimulus of gold, population increased over ninefold in 30 years to be 282 000 by 1911.

The history of the Western Australian gold industry is illustrated by Fig. 3.5, which shows the annual production from 1888, average grade for each year's production, and the dominant contribution of Kalgoorlie's Golden Mile to the total. The peak output of 72 600 kg was achieved in 1903. Total production to the end of

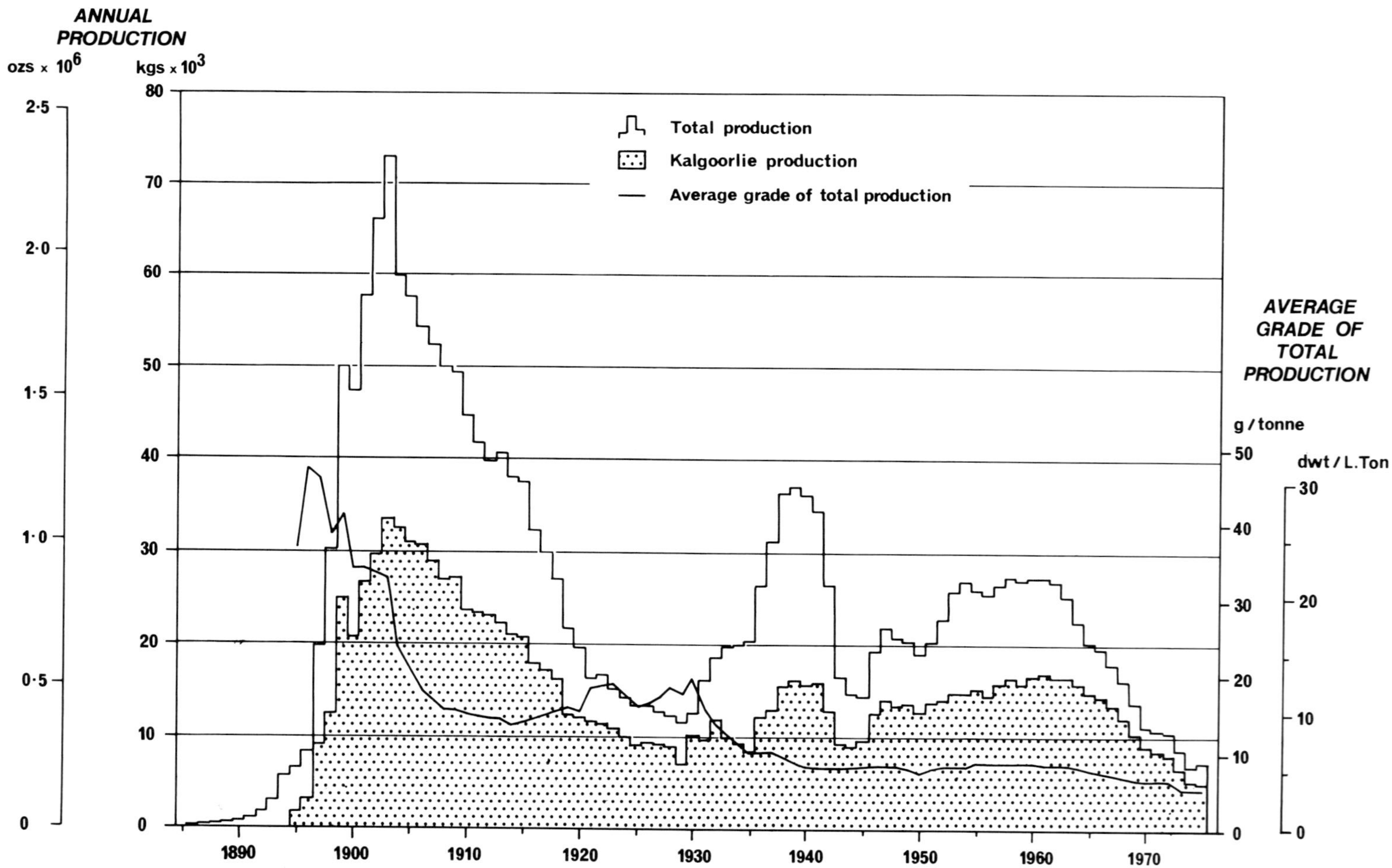

Fig. 3.5 History of gold production in Western Australia

1975 was 2.16 million kg, of which Kalgoorlie contributed 1.2 million kg, i.e. 52 per cent of the total.

Gold-mining declined after the 1914-18 war in the face of an essentially static gold price, a falling grade, and increased production costs, but when the price of gold almost doubled in the early 1930s, a resurgence took place in the industry. British capital again flowed into Australian gold-mines—to some extent for want of alternative investment opportunities in a period of world-wide depression—and the resulting revival in gold-mining once again eased economic hardship and reduced unemployment in the coastal cities. Many of Western Australia's larger gold-mines were re-equipped and production increased.

Output declined again during the 1939-45 war, but between 1946 and the early 1960s, a good deal of refurbishment of the gold-mines took place on the assumption that a rise in the gold price was imminent. Output began to increase once more. Unfortunately the price rise did not materialize until 1972, too late for many mines which were forced to close under the pressures of a fixed gold price and escalating costs.

Currently three important gold deposits are being mined with a combined annual production of 10 000 kg: The Mt Charlotte orebody at Kalgoorlie, the Norseman quartz reefs, and the Telfer deposits in the remote Paterson Range area east of Nullagine. These last deposits were located by surface prospecting in 1971 on the western edge of the Great Sandy Desert—one of the few outcropping gold occurrences missed by the early prospectors. Commissioned in 1976, this is the only new goldfield to be discovered since 1936.

Fig. 3.6 shows the history of the state's gold production with the influence of the Golden Mile removed. The gold production statistics essentially relate to primary gold production, alluvial gold production having been of little significance.

GENERAL GEOLOGY OF THE DEPOSITS

The state's gold deposits principally occur in Archaean rocks, in particular, in the greenstone belts of the Yilgarn Block (Fig. 3.7). Most of the deposits are transgressive, silicified lodes or quartz veins that have formed in structurally prepared sites in mafic volcanics and intrusives, although some important deposits also occur in siliceous, sedimentary units.

Sulphides may be a major constituent of the ore or extremely sparse, pyrite or pyrrhotite being the dominant sulphide. In general, the stratiform ores in sedimentary or tuffaceous rocks contain more sulphides than the transgressive quartz reefs and lodes. Gold is often found in association with arsenopyrite or galena; and chalcopyrite, sphalerite, tetrahedrite, and stibnite are often present in the ores. Gold tellurides are important at Kalgoorlie and have been observed in trace amounts at Norseman and Ora Banda.

Intrusive 'albite porphyries' are associated with many deposits, notably at Kalgoorlie, Norseman, Mt Magnet (Hill 50), Meekatharra, Coolgardie (Bayley's

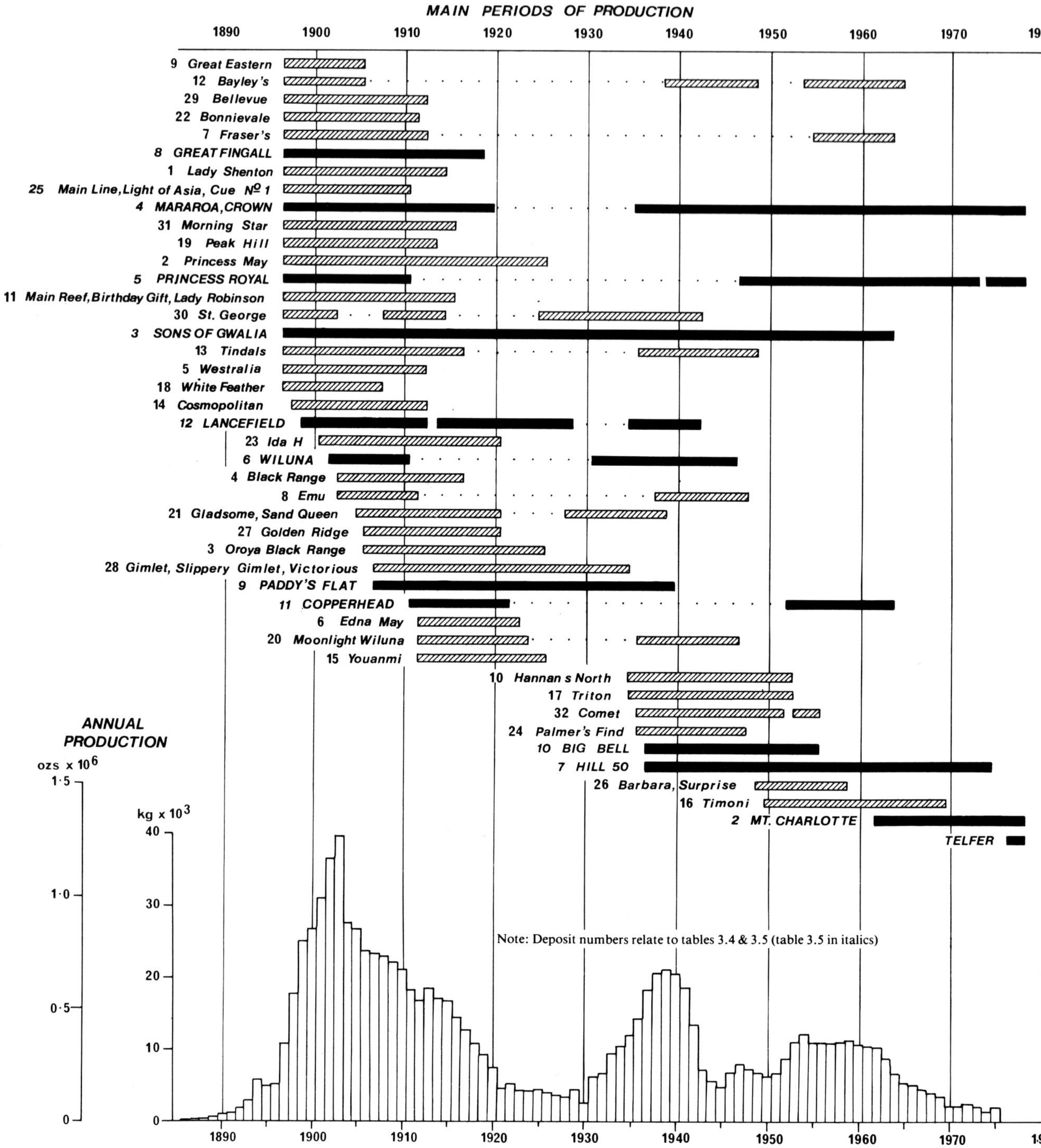

Fig. 3.6 History of gold production in Western Australia, excluding the Golden Mile

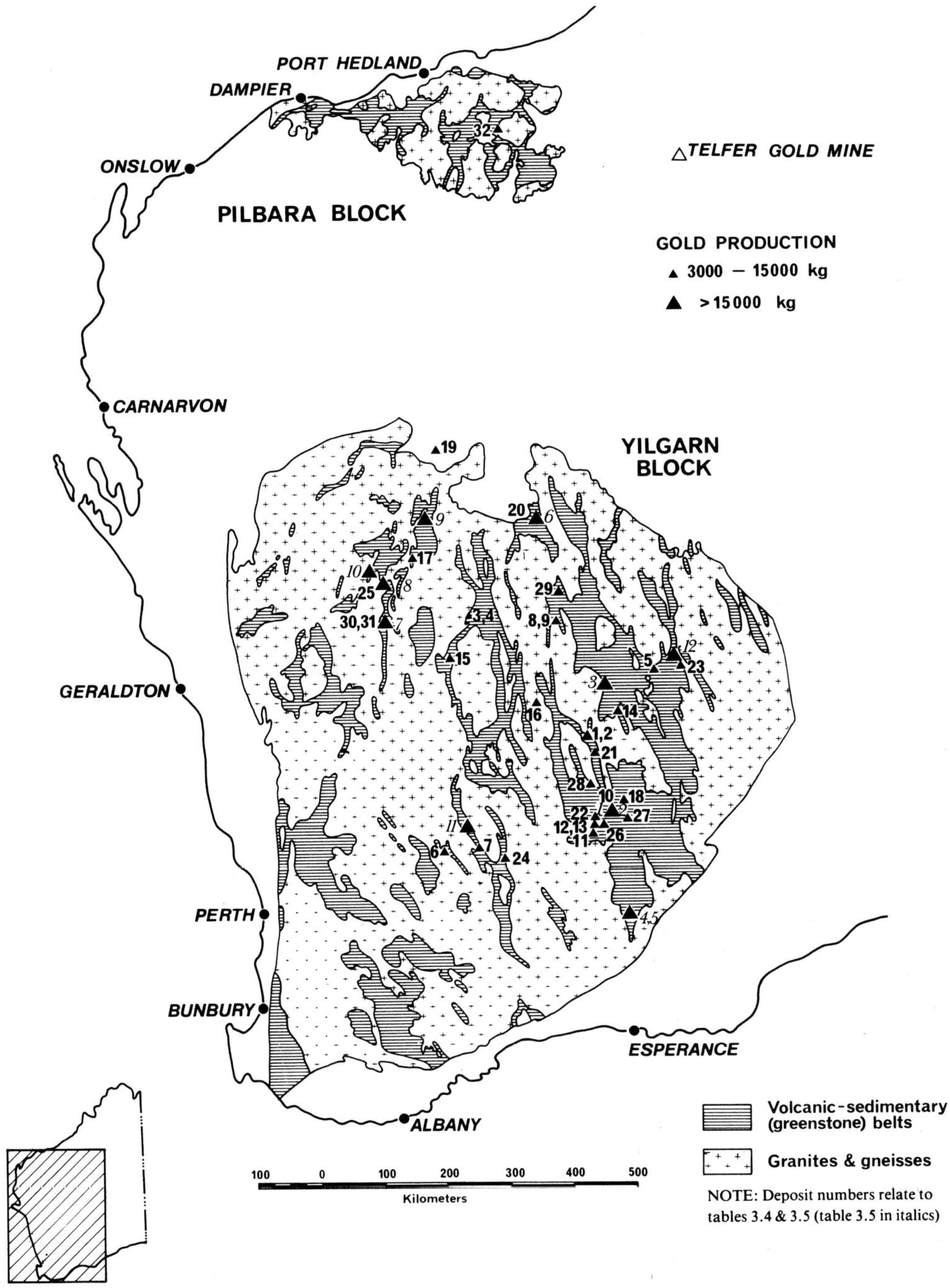

Fig. 3.7 Gold deposits in Western Australia

Table 3.4

Production statistics of major gold deposits of Western Australia—based on records of Western Australian Department of Mines to December 1975, except where otherwise indicated

Ref. No.	Mining centre	Main deposit	Tonnes ore	Kg gold	g/tonne	Discovered	Main producing period
1	Kalgoorlie	Golden Mile	94 487 192	1 120 556	11.9	1893	1895-1975
2	— —	Mt Charlotte	6 531 976	27 955	4.2	1893	1962-present
3	Leonora	Sons of Gwalia	7 254 395	82 025	11.3	1896	1897-1963
4	Norseman	Mararoa-Crown	5 187 211	63 431	12.2*	1892	1897-1919, 1936-present
5	— —	Princess Royal	1 395 079	31 380	22.5*	1892	1897-1910, 1947-1972, 1974-present
6	Wiluna	Wiluna	7 810 552	45 657	5.8	1904	1906-1910, 1931-1946
7	Mt Magnet	Hill 50	3 219 720	42 174	13.1	1898	1937-1974
8	Day Dawn	Great Fingall	1 927 219	38 550	20.0	1891	1897-1918
9	Meekatharra	Paddy's Flat	1 726 214	27 890	16.2	1894	1907-1939
10	Big Bell	Big Bell	5 094 357	23 077	4.1	1904	1937-1955
11	Bullfinch	Copperhead	3 935 524	20 552	5.2	1909	1911-1921, 1952-1963
12	Laverton	Lancefield	1 587 706	16 464	10.4	1897	1899-1912, 1914-1928, 1935-1942

* estimates by Central Norseman Gold Corporation N.L. to June 1975

and Tindals), Kanowna, Golden Ridge and Hampton Plains. Where petrological data are available, these rocks have been shown to be sodic dacites or sodic rhyolites.

The twelve gold deposits in Western Australia which have produced more than 15 000 kg of gold (Table 3.4 and Fig. 3.7) are briefly described in the following sections. Summary data on an additional 32 deposits which have produced in excess of 300 kg of gold are given in Table 3.5 and also located on Fig. 3.7.

Golden Mile, Kalgoorlie
(References: Campbell 1953*a*, Tomich 1959, Finucane 1965*a*, Woodall 1965, Travis *et al.* 1971).

More gold has been mined from this area then from any other in Western Australia and in terms of the world-wide occurrence of vein-lode gold mineralization it is second in size only to the Porcupine field in Canada. The mineralized zone, 4000 m long by 1200 m wide, and extending to depths of up to 1500 m, has long been known as the 'Golden Mile'. Most of the gold within this zone occurs in a complex system of silicified, pyritic shears and fractures in mafic rocks that occupy the central portion of a faulted syncline.

The bulk of production, perhaps 80 per cent, came from an altered mafic sill, the Golden Mile Dolerite. This is a complex differentiated intrusion of up to ten conformable units, which range in composition from an ultramafic layer adjacent to the lower chilled contact to a quartz-rich, granophyric unit in the upper section. Gold has also been mined from lodes in the underlying Paringa Basalt and from flatly-plunging fold- and fault-controlled oreshoots in sediments on the contact between the Paringa Basalt and Golden Mile Dolerite.

The Golden Mile Dolerite is overlain by black shales, which are part of a sedimentary-tuffaceous succession (Black Flag Beds). These rocks are barren.

Widespread and intense metasomatism is a feature of the mineralized area. The actinolite-albite-chlorite-epidote assemblage of the igneous host rocks is progressively altered to chlorite-calcite-albite assemblages and finally to sericite-ankerite-siderite adjacent to the lodes. Ore consists of carbonate-bearing, silicified zones containing 5-10 per cent fine-grained pyrite, plus minor tellurides, chalcopyrite, arsenopyrite, tetrahedrite, stibnite, sphalerite, and galena.

Most of the gold occurs as minute globules 0.5-20 microns in diameter in the pyrite, and about 15-20 per cent as gold tellurides.

Mt Charlotte, Kalgoorlie
(References: Haycraft 1965, Travis *et al.* 1971)

The deposit is a stockwork of quartz veins in the granophyric unit of the Golden Mile Dolerite. It occurs adjacent to an oblique fault similar in type to those which limit the Golden Mile deposit.

Although Mt Charlotte ore is different in appearance from the gold-telluride mineralization of the Golden Mile lodes, it is chemically and mineralogically similar except for its low tellurium and lower average gold content.

Sons of Gwalia, Leonora
(Reference: Finucane 1965*b*)

The deposit, much of which is mined out, is in a 150-m wide band of schist (chlorite-biotite-quartz-actinolite schist) in a thick sequence of basalts. The schist zone is probably a bedded tuffaceous zone in the volcanics.

The ore, which is quartz or quartz-impregnated schist with associated pyrite, chalcopyrite, pyrrhotite and arsenopyrite, occurs in a series of shoots over a strike length of 450-600 m and has been mined down-plunge for 1900 m. The oreshoots, most likely all in the one bed, are controlled by drag folds.

Mararoa-Crown and Princess Royal, Norseman
(References: Hall & Bekker 1965, Cameron, unpubl. data 1968, Christie unpubl. data 1969, Larking, unpubl. data 1969)

The Norseman gold belt is localized in the Wolyeenyer Group, a 10 000 m thick sequence of tholeiitic basalt, dolerite-gabbro intrusives and minor interbedded black shales.

The main deposits are quartz reefs in reverse faults that occur at about the same stratigraphic position in the succession over a strike length of 10 000 m. The three principal reefs are the Mararoa, Crown and Princess Royal.

The host rocks for mineralization are tholeiitic basalts and quartz gabbro. Wall-rock alteration consists of weak silica leaching and potash enrichment. Albite porphyries are abundant in the productive area.

Typical ore is laminated quartz containing collectively less than 1 per cent of sericite, green hydromica, ankerite or dolomite, pyrite, pyrrhotite, sphalerite and galena. The ore zones favour the flatter sections of the reef shears.

Wiluna
(References: Larcombe 1926, Edwards 1953)

The Wiluna gold deposits occur in a thick sequence of pillowed lavas which contains a quartz dolerite sill. Some flows are definitely basalts, but others are highly altered and there are insufficient data available for reliable classification. Rock alteration is similar but less intense than at Kalgoorlie. Most of the ore mined was adjacent to post-folding faults in the upper section of the quartz dolerite and in the overlying basalt.

The host rock for the mineralized lodes is silicified and carbonatized sheared basalt and dolerite. Pyrite and arsenopyrite are the dominant sulphides, each constituting 2-3 per cent of the ore. The close correlation between gold content and arsenic indicates that the gold is chiefly contained in the arsenopyrite.

Hill 50, Mt Magnet
(References: Forman 1960, Lewis 1965)

The Hill 50 deposit at Mt Magnet occurs in jaspilite interbedded with basic lavas and tuffs. Gold abundance closely follows pyrrhotite, the dominant sulphide. Pyrite is also common. The sulphides, which appear to replace the jaspilite constitute 50-80 per cent of the rock in the high-grade areas. There is also association between high-grade ore, faulting and albite porphyry.

Great Fingall, Day Dawn
(References: Woodward 1907, McLaren, unpubl. data 1910, Connolly & Cole, unpubl. data 1935)

The host rock at Great Fingall, near Day Dawn, is a quartz dolerite sill in a pillowed basalt sequence containing minor slate members, and the rock associations, although poorly documented, appear to resemble closely those at Kalgoorlie.

The ore mined was a quartz vein in a major fracture which, although at least 3500 m long, only contained appreciable quartz over a length of 600 m where it cut across the quartz dolerite. The auriferous quartz contained 1.5-2 per cent sulphides, of which pyrrhotite, galena, arsenopyrite, pyrite, and sphalerite were generally associated with the better values. Wall-rock alteration has not been studied, but old reports suggest that the country rocks in places are as intensely chloritized and carbonatized as at Kalgoorlie.

Paddy's Flat, Meekatharra
(References: Clarke 1916, Forman 1934)

The main production at Paddy's Flat, Meekatharra was from a series of quartz reefs intimately associated with a dyke of albite-quartz porphyry. The country rock is described as 'flecked schist' and peridotite, the 'flecked schist' being slightly schistose talc-chlorite rock with textures suggesting derivation from a medium-grained intrusive. Some ore has been found in the porphyry where brecciated by cross-faults. The ore contained only very minor sulphides, arsenopyrite and pyrite being recorded.

Big Bell
(Reference: Staff of Big Bell Mines Ltd 1953)

The Big Bell orebody is a zone of quartz-mica schist up to 560 m long and 37 m wide, which contains 6 per cent disseminated pyrite. The schist is bedded, with narrow, pyritic, quartz-feldspar-muscovite lenses semi-parallel to the bedding. The footwall rocks, which are probably basic volcanics, are separated from the orebody by a 3 m quartzite bed. The hanging wall rocks are fine-grained lavas and sandstone. Minor associated sulphides are marcasite, pyrrhotite, chalcopyrite, arsenopyrite, sphalerite, tetrahedrite, bornite, molybdenite, and stibnite. The gold is very fine-grained and associated with the sulphides.

Copperhead, Bullfinch
(References: Matheson & Miles 1947, Clappison & Zani 1953, Williamson & Barr 1965)

The ore mined at Copperhead, Bullfinch, occurred at two horizons in a sequence of basaltic lavas. The upper or 'northern series' is a dolomitic bed and the lower or 'southern series' a jaspilite. In both units gold was associated with quartz veins, pyrrhotite, pyrite and minor galena. Sulphides were minor in the dolomitic ore but pyrrhotite was fairly abundant in the jaspilite ore.

Lancefield, Laverton

(References: Matheson 1938, Miles 1938, Hobson & Miles 1951, Chenoweth, unpubl. data 1969)

The Lancefield deposit at Laverton is in jaspilite in a sequence of basalts and dolerites, and the ore contains carbonate, pyrite, marcasite, arsenopyrite, plus minor chalcopyrite and sphalerite. Much of the gold in the ore mined was fine-grained and included in the sulphides. Along strike the ore horizon is pyritic shale with some chert and carbonaceous material.

Other Significant Deposits

Data on thirty-two other significant deposits are presented in Table 3.5.

Peak Hill and Paterson Range are the only two important gold-producing centres that are not of Archaean age. The poorly-documented Peak Hill deposits are quartz-impregnated shear and facture zones in moderately-folded siliceous iron formations and micaceous sediments (MacLeod 1970). The host-rocks are probably Lower Proterozoic, but this has not yet been verified by radiometric dating. The Paterson Range deposits, with reserves in excess of 30 000 kg, occur in pyritic siltstones and shales of Proterozoic age. Assuming economic conditions will allow continuity of production, the Paterson Range should become one of the state's top ten gold-producing districts.

THE FUTURE OF GOLD EXPLORATION AND MINING IN WESTERN AUSTRALIA

With gold occurrences widespread in the state there are opportunities both for new discoveries and further development of the industry. However, these opportunities need qualification.

Currently the industry is at a low ebb, treading a precarious financial path between rising production costs and the uncertainties of the price of gold. The most desirable deposits for future development are those of high grade, or those which can be mined by open pit or highly-mechanized underground methods (e.g. Mt Charlotte). Location is equally important in such a large and sparsely-populated state, for the modern community facilities which are now essential to attract employees are expensive to establish and maintain in remote localities.

Of one thing we can be fairly sure: small orebodies requiring underground mining have never been and are never likely to be economically viable. Fig. 3.8 illustrates this point. Note that operations based on orebodies containing less than 60 kg of gold per vertical metre are particularly bad risks unless the grade is above 20 grams per tonne.

A study of the history of gold-mining in Australia also sounds its own note of caution. Although great personal wealth was created during the first 20 years of

Table 3.5

*Production statistics and geological characteristics of other significant gold deposits of Western Australia—based on records of Western Australian Department of Mines except where indicated by **

Mining centre	Deposit	Tonnes ore	Kg gold	g-tonne	Main producing period	Geological characteristics	Reference
1 Menzies	Lady Shenton-Crusoe Line	440 599	14 383	32.6	1897-1914	Quartz reefs and veins in a basic volcanic. Largest deposit, the Lady Shenton at top of 30 to 50 m thick dolerite. Associated sulphides: pyrite, galena, sphalerite	Woodward (1906) Miller, unpublished data (1935)
2 — —	Princess May	526 242	8 515	16.2	1897-1925	Siliceous-pyrrhotitic lode in basic volcanics close to a sedimentary member	Woodward (1906) Miller, unpublished data (1935)
3 Sandstone	Oroya Black Range	754 214	12 097	16.0	1906-1925	Quartz reefs in chloritized and carbonatized quartz-dolerite. Sparse pyrite, chalcopyrite, galena and sphalerite	Clarke (1914)
4 — —	Black Range	264 628	6 447	24.4	1903-1916	,, ,, ,,	
5 Mt Morgans	Westralia	815 793	11 705	14.3	1897-1912	Pyritic jaspilite with quartz stringers	Hobson and Miles (1951)
6 Westonia	Edna May	575 104	11 191	19.5	1912-1922	Folded quartz reef, probably stratiform, in hornblende gneiss and biotite schist. Abundant pyrite, marcasite plus galena and molybdenite	Matheson and Miles (1947)
7 Southern Cross	Fraser's	851 631	9 392	11.0	1897-1912 1955-1963	Quartz veins and pyrrhotitic lode in a basaltic tuff-bed	Williamson and Barr (1965)
8 Lawlers	Emu	1 107 025	8 677	7.8	1903-1911	Quartz reefs, probably stratiform, in silicified, actinolitic lode formation at the contact of an ultramafic unit and gneissic sediments. Minor arsenopyrite, chalcopyrite and pyrite	Ward (1950) Forman, unpublished data (1960)
9 — —	Great Eastern	222 083	4 151	18.7	1896-1905	Quartz reefs associated with mafic volcanic remnants in granite	Gibson (1907)
10 Kalgoorlie	Hannan's North	591 339	8 553*	14.5	1935-1952	Pyritic, siliceous replacement lodes in chloritized and carbonatized quartz-dolerite	Haycraft (1965)

Table 3.5 (continued)

11	Burbanks	Main Reef-Birthday Gift-Lady Robinson	386 684	8 544	22.1	1897-1915 Folded quartz reef in bedded horizon in basic volcanics	Blatchford (1913) McMath (1953)
12	Coolgardie	Bayley's	508 540	8 490	16.7	1897-1905 1938-1948 1954-1964 Quartz reefs and mineralized lodes in stratigraphic and semi-stratigraphic locations in ultramafic lavas. Ore occurs at two horizons, both associated with graphitic shale-chert members. One ore zone also coincides with contact of stratigraphic sodic dacite-porphyry. Moderate sulphides: pyrrhotite, pyrite and arsenopyrite. Sparse sphalerite, chalcopyrite, galena and scheelite	McMath (1953) Blockley, unpublished data (1956)
13	— —	Tindals	547 065	4 023	7.4	1897-1916 1936-1948 Sodic dacite, fairly heavily mineralized with pyrrhotite, pyrite and quartz veinlets. Country rock is probably basic volcanics	McMath (1953)
14	Kookynie	Cosmopolitan	558 881	8 360	15.0	1898-1912 Quartz reefs in biotite granite, sparse pyrite, galena, chlorite, calcite and ankerite	Jutson (1921*a*) Finucane and Miller, unpub. (1935)
15	Youanmi	Youanmi	713 330	8 265	11.6	1912-1925 1936-1943 Lodes in basalt. Abundant pyrite, arsenopyrite, and stibnite common	Feldtman (1923, 1924)
16	Mt Ida	Timoni	477 933	7 486	15.7	1950-1969 Quartz reef parallel to schistosity in basic tuffs, probably stratiform	Johnson (1950) Tomich (1956)
17	Reedys	Triton	730 471	7 116	9.7	1935-1952 Siliceous lode containing fine-grained auriferous pyrite and some pyrrhotite. The lode occupies a bed of probable tuffaceous origin between massive lavas and a predominantly tuffaceous sequence	Campbell (1953b)
18	Kanowna	White Feather	429 954	7 078	16.5	1897-1907 Quartz reef in agglomerate, the main constituent of which is porphyritic dacite. Minor galena and sparse pyrite associated with the gold	Blatchford and Jutson (1912) Williams (1970)
19	Peak Hill	Peak Hill	469 472	6 945	14.8	1897-1913 Network of interlacing auriferous quartz veins in quartzite and mica schist	Maitland (1904)
20	Wiluna	Moonlight Wiluna	899 332	6 775	7.5	1912-1923 1936-1946 Stibnite (1-1.25 per cent) plus subordinate pyrite and arsenopyrite in lode material in fractured, altered basalt	Edwards (1953)

21	Comet Vale	Gladsome-Sand Queen	249 017	5 662	22.7	1905-1920 1928-1938	Quartz veins in basalt. Minor associated pyrite, calcite and talc	Jutson (1921*b*)
22	Bonnievale	Bonnievale	343 098	5 482	16.0	1897-1911	Quartz reefs in biotite granite. Reefs laminated and carry minor galena	McMath (1953)
23	Laverton	Ida H.	235 214	5 347	22.7	1901-1920	Quartz reef parallel to schistosity in sedimentary schists. Considerable pyrite in schists, minor pyrite and sphalerite in quartz	Hobson and Miles (1951)
24	Mt Palmer	Palmer's Find	310 728	4 928	15.9	1936-1947	Bedded quartz reef in basic volcanics	Matheson and Hobson (1940)
25	Cue	Main Line-Light of Asia-Cue No. 1	194 131	4 593	23.7	1897-1910	Quartz reefs in granodiorite	Woodward (1907)
26	Hampton Plains	Barbara-Surprise	284 640	4 485	15.8	1949-1958	Mineralized shears in ultramafic volcanics and folded felsic porphyry. Minor pyrite and pyrrhotite	McMath (1953)
27	Golden Ridge	Golden Ridge	254 847	4 368	17.1	1901-1920	Quartz reefs in carbonatized and sericitized zone in a stratiform chloritic quartz-porphyry. Sparse sulphides: pyrite, galena and sphalerite	Honman (1916)
28	Ora Banda	Gimlet-Slippery Gimlet-Victorious	409 967	4 328	10.6	1907-1934	Fractures in basalt mineralized with quartz, calcite, pyrite, pyrrhotite plus minor arsenopyrite and tourmaline	Jutson (1914)
29	Sir Samuel	Bellevue	252 340	4 033	16.0	1897-1912	Quartz in pyritic basic schist containing 1-2 per cent chalcopyrite. The gold is contained almost entirely in the sulphides	Gibson (1907)
30	Mt Magnet	St George	477 190	3 663	7.7	1897-1902 1908-1914 1925-1942	Siliceous, pyritic, jaspilite lode and 'laminated quartzite' in sedimentary —tuffaceous basic schists	Le Grange, unpublished data (1934) Berliat (1958) de la Hunty (1958)
31	— —	Morning Star	185 300	3 087	16.7	1897-1915	Quartz veins in basic schists, close to siliceous jaspilite. Probably same horizon as St George	Do. do.
32	Marble Bar	Comet	125 281	3 513	28.0	1936-1951 1953-1955	Quartz-carbonate-pyrite replacement lodes in basic volcanics	Noldart & Wyatt (1962)

* Production statistics from Haycraft (1965).

gold-mining in Australia (1851-1870), when 1.1 million kg of gold (37 million oz) was produced from alluvials and shallow diggings on the rich Victorian and New South Wales goldfields, since then the number of prospectors and investors to prosper from gold-mining has steadily declined. Over the last 95 years the main beneficiaries appear to have been the state and the nation, through the employment opportunities gold-mining has created, the taxes paid and the demand that gold-mining has generated for goods and services.

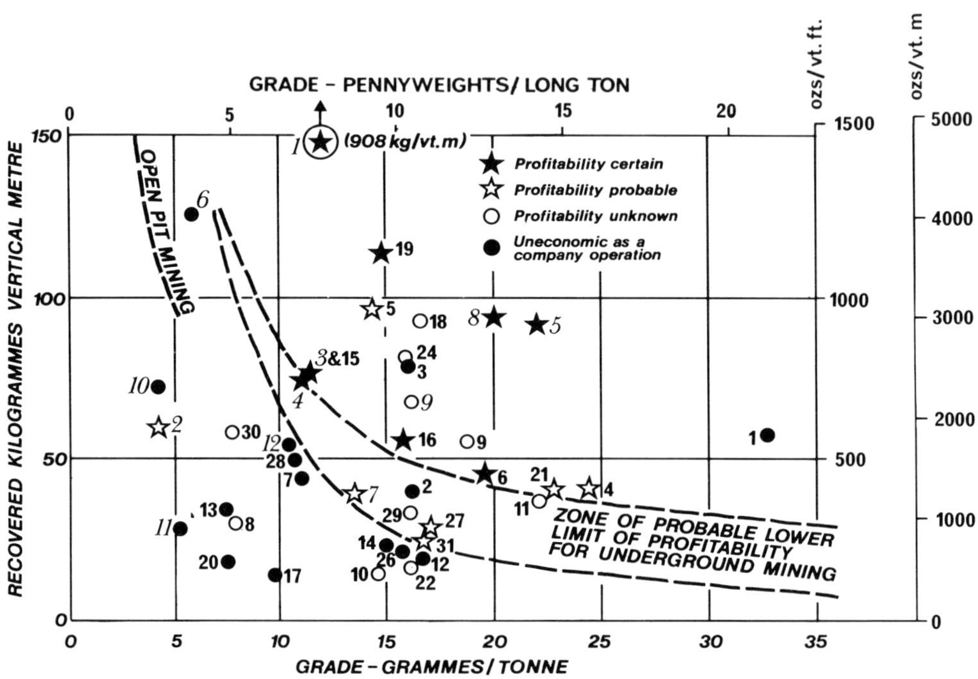

Note: Deposit index numbers relate to tables 3.4 and 3.5 (table 3.5 in italics)

Fig. 3.8 Profitability of gold-mining

Tens of millions of dollars were spent by hundreds of gold-mining companies during the early part of this century to develop Western Australia's gold deposits, but up to 1920 only $26 million was paid in dividends, and of this amount, $22 million was paid by a handful of companies working deposits on the 'Golden Mile' at Kalgoorlie. To 1919, of companies formed to work Archaean gold deposits outside the Golden Mile, only fourteen returned their capital and only two paid dividends of more than $1 million. The situation has not changed significantly since.

One of the few success stories in recent years is that of Western Mining Corporation Limited, which was formed in 1933 to use the geological sciences to find extensions and repetitions of the known Archaean gold deposits. Although the Company investigated hundreds of uneconomic prospects and made several unsuccessful attempts to re-open abandoned gold-mines, it succeeded, after many difficulties, on the Norseman field. Here, discoveries have resulted in 40 years of profitable gold-mining, the production of 85 000 kg of gold (2.7 million oz) and the payment of $18 million in dividends on the $1.3 million of capital invested. These dividends in turn helped to finance the successful mineral exploration which lead to Western Mining Corporation Limited pioneering the development of the state's nickel industry and becoming one of Australia's largest companies.

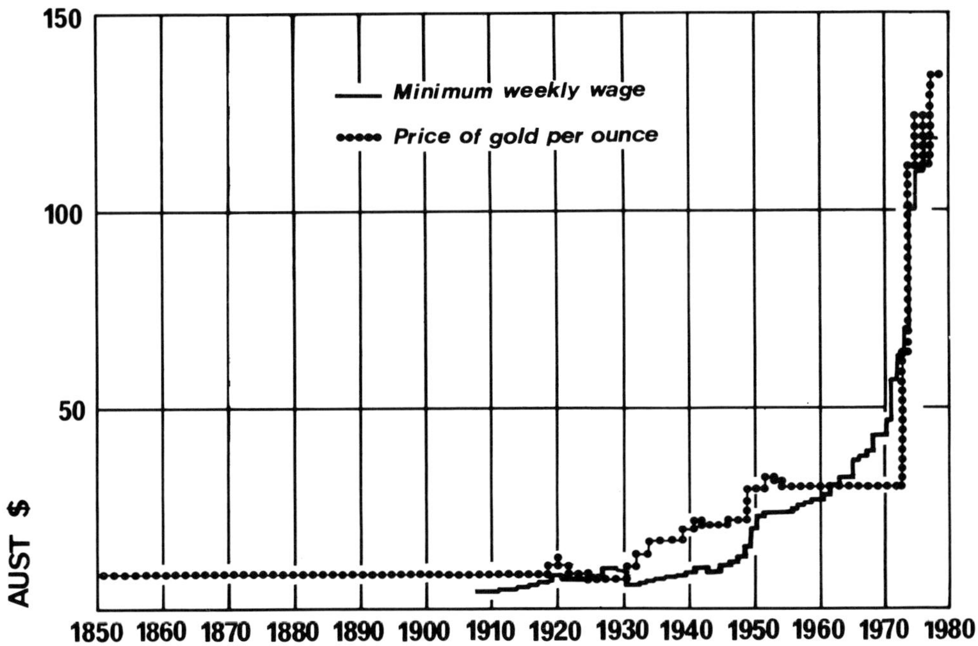

Fig. 3.9 Price of gold and Australian minimum weekly wage

There are many unknowns which will affect the future profitability of Australian gold-mining. For how long will underground mine labour be available? To what extent will governments once again provide some of the services and community facilities required in new inland mining towns? And, most important of all, what will be the future price of gold in Australian currency relative to the cost of labour? Fig. 3.9 illustrates the historical situation in Australia with respect to the cost of labour and the price of gold. It would seem that gold-mining is unlikely to be generally profitable unless the price of gold per ounce in Australian dollars is at least double the Australian minimum weekly wage. Given sufficient economic incentive, where might gold exploration and gold mining revive in Western Australia?

Gold deposits undoubtedly remain to be discovered in the Archaean, where 80 per cent of the terrain is covered by soil, shallow sediment or deeply weathered zones.

However, this concealed ground is expensive and difficult to prospect, and despite considerable exploration over the last 70 years no new Archaean goldfields have been found. Improved geophysical and geochemical exploration, coupled with an increased understanding of the structural control of mineralization will improve the effectiveness of future exploration, but the magnitude of the task should not be underestimated.

The areas of Proterozoic rocks in Western Australia are vast and the recent discoveries at Paterson Range suggest that there is still potential for new finds of Proterozoic gold. But much of the Proterozoic is in particularly remote areas and discoveries will not be easy or cheap to develop.

There is still scope for gold exploration in Western Australia and a continuing exploration effort is justified. Enthusiasm, however, should be tempered with caution as a further substantial increase in the price of gold relative to costs is required before vigorous gold exploration and mine development can be expected on a large scale.

REFERENCES

Berliat, K. 1958. A summary report on the Mt Magnet district, Murchison Goldfield, W. A., Part I: Regional Geology. *Bull. geol. Surv. W. Aust.* **109:** 21-9.

Blainey, G. 1963. *The rush that never ended.* Melbourne: Melbourne University Press.

Blatchford, T. 1913. Geological investigations in the area embracing the Burbanks and Londonderry mining centres with special reference to the ore deposits and their future prospects. *Bull. geol. Surv. W. Aust.* **53.**

Blatchford, T. & Jutson, J. T. 1912. Mining geology of Kanowna main reef line, Kanowna, North East Coolgardie Goldfield. *Bull. geol. Surv. W. Aust.* **47.**

Blockley, J. G. 1974. Notes on the Paterson Range gold prospects. *Ann. Rep. geol. Surv. W. Aust. for 1973,* 71-3.

Campbell, J. D. 1953*a*. The structure of the Kalgoorlie Goldfield. In *Geology of Australian Ore Deposits* (ed. A. B. Edwards), 79-93. 5th Empire Min. Metall. Congr., Melbourne.

——— 1953*b*. Triton Gold Mine. In *Geology of Australian Ore Deposits* (ed. A. B. Edwards), 195-207. 5th Empire Min. Metall. Congr., Melbourne.

——— 1965. Gold deposits of Australia. In *Geology of Australian Ore Deposits* (ed. J. McAndrew), 31-8. 8th Common. Min. Metall. Congr., Melbourne.

Clappison, R. J. S. & Zani, J. A. 1953. The structures of the Southern Cross-Bullfinch belt. Yilgarn Goldfield. In *Geology of Australian Ore Deposits* (ed. A. B. Edwards), 128-37. 5th Empire Min. Metall. Congr., Melbourne.

Clarke, E. de C. 1914. Notes on the geology and mining at Sandstone and Hancock's, East Murchison Goldfield. *Bull. geol. Surv. W. Aust.* **62.**

——— 1916. The geology and ore deposits of Meekatharra, Murchison Goldfield. *Bull. geol. Surv. W. Aust.* **68.**

de la Hunty, L. E. 1958. A summary report on the Mt Magnet district, Murchison Goldfield, W. A. Part II: Report on selected mining groups in the Mt Magnet district, Murchison Goldfield, *Bull. geol. Surv. W. Aust.* **109**: 34-76.

Edwards, A. B. 1953. Gold deposits of Wiluna. In *Geology of Australian Ore Deposits* (ed. A. B. Edwards), 215-23. 5th Empire Min. Metall. Congr., Melbourne.

Feldtman, F. R. 1923. The Youanmi goldmining centre, East Murchison Goldfield. *Ann Rep. geol. Surv. W. Aust. 1922:* 5-8.

——— 1924. The Youanmi goldmining centre, East Murchison Goldfield. *Ann. Rep. geol. Surv. W. Aust. 1923: 8-12.*

Finucane, K. J. 1965*a*. Ore distribution and lode structures in the Kalgoorlie Goldfield. In *Geology of Australian Ore Deposits* (ed. J. McAndrew), 80-6. 8th Common. Min. Metall. Congr., Melbourne.

——— 1965*b*. Geology of the Sons of Gwalia Gold Mine. In *Geology of Australian Ore Deposits* (ed. J. McAndrew), 95-7. 8th Common. Min. Metall. Congr., Melbourne.

Forman, F. G. 1934. Ingleston Alberts gold mine, Meekatharra, *Ann. Rep. geol. Surv. W. Aust. 1933:* 8-13.

Forman, D. J. 1960. Stratigraphy of the Boogardie Group, *J. roy. Soc. W. Aust.* **43**: 83-96.

Gibson, C. G. 1907. The geology and mineral resources of the Lawler's, Sir Samuel and Darlot (East Murchison Goldfield), Mt Ida (North Coolgardie Goldfield and a portion of the Mt Margaret Goldfield). *Bull. geol. Surv. W. Aust.* **28.**

Hall. H. I. E. & Bekker, C. 1965. Gold Deposits of Norseman. In *Geology of Australian Ore Deposits* (ed. J. McAndrew.), 101-6. 8th Common. Min. Metall. Congr., Melbourne.

Haycraft, J. A. 1965. Orebodies in the Mt Charlotte-Hannan's North area, Kalgoorlie. *Proc. Australas. Inst. Min. Metall.* **213**: 49-64.

Hobson, R. A. & Miles, K. R. 1951. Geology of portion of the Mt Margaret Goldfield. *Bull. geol. Surv. W. Aust.* **103.**

Honman, C. S. 1916. The geology of the country to the south of Kalgoorlie, Coolgardie and East Coolgardie Goldfields. *Bull. geol. Surv. W. Aust.* **66.**

Johnson, W. 1950. Report on the Timoni workings of Mt Ida Gold Mines, Copperfield, Menzies district, North Coolgardie Goldfield. *Ann. Rep. geol. Surv. W. Aust. 1948:* 53-7.

Jutson, J. T. 1914. The mining geology of Ora Banda, Broad Arrow Goldfield. *Bull. geol. Surv. W. Aust.* **54.**

——— 1921*a*. The mining geology of Kookynie, Niagara and Tampa, North Coolgardie Goldfield. *Bull. geol. Surv. W. Aust.* **78.**

——— 1921*b*. The mining and geology of Comet Vale and Goongarrie, North Coolgardie Goldfield. *Bull. geol. Surv. W. Aust.* **79.**

Larcombe, C. O. G. 1913. *The Geology of Kalgoorlie with special reference to the ore deposits.* Melbourne: Australas. Inst. Min. Engineers.

——— 1926. A petrological investigation into the cores from bores put down by the Wiluna Development Syndicate, Wiluna, East Murchison G.F. *Ann. Rep. geol. Surv. W. Aust. 1925:* 23-30.

Lewis, B. R. 1965. Gold Deposit of Hill 50 Mine. In *Geology of Australian Ore Deposits* (ed. J. McAndrew), 98-100. 8th Common. Min. Metall. Congr., Melbourne.

MacLeod, W. N. 1970. Peak Hill, W. A. *W. Aust. geol. Surv. 1:250 000 Geol. Series Explan. Notes.*

Maitland, A. G. 1904. The mineral production of Western Australia up to the end of the year 1903. *Bull. geol. Surv. W. Aust.* **16.**

Matheson, R. S. 1938. Report on 'Beria Main Lode' G.M.L. 2416T. Mt Margaret Goldfield. *Ann. Rep. geol. Surv. W. Aust. 1937:* 14-18.

Matheson, R. S. & Hobson, R. A. 1940. Mining groups of the Yilgarn Goldfield south of the Great Eastern Railway. *Bull. geol. Surv. W. Aust.* **98.**

Matheson R. S. & Miles K. R. 1947. The mining groups of the Yilgarn Goldfield, north of the Great Eastern Railway. *Bull. geol. Surv. W. Aust.* **101.**

Miles, K. R. 1938. Report on 'Beria Main Lode', G.M.L. 2416T, Mt Margaret Goldfield. *Ann. Rep. geol. Surv. W. Aust. 1937:* 20-1.

McMath, J. C. 1953. The geology of the country about Coolgardie, Coolgardie Goldfield, W. A. *Bull. geol. Surv. W. Aust.* **107.**

Noldart, A. J. & Wyatt, J. D. 1962. The geology of portion of the Pilbara Goldfield. *Bull. geol. Surv. W. Aust.* **115.**

Prider, R. T. 1965. Geology and mineralization of the Western Australian Shield. In *Geology of Australian Ore Deposits* (ed. J. McAndrew), 56-65. 8th Common. Min. Metall. Congr., Melbourne.

Staff of Big Bell Mines Limited, 1953. The Big Bell Gold Mine. In *Geology of Australian Ore Deposits* (ed. A. B. Edwards), 224-230. 5th Empire Min. Metall. Congr., Melbourne.

Tomich, S. A. 1956. Summary report on the geology of portion of the Mt Ida district, North Coolgardie Goldfield. *Ann. Rep. geol. Surv. W. Aust. 1953:* 18-20.

——— 1959. The Oroya Shoot and its relationship to other flatly plunging ore pipes at Kalgoorlie. *Proc. Australas. Inst. Min. Metall. 190:* 113-24.

Travis, G. A., Woodall, R. & Bartram, G. D. 1971. The geology of the Kalgoorlie Goldfield. *Spec. Publ. Geol. Soc. Aust.* 3: 175-90.

Ward, H. J. 1950. Notes on Emu gold mine, Agnew, East Murchison Goldfield. *Ann. Rep. geol. Surv. W. Aust. 1948:* 60-2.

Williams, I. 1970. Explanatory notes on the Kurnalpi 1:250 000 geological sheet, W.A. *Rec. geol. Surv. W. Aust.* 1970/1.

Williamson, H. C. & Barr D. J. 1965. Gold mineralization in the Yilgarn Goldfield. In *Geology of Australian Ore Deposits* (ed. J. McAndrew), 87-94. 8th Common. Min. Metall. Congr., Melbourne.

Woodall, R. 1965. Structure of the Kalgoorlie Goldfield. In *Geology of Australian Ore Deposits* (ed. J. McAndrew), 71-9. 8th Common. Min. Metall. Congr., Melbourne.

——— 1975*a*. Gold in the Precambrian shield of Western Australia. In *Economic Geology of Australia and Papua New Guinea—1. Metals* (ed. C. L. Knight). Australas. Inst. Min. Metall., Melbourne, Monogr. Ser. 5: 175-84.

——— 1975*b*. Gold resources of Australia. *Proc. 5th Gold & Money Session and Gold Technical Session.* Dept Geology and Mineral Industries, Portland, Oregon, 106-38.

Woodward, H. P. 1906. The auriferous deposits and mines of Menzies, North Coolgardie Goldfield. *Bull. geol. Surv. W. Aust.* **22.**

——— 1907. A report on the geology, together with a description of the productive mines of the Cue and Day Dawn districts, Murchison Goldfield. *Bull. geol. Surv. W. Aust.* **29.**

3C

Iron

A. F. Trendall
Geological Survey of Western Australia

General and historical introduction

If one mineral product had to be singled out as the most indispensable to ordinary daily life, iron would be the obvious choice. Not only is iron the structural basis for all modern technology, but it also plays an essential part in the mining and extraction of all other mineral products and in subsequent manufacturing processes.

In 1976 over 85 million tonnes of iron-ore were mined in Western Australia, of which over 77 million tonnes were shipped by bulk carrier for use outside Australia; this was about 18 per cent of all iron-ore transported across international boundaries throughout the world in that year. The $A776 million value of this Western Australian export tonnage represented about 7 per cent of the total value of all Australian exports for that year.

As shown in Fig. 3.14, export began in 1966, so that this position of national and international significance was achieved over a period of only 10 years. This almost explosively abrupt emergence of a new mineral field had no parallels in scale in the entire previous history of the minerals industry, and it can only be fully understood by looking briefly at the world history of man's use of iron.

As discussed further below, most iron-ores are oxides in which metallic iron (Fe) is tightly bound chemically with oxygen. To obtain the metal such ores must first be smelted, a process which involves the removal of the oxygen by placing them in contact with a 'reductant', which is commonly carbon in the form of coke, but which in some modern processes may be hydrogen gas. Such reactions can only proceed at high temperatures, which are achieved by burning a fuel. It is traditional and convenient, but not essential, to use the same substance as both fuel and reductant; thus part of the coke used in smelting first burns to produce the high temperature needed for the remainder to reduce the ore.

After metallic iron was first systematically smelted in the Middle East in about 1400 BC, its development became a story of continuous growth of output, improvement of production methods, and extension of its uses. At first, wood (as charcoal) was used both as a fuel and reductant to produce spongy 'blooms' of iron from small earth or brick furnaces close to sources of iron-ore; after each bloom was removed, the furnace was recharged and lighted for the process to be repeated. By the seventeenth century depletion of forests by the ever-increasing iron production of western European countries was of serious concern, and contributed to the switch to the use of coal. When coal, and then coke, became established as the fuel and reductant for the new larger blast furnaces, which could be kept in continuous operation and from which liquid metal could be tapped at regular intervals, the industrial revolution gathered momentum, and the manufacturing areas of the western industrial world were established on or close to coalfields, to which iron-ore was carried, mainly by rail, from the nearest sources.

Throughout the nineteenth and early twentieth centuries, as the world's manufacturing industries steadily expanded, few if any iron producers were interested in iron-ore deposits that were remote from coalfields. Thus the Western Australian Government Geologist of 1887-92 was able to write (Woodward 1890, pp. 25-6): 'This is essentially an iron country . . . . There is enough to supply the whole world, should the present sources be worked out'; but he also later continued (Woodward 1894, p. 48): 'but as iron ores are of no value, it is useless to trouble about them'!

In 1919 the Government Geologist wrote in a vein similar to that used 30 years previously (Maitland 1919, p. 3): 'The large iron deposits of the State extend over an area from Kimberley to Cape Leeuwin . . . and are . . . under present conditions, practically valueless.' It was not until after the Second World War that the blast furnaces of the industrialized world had so depleted their closer sources of iron-ore that its large-scale long-distance transport became both feasible and essential. However, when overseas interest in Western Australian iron-ore first began to appear in 1938, the Commonwealth Government placed an embargo on its export until such time as Australia's resources of iron-ore were known with certainty to be sufficient for future domestic needs. This embargo remained until 1960. During those 22 critical years long-distance international trade in iron-ore continued to grow exponentially, so that when the embargo was lifted Western Australia was able to establish, almost instantaneously, a position consistent with its large supply potential. Thus, in historical perspective, the breathtaking growth of the state's iron-ore mining industry was due to the ending of an unnecessary and artificial restraint which had been applied during a period of enormously-increasing world demand.

By contrast with its important international position in iron-ore supply, Western Australia has never established a significant iron and steel industry based on the processing of its own ore; this was to be expected in a state whose population growth has lagged well behind the market requirements of an industry always dominated by the need for economies of scale. A small charcoal blast furnace plant was established by the state government at Wundowie, 60 km east-northeast of Perth, in 1945; it used ore from Koolyanobbing (Fig. 3.10) after local ore proved unsuitable. In 1976 it was sold to a private company. Its greatest production of pig-iron in one year was 63 000 tonnes in 1975. Ore from Koolyanobbing has also supplied a blast furnace at Kwinana since 1968, the greatest annual production so far being 775 000 tonnes in 1973-74. No steel has yet been produced within Western Australia.

The nature of iron-ore

The main ore minerals of iron are its two oxides, magnetite (Fe_3O_4) and hematite (Fe_2O_3); its hydroxides, goethite ($Fe_2O_3.H_2O$) and limonite ($Fe_2O_3.nH_2O$) are also important, while small amounts of iron are obtained from the carbonate, siderite ($FeCO_3$), the sulphide, pyrite (FeS_2), and the complex silicates chamosite and greenalite. About one quarter of the world's iron is obtained from an iron-ore known as taconite. This is a rock consisting of about 45 per cent by weight magnetite, the remainder being mainly quartz. As magnetite contains about 70 per cent of iron (Fe) the taconite itself contains about 30 per cent Fe. To use such ore it is necessary to first grind the taconite to a fine powder, after which black magnetic magnetite can be separated with special processing equipment.

The iron-ores of Western Australia differ from taconite in that they consist almost entirely of non-magnetic oxides of iron. Like the similar ores of Brazil, Africa, and India, they consist almost entirely of mixtures of hematite, goethite and limonite. Their natural iron content, as mined, is much higher and lies in the range 55-65 per cent Fe. The high-grade (more than 60 per cent Fe) ores in particular can potentially be fed directly into a blast furnace after crushing. From a chemical viewpoint, they do not require any special processing between mining and reduction, although if they are too soft and powdery they may need to be formed artificially into lumps (pellets) before introduction to the blast furnace.

The physical appearance of the Western Australian ores varies widely according to the relative proportions of hematite, goethite and limonite, and the type of crystallinity developed in these minerals. Pure, coarsely crystalline hematite forms bladed black crystals with such brilliantly reflecting surfaces that it is then called specularite (Latin *speculum* = mirror). Specularite is developed in some Koolyanobbing material. Most of the deeper-level Hamersley Range hematite ore is finely crystalline, and varies from hard, heavy, massive purplish black or dark-red rock to porous red, brownish-red, or brownish-yellow ore, which may be quite friable and powdery, or full of cavities. The brown and yellow colours are related to increasing amounts of goethite and limonite.

Occurrence, distribution and origin of Western Australian iron-ore

The distribution of iron-ore deposits in relation to the broad geological structure of Western Australia is shown in Fig. 3.10. Three types of ore are there distinguished: hematite enrichment ore, pisolitic limonite ore, and sedimentary ore. These three main types differ markedly, not only in their geological environment, age and origin, but also in their physical and chemical properties. They are described in the listed order below. A more detailed account of these types has been given by Trendall (1975) in a volume which includes full descriptions of many individual deposits.

Hematite enrichment ore is the most widespread and important type; it forms over 80 per cent of the currently mined ore, and about 75 per cent of the known reserves of the state. Bodies of hematite enrichment ore invariably occur in association with a distinctive host rock known as banded iron-formation, usually abbreviated to BIF. For a complete understanding of these ores we need first to look more closely at the nature of this host.

As its name implies, BIF is an iron-rich rock with a conspicuous banded structure. It is made up of thin layers alternately very rich and relatively poor in iron; these layers usually range from less than a millimetre up to several centimetres in thickness. In unweathered rock the iron-rich layers, or bands, are usually black or dark red and consist mainly of magnetite, with subordinate hematite, siderite, iron silicates (grunerite), and quartz; the iron-poor bands (often called chert) may be white, green, red or grey, or clear and glassy, and consist mainly of fine-grained quartz, usually with a small amount of the minerals which form the iron-rich bands. Near the surface most BIFs are affected by weathering to some extent, so that instead of the sharply contrasting colour-banding of the fresh rock there is a general brownish or reddish colour due to oxidation and hydration of the magnetite to hematite and goethite.

In total chemical composition, most BIFs contain about 50 per cent iron oxides and 45 per cent silica, with all other constituents together forming the remaining 5 per cent. Their total iron content is thus about 30 per cent.

BIFs occur within the older (Precambrian) rocks of the earth's crust, throughout all continents. They occur consistently in rock successions older than about 2500 million years (Archaean), but are particularly abundant in successions between 2500 and 1800 million years old (Early Proterozoic). The present consensus of research opinion is that they represent chemical sediments precipitated locally in lakes or restricted seas under very quiet conditions in which there was no contaminating mud or sand washed in by rivers. There are no rocks exactly like them being formed at the present day.

In Western Australia numerous BIF units occur within the Archaean Pilbara and Yilgarn Blocks (Fig. 3.10); these BIFs were laid down more than 2700 million years ago. They have usual stratigraphic thicknesses (across the banding) of 10 to 70 m, and normally dip steeply, with outcrop lengths (along the banding) of many kilometres.

The Hamersley Basin (Fig. 3.10) developed in the north-western part of the state at a much later period, probably between 2300 and 2000 million years ago. During its development, several different sheets of BIF were formed, separated by other types of sedimentary rocks, and by intrusive igneous rocks. The whole succession was later folded and eroded, so that the Early Proterozoic BIFs of the Hamersley area, unlike those of the Archaean blocks, cover extensive areas, and often have low dips. They are also thicker than those of the Archaean, and the three which are the host rocks for the main iron-ore deposits are underlined in the stratigraphic table below:

Brockman Iron Formation	Yandicoogina Shale Member <u>Joffre Member</u> Whaleback Shale Member <u>Dales Gorge Member</u>
Shales and carbonates	
<u>Marra Mamba Iron Formation</u>	

The hematite enrichment ores of both the Hamersley Basin area and the Pilbara and Yilgarn Blocks formed, long after their deposition, within these BIFs. This may

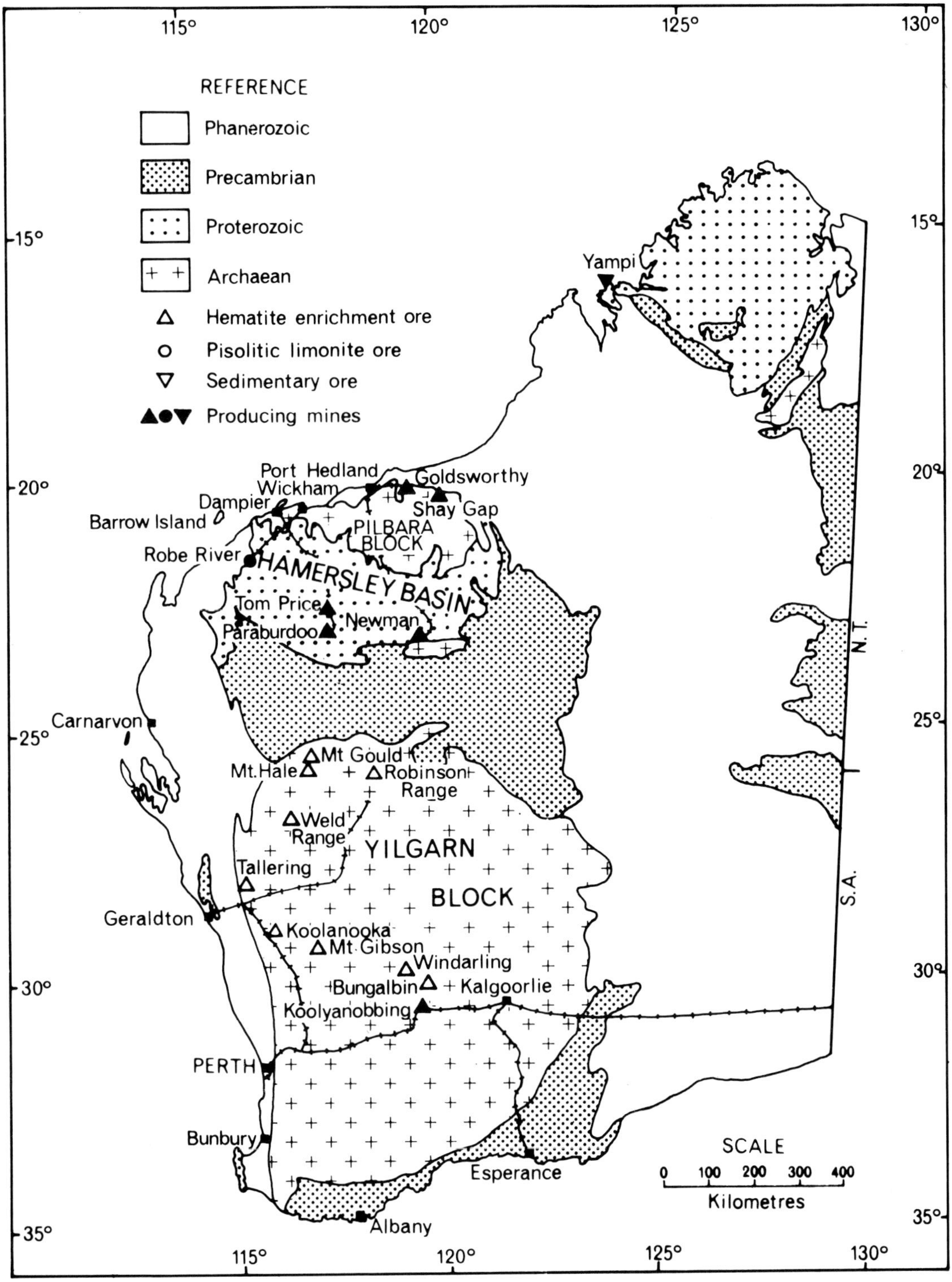

Fig. 3.10 Major iron-ore deposits of Western Australia in relation to the regional geology

have taken place about 1500 to 1300 million years ago, but the evidence for this is not definite. The ores formed in local pockets by a process of 'enrichment' (hence their name) of the iron content of the parent BIF from its initial 30 per cent to a final value of over 60 per cent, by which time almost all the minerals of the original BIF had been replaced by hematite.

The details of this process are not well understood, but it must have been effected by circulating hot water, which gradually dissolved out and carried away the silica (quartz) and other minor constituents, leaving the iron of the magnetite of the iron-rich bands behind, but altered to the more oxidized form of hematite. The process must have occupied many millions of years, and probably took place at a considerable depth. After its completion the surface of all the Precambrian areas of the state were further lowered by erosion, and as the ore bodies were exposed, a proportion of the hematite was further modified by weathering to goethite. This type of ore is therefore sometimes referred to as 'hematite-goethite ore', a term which is mineralogically accurate. Like the BIF itself, the iron-ore tends to be more resistant to erosion than its neighbouring rocks, so that the larger bodies form rounded tops of conspicuous hills, as at Mount Tom Price and Mount Whaleback.

In general, the size, shape and attitude of bodies of hematite enrichment ore are closely controlled by the thickness and local structure of their parent BIFs. Thus the largest bodies are those of the Hamersley Basin (Fig. 3.10), or the Hamersley Iron Province, as the area is called by economic geologists.

The Mount Tom Price orebody is described briefly below as a typical example of the larger deposits of the Hamersley Iron Province. As shown in Figs. 3.11 and 3.12 it lies at the eastern end of an extensive downfold of the BIF layers, known as the Turner Syncline. The ore occurs mainly within the Dales Gorge Member, but to a lesser extent also in the Joffre Member, of the Brockman Iron Formation. The main orebody is about 7 km long (east-west) with an average width of 0.6 km, and forms an irregular hill-top sheet with an average thickness of about 50 m. The base of the orebody is marked by underlying shale, but the shale which formerly marked the top has largely been stripped off by erosion. Wherever the lateral margins of the orebody (against unenriched BIF) are visible, the contact between BIF and ore is fairly sharp, transition from one material to the other taking only about 10-20 m, in spite of the wide difference in chemical composition.

The orebody initially contained about 750 million tonnes of hematite. The best quality ore contains about 68.5 per cent Fe, 1.07 per cent SiO_2 (silica), 0.47 per cent Al_2O_3 (alumina), 0.024 per cent P (phosphorus), and 0.94 per cent combined H_2O (water). However, owing to the presence of thin shales within the Dales Gorge Member, which persist in an altered form within the ore body after its conversion to hematite, a proportion of the ore is of lower grade, with a higher proportion of the undesirable contaminants listed. The best ore is a hard, massive, fine-grained, bluish-black rock with a density of about 4.0 g/cm^3. However this hematite-rich ore is capped by about 20 m of ore rich in goethite, which is less massive, and generally of a brownish colour.

The orebodies at Paraburdoo (700 million tonnes) and Mount Whaleback (over

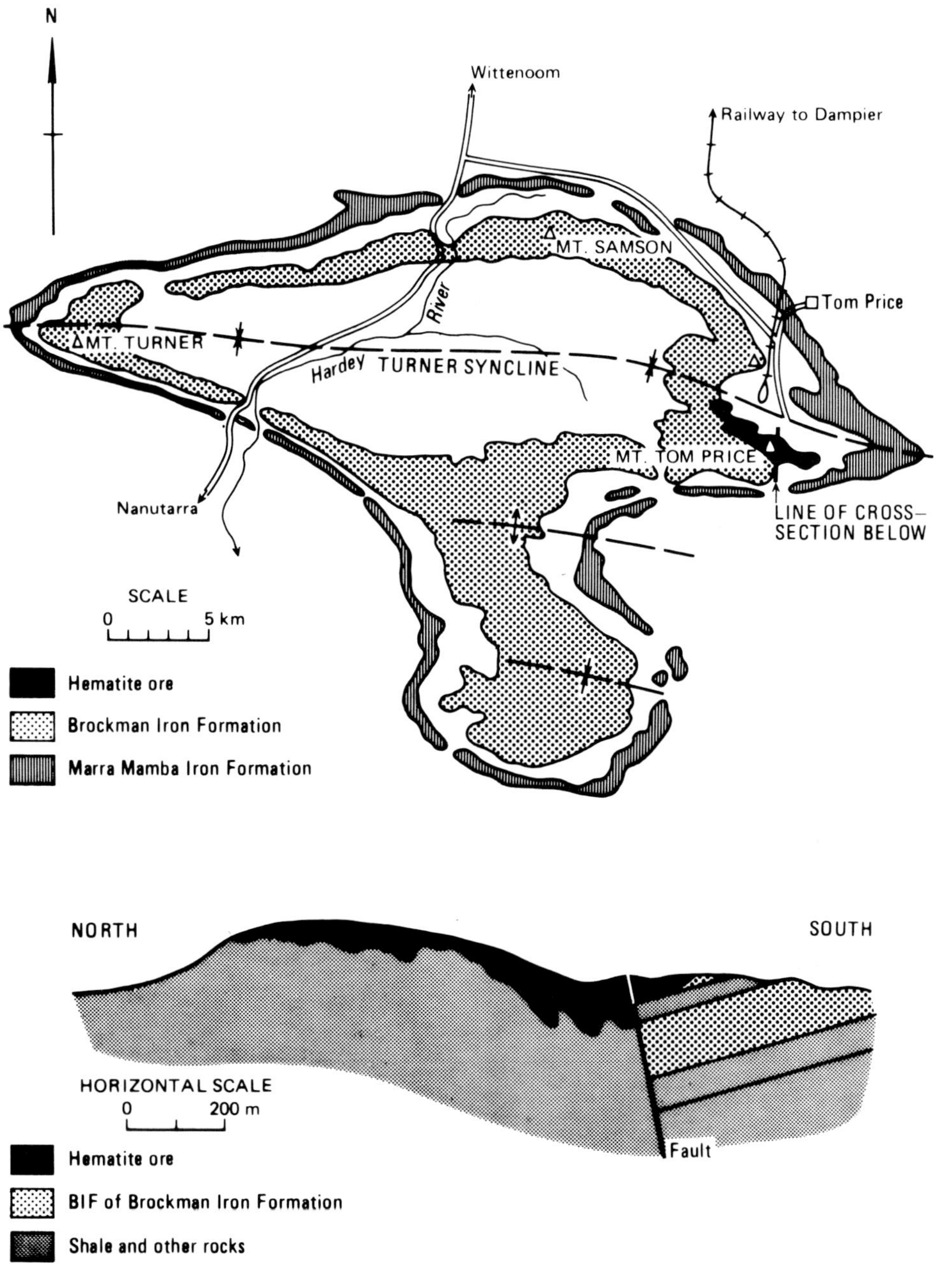

Fig. 3.11 The Turner Syncline, Hamersley Iron Province

1000 million tonnes) are essentially similar to that at Mount Tom Price, and like it lie within the Dales Gorge Member of the Brockman Iron Formation, although they also extend upwards into the Joffre Member. Other large deposits within the Brockman Iron Formation include Koodaideri (700 million tonnes), Marillana, Rhodes Ridge and Jimblebar. Unfortunately all these have a higher phosphorus content (mainly above 0.1 per cent P), which decreases their marketability. The bulk chemical composition of most of the large orebodies within the Brockman Iron Formation corresponds to a mineral mixture of 70 per cent hematite, 23 per cent goethite, 5 per cent kaolinite and 2 per cent quartz, the latter two minerals being derived from the associated shale. This gives a composition of about 63.5 per cent Fe, 4.3 per cent SiO_2, 2.0 per cent Al_2O_3 and 3.0 per cent combined water.

Hematite enrichment ore within the Marra Mamba Iron Formation of the Hamersley Iron Province tends to be developed in thinner bodies, corresponding to the smaller thickness of the parent BIF. Such bodies also tend to be less conspicuous topographically, so that instead of forming prominent hilltop masses they may lie along comparatively low ridges or even at general plain level. The largest prospective bodies so far explored within the Marra Mamba Iron Formation are those of Marandoo and 'MAC' (Mining Area C).

Hematite enrichment orebodies within the older BIFs of the Archaean Pilbara and Yilgarn Blocks include (Fig. 3.10): those mined at Mount Goldsworthy (55 million tonnes) and Shay Gap, that formerly mined at Koolanooka (7 million tonnes), and that mined at Koolyanobbing. Other deposits which have been investigated include those at Mount Gould, Robinson Range, Weld Range, Tallering Peak, Mount Gibson, Windarling, and Bungalbin; and there are many other scattered deposits. In general these orebodies are much smaller than those of the Hamersley Iron Province, and they are narrow and steeply dipping, so that more of the neighbouring waste rock must be removed in the mining processes. However, some of the bodies contain ore of very high quality, which in places attains a grade of almost 69 per cent Fe, with less than 1 per cent SiO_2 and 0.01 per cent P.

Pisolitic limonite ore is the next most important type of iron-ore from an economic viewpoint. It formed in Tertiary times, that is about 20 million years ago, in the general area of the Hamersley Iron Province. By that time, the hematite enrichment ore deposits (which had formed at depth well over 1000 million years earlier within their even older BIF host rocks) had been exposed at the surface by steadily continuing erosion. The landscape had developed to a point of broad similarity to that of the present day: the present drainage lines had been established, and the general distribution of ranges and plains would have been discernible. However, the now rugged topography of the Hamersley and Ophthalmia Ranges then had a more subdued and rounded outline, and there was a good cover of forest owing to a warm, wet climate. The rivers which drained this forested landscape had broad, meandering, permanently swampy channels. In the beds of these rivers iron dissolved from all the iron-rich rocks of the Hamersley Basin was reprecipitated and gradually replaced the associated alluvial muds and sands, to form thick valley-floor sheets of limonitic iron-ore. This material became harder as time passed, and

Fig. 3.12 Mt Tom Price, Hamersley Iron Province

Fig. 3.13 Pannawonica mine, Hamersley Iron Province

developed internally a nodular structure as though composed of closely-packed peas, each a few millimetres in diameter. This nodular arrangement is referred to as pisolitic texture and is commonly, but not always, present in pisolitic limonite ore. Small fragments of fossil wood, preserved in limonite, are also not uncommon. As the climate later became gradually more arid and closer to that of the present day, the whole landscape was further lowered by erosion. At that stage the sheets of limonitic iron-ore in the valleys proved resistant to erosion, and as the rivers cut down further, the ore was preserved as cappings on discontinuous lines of flat-topped hills known as mesas, standing up above, and running sinuously alongside the new drainage lines.

Thus at present the pisolitic limonite ores occur on mesas along all the major lines of drainage of the Hamersley Iron Province. They are best developed along the Robe River, in the northwestern part (Fig. 3.13). There the mesas stand about 75 m above the present valley floors and are capped by limonite with a thickness usually less than 30 m, but locally up to 60 m. As yet this type of ore is mined only at Pannawonica, but extensive deposits also occur at Deepdale, further down-stream. It is a tough but fairly brittle brown or reddish-brown material with a density between 2.5 and 3.5 g/cm^3. It is of lower grade than hematite enrichment ore, but this is compensated for by its low content of impurities, especially phosphorus. This ore has a compositional range as follows: 40-60 per cent Fe, 7.5-11 per cent H_2O+, 3-7 per cent SiO_2, 1.8-4.5 per cent Al_2O_3, 0.015-0.05 per cent P.

Sedimentary ore occurs only in the Kimberley area. About 1800 million years ago, at much the same time as the Hamersley Basin existed, the Kimberley area was covered by a somewhat similar sedimentary basin, which differed in that it was filled in by mainly clastic sediments : sandstones, siltstones and shale. Locally in the basin, beds of sand-sized and silt-sized grains of hematite were deposited. Subsequent lithification, folding and erosion of this sedimentary sequence has produced bodies of high-grade hematitic ore not unlike (in shape and size as well as in composition) many bodies of hematite enrichment ore, although their origin is, as described, very different.

This type of ore is mined only at Cockatoo and Koolan Islands, in Yampi Sound. The mined orebodies have a width of up to 40 m (but usually about 25 m), and reach over 2 km in length. The material is soft at depth, thereby differing from the Hamersley ores, but a hard surface crust is locally present.

The Yampi Sound ore is of excellent quality: the main orebody on Koolan Island has an average composition of 67.0 per cent Fe, 1.5 per cent SiO_2, 1.2 per cent Al_2O_3, 0.6 per cent H_2O+, and 0.02 per cent P.

Production and reserves

Before export began in 1966, production of iron-ore in Western Australia was comparatively small, most of Australia's needs being supplied by the Middleback Range Mines of South Australia. The abrupt expansion initiated by export is clearly shown in Fig. 3.14, in which Australian and Western Australian productions from

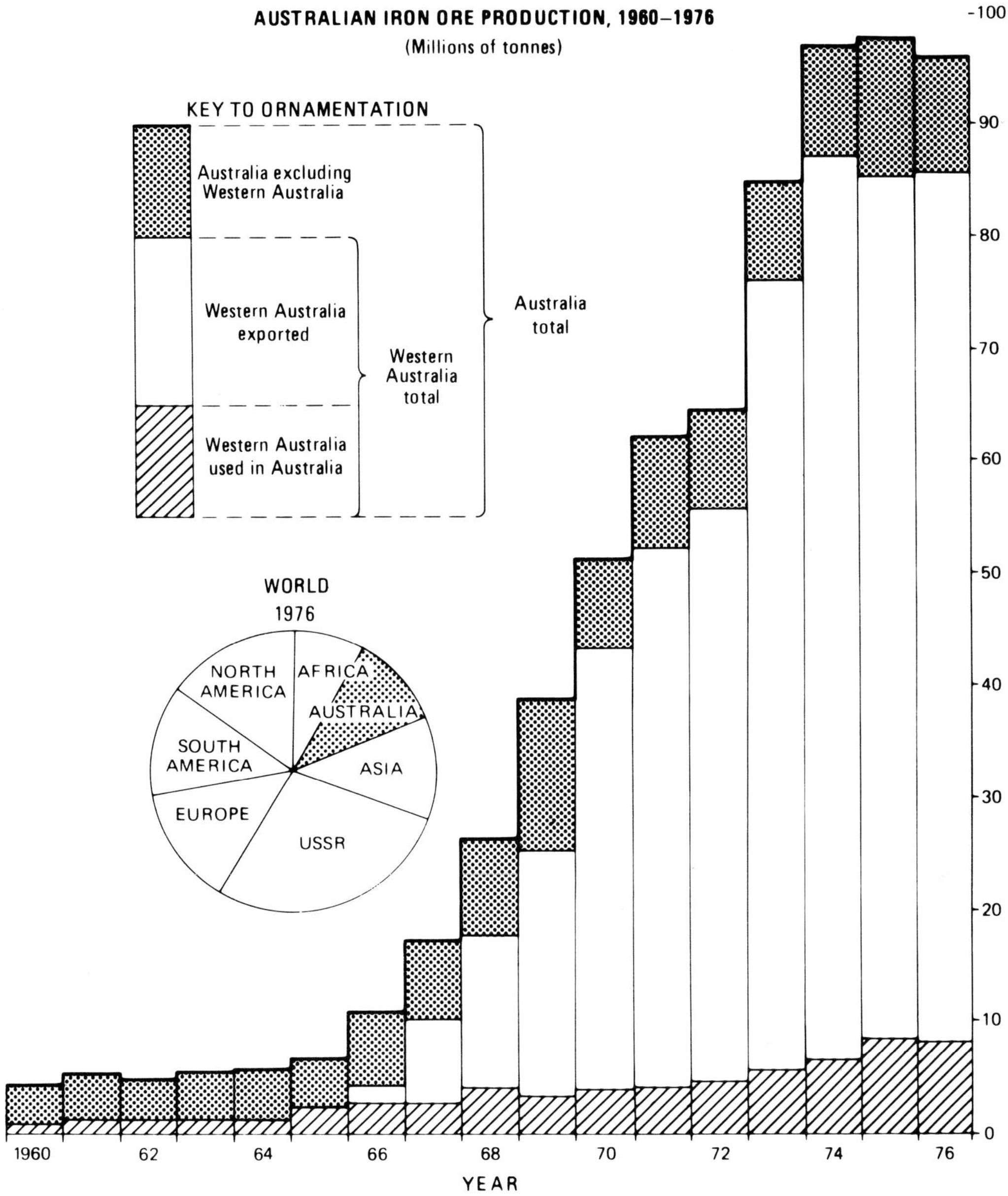

Fig. 3.14 Australian iron-ore production, 1960-76 (millions of tonnes)

1960 to 1976 inclusive are illustrated, as well as the contribution of Australia in relation to world production for a representative recent year. Insofar as it is possible to draw general conclusions from recent years, the histogram suggests development of a 'plateau' after the initial period of rapid growth following Western Australia's entry into the international market.

Iron is one of the few mineral commodities which appear to be in abundant supply in the long-term future. For Western Australia, reserves of hematite ore of 55 per cent Fe or above stand at about 30 billion tonnes, which includes more than 8 billion tonnes of high-grade (60 per cent Fe or better) ore reserves. At present mining levels it appears that almost a century of mining remains before ores of lower grade need to be mined. However, the real position is more complicated than this, since the ability to sell iron-ore in the export market depends on very small details of its chemical and physical characteristics. For example, a high proportion of Western Australian high-grade ore has a comparatively high phosphorus content, which makes it unsaleable under existing conditions. Therefore, while it can safely be said that Western Australia has sufficient reserves of iron-ore to supply the domestic needs of Australia for many centuries, it is less certain that the state will be able to maintain a competitive position in the international export market.

REFERENCES

Maitland, A. G. 1919. The iron deposits of Western Australia. In *The Mining Handbook*. W. Aust. geol. Surv. Mem. **1**, chap. II: 3-12.

Trendall, A. F. 1975. Geology of Western Australian iron ore. In *Economic Geology of Australia and Papua New Guinea—1. Metals* (ed. C. L. Knight). Australas. Inst. Min. Metall. Monogr. Ser. 5: 883-92.

Woodward, H. P. 1890. Mineral Wealth. In *Annual General Report of Government Geologist for 1888-1889*. Perth: Govt Printer: 21-7.

——— 1894. Metallic Minerals. In *Mining Handbook to the Colony of Western Australia*. Perth: Govt Printer: 39-50.

3D

Nickel

R. Woodall and G. A. Travis
Western Mining Corporation Ltd

HISTORY

Western Australia's small population of 1.2 million is concentrated in Perth and the nearby agricultural areas, leaving the semi-arid to arid interior only sparsely settled. The interior is largely flat, deeply weathered and soil-covered, so that, despite the activities of several generations of prospectors, it is still only partly explored in terms of near-surface mineral deposits. Extensive surface concealment, leached outcrops and deep overburden combine to make this area some of the most difficult and costly to explore—which helps to explain why Western Australia's nickel potential remained unknown for so long.

Exploration for nickel in Western Australia was first undertaken in 1955, in the remote Wingelinna area near the border of South Australia and the Northern Territory. Intermittent drilling programmes between 1955 and 1970 outlined reserves of 56 million tonnes of nickeliferous laterite averaging 1.24 per cent Ni in a number of deposits developed in sheared ultramafic rocks (Sprigg & Rochow 1975). Although the activities of the 1950s received considerable publicity, they did little to stimulate exploration in the more accessible Archaean areas where major nickel sulphide deposits were to be discovered a decade later.

The discovery of high-grade nickel sulphide ore at Kambalda in January 1966, which triggered-off five years of intensive prospecting and set Australia on the path to becoming the Western world's third largest nickel producer, was the culmination of events which began twelve years earlier. In late 1954 prospector-farmer George Cowcill submitted to the Western Australian School of Mines in Kalgoorlie some green-stained ferruginous samples from Kambalda, after being struck by the similarity of this material to some green, secondary uranium-bearing samples from the Northern Territory on display at the Mines Department in Coolgardie. W. H. Cleverly, then head of the Geology Department at the School of Mines, established that there was no uranium present, and in terms of the request from Cowcill, there was no obligation to carry out any further investigations. However, Cleverly decided to identify the green staining, and when chemical tests established the presence of significant amounts of nickel he wrote to Cowcill as follows: 'In view of the very meagre Australian resources of nickel and the relatively high price, it might be worth while to explore the extent of the occurrence and to submit representative samples for nickel assay.' On a further sample submitted by Cowcill early in 1955, Cleverly reported: 'It is an ironstone of sulphide derivation with a few thin greenish stains which gave a positive test for nickel.' Mr Cowcill has stated that had he not been supplied with the information that the samples contained nickel, they would have been discarded and forgotten.

In 1964 Cowcill arranged for his prospecting mate, John Morgan, to collect further samples and to show them to Western Mining Corporation. The samples were brought to Roy Woodall, then assistant chief geologist of Western Mining Corporation in Kalgoorlie, but without the earlier reports by Mr Cleverly.

After confirming that significant nickel was present in the samples and after checking their field occurrence, Woodall also considered that the samples were probable nickel gossans, and from their chemical composition and location on a serpentinite-basalt contact recognized their possible economic significance. An exploration programme was initiated in late 1964, high grade nickel sulphides were discovered in January 1966, and a nickel exploration boom in Western Australia was born.

In looking back on the Kambalda discovery, it is pertinent to remember that the decision to investigate the origin of the suspected gossans was made at a time when no nickel sulphides were known in the Archaean rocks of Western Australia. Moreover, this geological environment was not known to host major nickel deposits anywhere in the world. Geological attitudes to the nickel potential of Western Australia were then very different to those which now prevail, a fact borne out by the attitude of several major mining companies, which at the time declined to participate with Western Mining Corporation in the exploration of areas such as Kambalda.

Although several other Australian and International mining companies were actively exploring the Archaean rocks of Western Australia at the time of the Kambalda discovery, their efforts were directed towards copper-zinc exploration. However, they were the first to react to the Kambalda discovery and, in 1966 and 1967 large areas were taken up for exploration, principally as temporary reserves. With the extremely cheap rental of \$60 per reserve and no real restriction on the area which could be applied for, most of the greenstone belts were held before a ban on the granting of reserves was introduced in June 1967. By this time exploration rights had been granted over approximately 90 000 km^2, and as holders of these large tenement blocks were required to halve each reserve every year, exploration commenced in earnest.

Continued success at Kambalda and the discovery of several other deposits in the Kalgoorlie region gave considerable encouragement to exploration companies and prospectors. The excitement of discovery and a buoyant stock market attracted prospectors and entrepeneurs—a trickle in 1967, a steady stream in 1968, and a flood in 1969 and 1970 (notably the Poseidon boom following the discovery at Mt Windarra in 1969). Old companies were resurrected and over 100 new companies formed during those years to join the ranks of the established explorers.

It is estimated that exploration expenditure for nickel in Western Australia was \$2.6 million in 1966, \$6.1 million in 1967, \$15 million in 1968, \$33 million in 1969 and \$46.5 million (the peak) in 1970. Since then expenditure has declined to current levels of \$5-10 million per year. Significant events during this decade are summarized below:

1966: High-grade nickel sulphides discovered at Kambalda.

1967: Western Mining Corporation commences production of nickel concentrates from the Lunnon Shoot at Kambalda at a rate of 5000 tonnes of contained nickel per year.

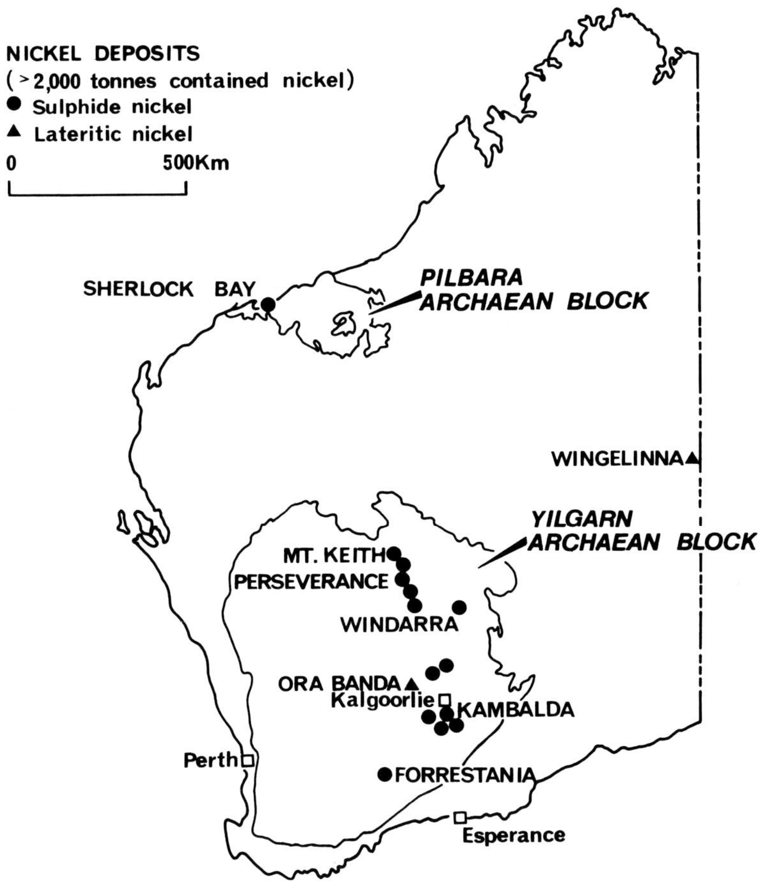

Fig. 3.15 Major nickel deposits of Western Australia

1968: Several new deposits discovered: Scotia, Nepean, Redross and others in the Widgiemooltha-Spargoville belt. They prove to be small to medium-size orebodies of similar type to those at Kambalda.

1969: New discoveries at Kambalda and St Ives by Western Mining Corporation since 1966 allow the production rate at Kambalda to double. The Mt Windarra deposit is discovered near Laverton and a very large, low-grade deposit is discovered at Mt Keith, 75 km south of Wiluna, extending the known nickel belts by several hundred kilometres. The Great Boulder-North Kalgurli consortium commences production from the Scotia nickel mine.

1970: Additional low-grade deposits are located in the Wiluna-Agnew greenstone belt and Australia's first nickel refinery is commissioned by Western Mining Corporation at Kwinana. The Nepean nickel mine is brought into production by Metals Exploration Ltd.

1971: The large Perseverance deposit is discovered by the Selection Trust group in the Agnew district, 330 km north of Kalgoorlie. Several new nickel deposits are discovered in the Forrestania belt on the eastern fringes of the wheat belt, 260 km southwest of Kalgoorlie. Kambalda production reaches 35 000 tonnes of nickel metal per annum.

1972: The Sherlock Bay deposit in the West Pilbara is discovered.

1973: Western Mining Corporation's nickel smelter at Kalgoorlie is commissioned.

1974: Production commences at Windarra by Windarra Nickel Mines and by Anaconda Australia at Redross.

1976: Plans announced for the mining of the Agnew deposit and expansion of Western Mining Corporation's nickel smelter to treat the Agnew nickel concentrates.

The deposits found during the period 1966-76, the geological ore reserves, and production to June 1976 are tabulated in Tables 3.6-3.8 and are located on Fig. 3.15.

Table 3.6

Nickel sulphide deposits in Western Australia
Higher-grade deposits (cut-off 1% Ni):
reserve estimates, June 1976

Location	*Million tonnes*	*Grade % Ni*	*Contained nickel '000 tonnes*
Kambalda—St Ives	24.3	3.2	779
Agnew	45.0	2.2	990
Mt Windarra	5.3	1.7	92
South Windarra	3.8	1.1	43
Spargoville	0.4	2.7	12
Widgiemooltha		Not available for publication	
Forrestania Belt	6.0	2.8	168
Scotia	0.1	1.6	4
Nepean	0.1	3.7	3
Carr Boyd	0.5	1.5	8
Total	85.6	2.4	2099

Table 3.7

Nickel sulphide deposits in Western Australia
Lower-grade deposits (cut-off 0.5% Ni):
reserve estimates June, 1976

Location	*Million tonnes*	*Grade % Ni*	*Contained nickel '000 tonnes*
Mt Keith	290	0.6	1 740
Yakabindie	200	0.6	1 200
Honeymoon Well	80	0.9	720
Sherlock Bay	75	0.5	375
Others (including Kambalda, Agnew, Windarra)	110	0.7	770
Total	755	0.6	4 805

In addition to the major exploration effort directed towards the search for sulphide nickel deposits, some exploration was carried out for nickeliferous laterites. Several discoveries were made close to the established facilities in the Kalgoorlie area, and these may become of some economic importance. The best known is at Ora Banda northwest of Kalgoorlie, where silcrete overlying the nickeliferous clays carries small amounts of nickel, and is already being used as a silica flux at the Kalgoorlie nickel smelter.

GEOLOGY AND REGIONAL SETTING OF THE DEPOSITS

The bulk of the Western Australian nickel sulphide deposits occur in, or adjacent to magnesium-rich ultramafic rocks in the volcanic 'greenstone' belts of the Archaean Yilgarn Block. These belts constitute about 20 per cent of the 600 000 km^2 area of the Yilgarn Block. Only a small part of the greenstone belts are exposed (approximately 15 per cent), the remainder being masked by soils, laterite and salt-lake sediments.

A typical greenstone belt consists of mafic to ultramafic volcanics and associated intrusives, felsic volcanic rocks (largely pyroclastics) with related volcanogenic sediments, and fine to coarse clastics. In the eastern half of the Yilgarn Block, where most of the nickel occurs, these belts are elongated NNW parallel to the dominant direction of strike faulting. Gemuts and Theron (1975) describe the stratigraphy in the Kalgoorlie-Norseman region and present evidence for the regional association of nickel sulphides with a time-equivalent ultramafic sequence. O'Driscoll (1971*a* & *b*) developed a tectonic pattern for the nickel sulphide province which relates the deposits to northeast and northwest trending intracontinental lineaments.

The economically significant nickel sulphide deposits can be classified into three environmental associations:

(1) volcanic association,
(2) dunitic association,
(3) mafic association.

The volcanic and dunitic associations are thought likely to be end-members of a genetically linked series, but it is convenient to discuss them as separate entities. Those with a mafic association form a more discrete family of deposits.

Volcanic association

Volcanic or subvolcanic ultramafic rocks host all the currently economic nickel sulphide orebodies, the more important being the Kambalda-St Ives deposits (Ross & Hopkins 1975), Nepean (Sheppy & Rowe 1975), Windarra (Roberts 1975), Scotia (Christie 1975) and those around the Widgiemooltha Dome (Dalgarno 1975) and deposits in the extensions of the Widgiemooltha ultramafic rocks north to the Spargoville area (Andrews 1975). The deposits in the Kambalda-Widgiemooltha area are shown in Fig. 3.16. About half of the known sulphide nickel reserves with grades in excess of 1 per cent Ni belong to this association and the deposits tend to be higher in grade than those of other associations.

Typically, the sulphides occur as pyrrhotite-pentlandite* layers in structural flexures and embayments at the base of small, altered, olivine-rich peridotite lenses up to 45 m thick. These are interpreted as the basal portions of ultramafic lava flows or shallow subvolcanic sills. The mineralized units are usually overlain by multiple thin ultramafic flows with komatiitic affinities which are lower in MgO and higher in CaO, Al_2O_3 and TiO_2 than the ore-bearing peridotites. Both the mineralized and overlying units may be extensively altered to talc-carbonate rocks (Kambalda, Redross, Windarra), serpentinites (Kambalda, Wannaway, Scotia), or metaperidotites (Nepean). Metasediments rich in albite and quartz are common on the footwall contact of the ultramafic sequence and at interflow positions within the sequence.

The deposits may contain massive, matrix, or disseminated sulphide concentrations. Where all three are present, as at Kambalda, massive ore underlies matrix ore, and both are overlain by disseminated sulphides. In some deposits matrix to disseminated sulphides are dominant and massive sulphides rare (Nepean, Windarra, South Windarra) but in others remobilized breccia ores are common (Redross). Although normally confined to the basal contact of the basal peridotite unit, ore may also occur at the base of overlying ultramafic units, as in the hanging wall orebodies at Kambalda, or as remobilized massive veins within rocks which form the footwall of the host ultramafic. In the volcanic association the Ni:Cu ratio is usually in the range 10:1 to 15:1.

The ore mineralogy is dominated by pyrrhotite-pentlandite, with subordinate or local concentrations of pyrite, chalcopyrite, magnetite, and ferroan chromite. They

* *Pyrrhotite* is $Fe_{1-x}S$ (where x is 0-0.2), containing a variable amount (up to 0.6 per cent) of nickel and *pentlandite* is (Fe,Ni)S containing up to 40 per cent Ni (36-38% at Kambalda).

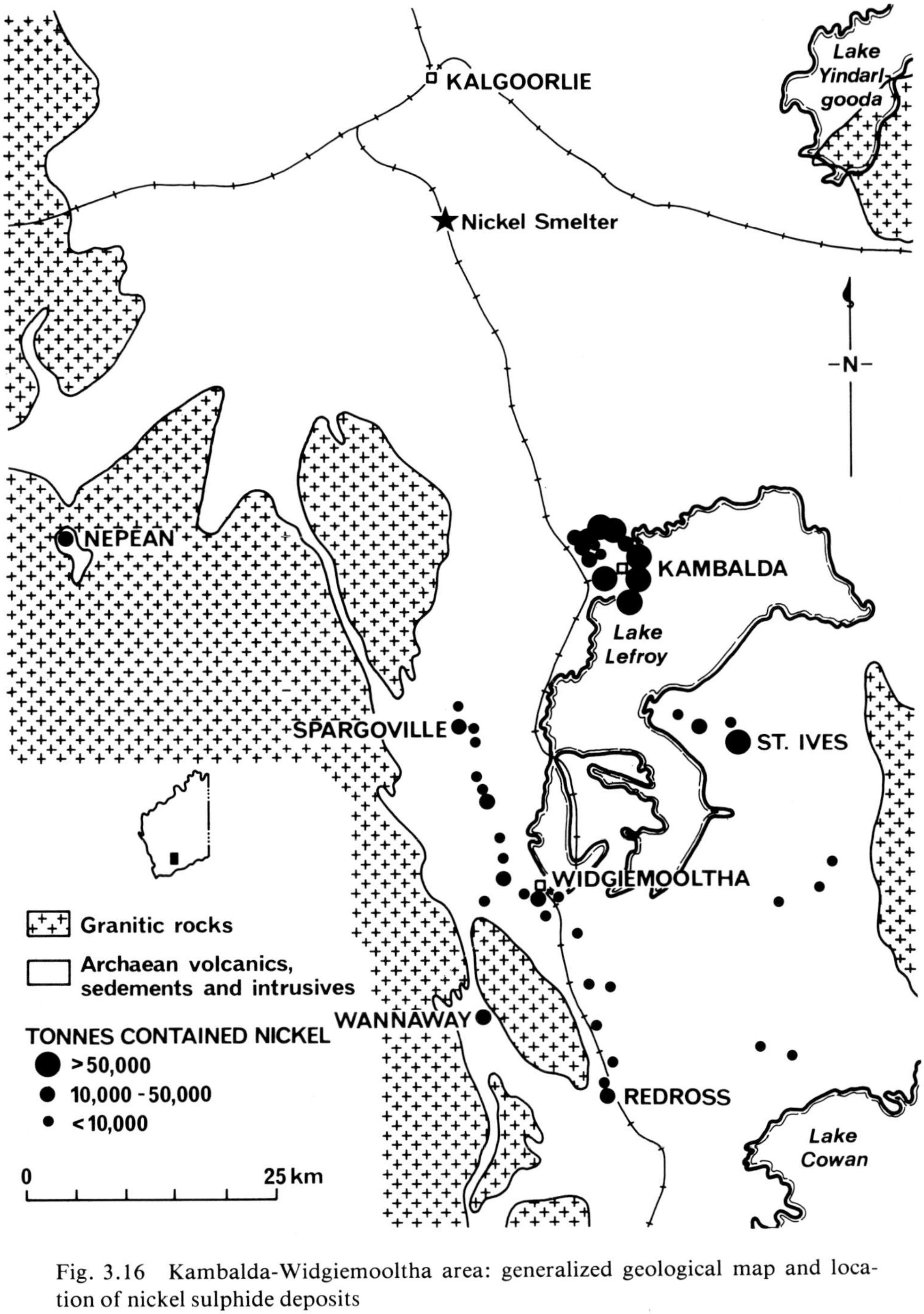

Fig. 3.16 Kambalda-Widgiemooltha area: generalized geological map and location of nickel sulphide deposits

are intergrown with silicates and carbonates and have undergone the same greenschist to amphibolite facies metamorphism as the enclosing rocks.

The ores have tectonite fabrics, best displayed by mineralogical layering in some of the massive ores. Pyrrhotite often exhibits strong physical and optical preferred orientation, together with kink bends and deformation twins.

The orebodies exhibit a significant structural control. In linear orebodies, the long axis is sub-parallel to the dominant regional fold axis, or to linear elements and shear zones in the enclosing rocks (Windarra, South Windarra, Kambalda).

Dunitic association

Collectively the deposits in this association constitute the largest sulphide nickel reserve in Western Australia, but many are low-grade (about 0.6 per cent Ni) and uneconomic under present conditions, e.g. Mt Keith (Burt & Sheppy 1975) and Yakabindie deposits (Turner & Ranford 1975, Naldrett & Turner 1977). Several important higher-grade finds have been made, principally the Perseverance deposit (Martin & Allchurch 1975) and several deposits in the Forrestania greenstone belt.

The bulk of the sulphide nickel in this association occurs in serpentinized and talc-carbonate-altered dunites in the northern section of the Leonora-Wiluna greenstone belt, encompassing the Weebo, Perseverance, Yakabindie, Mt Keith and Honeymoon Well deposits. Nickel sulphides are widely distributed in the high-magnesium rocks in this portion of the belt. Most of the sulphides occur as low-grade ($<$ 1.0 per cent Ni) disseminations within large bodies of altered dunite, and in a few places are distributed uniformly through substantial volumes of rock, e.g. Mt Keith (Burt & Sheppy 1975). Where the sulphide accumulations approach contact positions, the nickel grade tends to increase in spite of a parallel decrease in the Ni:S ratio. The most important contact deposit found to date is at Perseverance. The dunites in this belt occur in long sinuous bodies 300-500 m wide, and as discrete lenses, often overlain or linked by extrusive ultramafics. There appears to be a stratigraphic and tectonic control to the present distribution of the ultramafic rocks. The sulphides have Ni-Cu ratios in the range 30:1 to 60:1—higher than other associations.

The deposits in the Forrestania greenstone belt belong to this association, the more important being Cosmic Boy, Digger Rocks and Flying Fox.

Mafic association

Sulphide concentrations in this association characteristically occur in breccia pipes within layered intrusives or disseminated in certain members of mafic-ultramafic complexes. The best known example is at Carr Boyd, about 80 km north-northeast of Kalgoorlie, where several bronzite pegmatoid breccia pipes cut a troctolite-anorthosite sequence within the Carr Boyd complex. Others are known in the Pilbara and Kimberley regions.

Pyrrhotite usually dominates the sulphide assemblage and chalcopyrite is relatively more abundant than in other associations. The combined grade rarely exceeds 3.0 per cent Ni + Cu, and the Ni:Cu ratio is low, usually in the range 1:1 to 2:1.

Carr Boyd is the only deposit in this association that has been mined (Purvis *et al.* 1972, Schultz 1975.). Mining commenced in 1973 and has continued intermittently since then.

Several nickel deposits have been located which do not readily fall into these associations. One such example—the Sherlock Bay deposit in the West Pilbara—is in a quartz-magnetite-amphibole-sulphide schist, interpreted by Miller and Smith (1975) as a major fracture zone into which sulphides have been introduced from nearby ultramafic rocks.

There are also some small deposits in volcanic ultramafic environments in the Kalgoorlie district, where the nickel sulphides appear to be an intrinsic part of narrow volcanogenic sedimentary horizons rather than the overlying ultramafic unit.

Deep oxidation of the nickel deposits has given rise to a characteristic supergene alteration profile which may extend as much as 200 m below the surface. In most deposits the primary assemblage is pyrrhotite-pentlandite, and the initial affect of weathering is the development of violarite flecks in the pentlandite. Upward in the profile, pentlandite and then pyrrhotite are progressively replaced by violarite, and on complete replacement of pentlandite the remaining pyrrhotite is replaced by secondary pyrite. Near the water-table this pyrite-violarite zone grades into an oxide zone on decomposition of all sulphides to goethite. Leaching in the oxide zone results in progressive depletion of nickel and enrichment in silica towards the surface gossans.

DEVELOPMENT OF THE WESTERN AUSTRALIAN NICKEL INDUSTRY AND FUTURE PROSPECTS

Following the discovery of nickel sulphides at Kambalda in January 1966, Western Mining Corporation (WMC) Limited announced plans on 4 April 1966 for development of this orebody into Australia's first nickel mine. By June 1967, just 17 months after the discovery, the first nickel concentrates were produced, and in August 1967 the first concentrates were shipped overseas.

Rapid development was possible primarily because of the high demand for nickel at the time, the high grade of the ore and the proximity to an established pool of labour in the economically-depressed gold-mining centre of Kalgoorlie.

The WMC nickel refinery at Kwinana was commissioned early in 1970, and several increases in capacity have raised the original throughput of 16 000 tonnes of nickel per year to 30 000 tonnes per year in 1976. The refinery uses the Sherritt-Gordon Mines Ltd ammonia-leach process and can treat both nickel sulphide concentrates and nickel matte. Construction of the refinery prior to the smelter provided a cheaper and faster route to nickel-metal sales products. The products from the refinery are high-purity nickel metal powder and briquettes, mixed nickel-sulphides, and ammonium sulphate.

The WMC nickel smelter in Kalgoorlie, incorporating an Outokumpu-designed flash furnace, with an annual capacity of 200 000 tonnes of concentrate was commissioned in 1973.* The capacity was increased in 1975 to 350 000 tonnes per annum, and in October 1976 the company announced that agreement had been reached to toll-smelt up to 100 000 tonnes per annum of concentrate from the Perseverance Mine, near Agnew, when production commences in 1978. Expansion of the smelting facilities to cater for this additional feed is now in progress.

Table 3.8

Nickel sulphide deposits in Western Australia
Cumulative ore production, June 1976

Location	*Million tonnes*	*Grade % Ni*	*Contained nickel tonnes*
Kambalda	8.6	3.2	276 826
Scotia	0.7	2.3	14 171
Nepean	0.4	3.7	14 466
Carr Boyd	0.2	1.4	2 323
Windarra	1.7	1.6	27 836
Redross		Not available for publication	
Spargoville	0.2	2.3	4 012

Nickel output from the Kambalda mines, which has remained relatively constant at around 35 000 tonnes per year since 1971, has been supplemented by production from a number of small nickel mines in the Kalgoorlie area. Ore from the Nepean mine is milled at Kambalda, and that from Scotia, Carr Boyd, Spargoville and Redross is milled at Kalgoorlie in plants previously used for milling gold ores. Ore from the Mt Windarra and South Windarra mines is milled on site, and the concentrates are transported by road and rail to the Kalgoorlie nickel smelter. Table 3.8 shows the cumulative ore production to June 1976, and Fig. 3.17 the annual production of nickel in Western Australia since 1967. This is presently equivalent to approximately two-thirds of Australia's total production, the balance being supplied by the Greenvale laterite deposit in Queensland.

Apart from the planned production from the Perseverance mine in late 1978 or early 1979, no new developments or expansions of existing operations have been announced. This is due to a number of factors:

(1) In line with the stagnation which has affected most western economies, the world consumption of nickel has been unable to sustain its historical 6.5 per cent per annum compound growth rate.

(2) With recent technology advances, the steel industry has placed greater emphasis on lower-quality nickel products (e.g. ferro-nickels derived from

* Actually the first tonne of concentrate was put through on 5/12/72 but the smelter was not officially opened until April 1973.

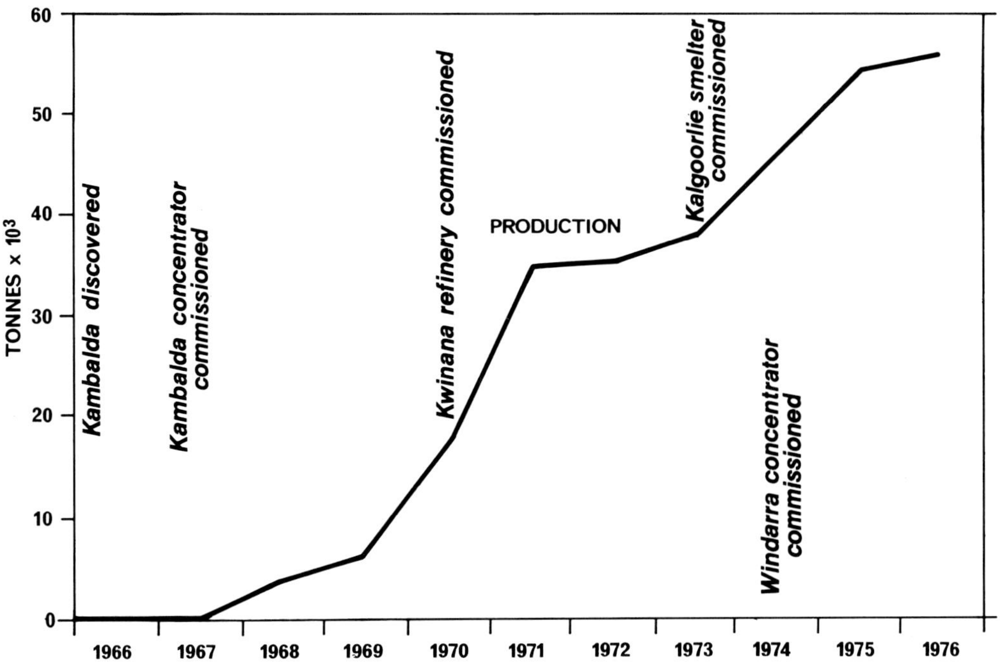

Fig. 3.17 Western Australian annual nickel production

laterites), thus causing consequential depressant effects on high-purity nickel products derived from sulphides. Since 1971 nickel price rises have not kept pace with the increased costs of production, which for the Australian mining industry have been much greater than the inflation figures based on the consumer price index. Since more than 90 per cent of Australia's nickel is exported and is sold in United States dollars, these problems have been compounded for the domestic producer by higher values of the Australian dollar relative to the United States dollar. The high value of the Australian dollar over most of the last seven years has seriously reduced the returns to Australian producers and put them at a considerable disadvantage compared with their North American counterparts. Fig. 3.18 shows the producer price of nickel metal from 1969-76 in both United States and Australian currency. Owing to the revaluation of the Australian dollar, the producer price for refined nickel in Australian currency during most of 1972, 1973 and the first half of 1974 was actually lower than the 1970 price.

(3) The nickel-mining industry in Western Australia has provided most of its own infrastructure: housing, community facilities, roads, water and power supplies, and in part its own railways. The cost of these basic facilities at Kambalda was $43 million and at Windarra $20 million, and to duplicate them today for new mines would cost substantially more.

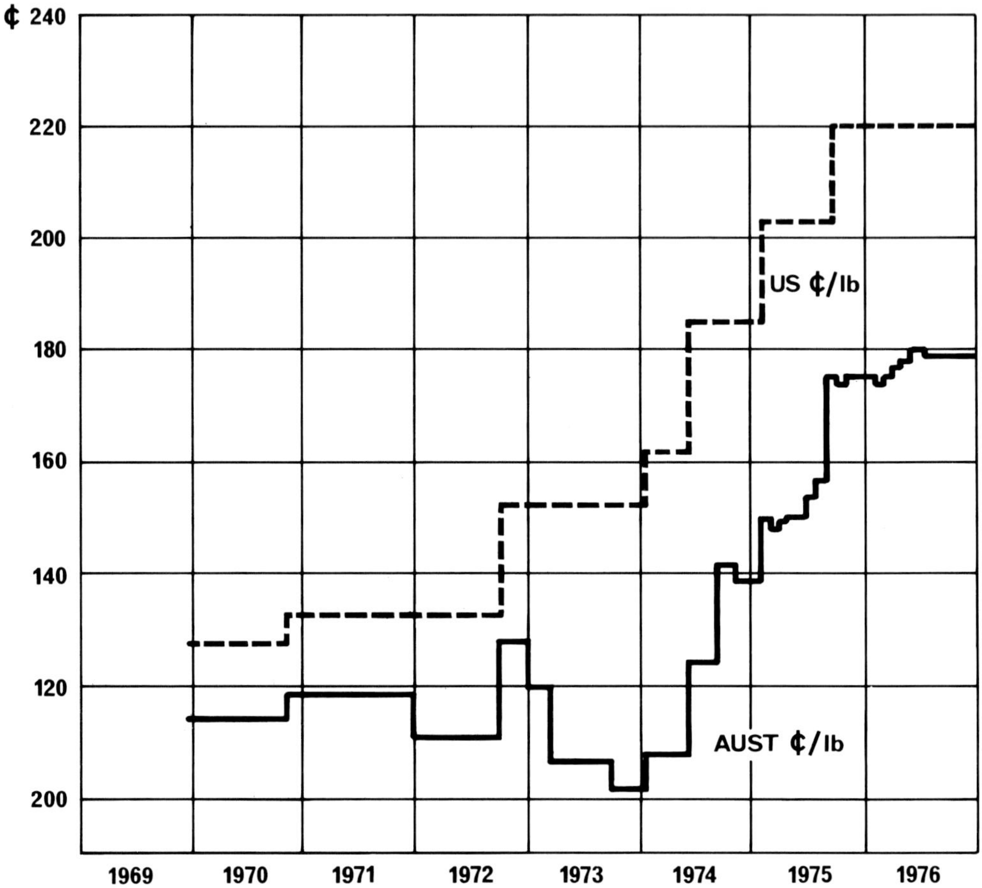

Fig. 3.18 Producer price of nickel metal, 1969-76

The combined effect of these adverse factors has been that investment in Australia's nickel-mining industry has to date yielded a relatively poor financial return.

Exploration expenditure of approximately $200 million in the last 10 years in this state has resulted in the discovery of approximately 100 million tonnes of mineralization averaging 2.3 per cent Ni and at least 600 million tonnes averaging 0.6 per cent Ni (Fig. 3.19). Future exploration will be more difficult and expensive, for it will be concerned with exploring the more concealed ground, and the unit cost of such exploration could rise exponentially as the depth of cover increases.

For future exploration to have the best chance of being economically viable, geophysical and geochemical exploration techniques need to be improved and the nature of concealment needs to be better understood. The concealment may be shallow residual soils, wind-blown sands, alluvial sheet-wash, thick variably stripped

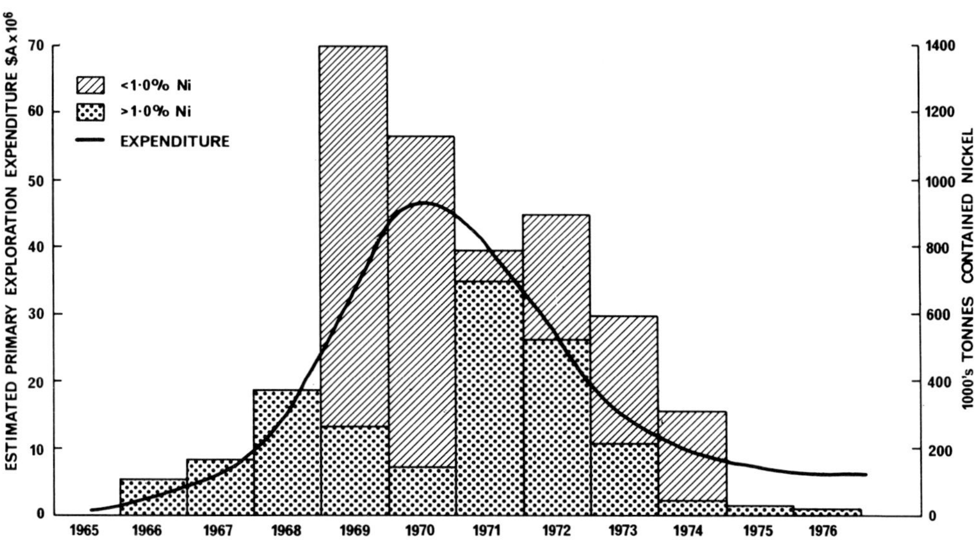

Fig. 3.19 Nickel exploration expenditure and nickel discovered in Western Australia

laterite profiles, alluvium in fossil or current drainage systems, salinas, or Tertiary marine sediments. Each type of concealment presents its own particular problems and several types of cover may be superimposed in any given area. Of special relevance to exploration is an understanding of all the parameters influencing the mobilization, transportation and precipitation of metals in the near-surface zone, so that geochemical surveys can be made more cost-effective and results more confidently interpreted. Data on the physical characteristics of the near-surface environment are relevant to the development of improved geophysical techniques and interpretation of the results.

If the frequency of nickel distribution in the covered areas is equivalent to that in exposed areas there are many deposits still to be discovered. The technical problems and challenges to future exploration are considerable, but these can be overcome, and the high cost and high risk can be justified, if there is confidence that reasonable returns will be obtained by the few who are fortunate enough to make discoveries.

REFERENCES

Andrews, P. B. 1975. Spargoville nickel deposits. In *Economic Geology of Australia and Papua New Guinea—1. Metals* (ed. C. L. Knight). Australas. Inst. Min. Metall., Melbourne, Monogr. Ser. 5: 89-91.

Burt, D. R. L. & Sheppy, N. R. 1975. Mt Keith nickel sulphide deposit. In *Economic Geology of Australia and Papua New Guinea—1. Metals* (ed. C. L. Knight). Australas. Inst. Min. Metall., Melbourne, Monogr. Ser. 5: 159-68.

Christie, D. 1975. Scotia nickel sulphide deposit. In *Economic Geology of Australia and Papua New Guinea—1. Metals* (ed. C. L. Knight). Australas. Inst. Min. Metall., Melbourne, Monogr. Ser. 5: 121-25.

Dalgarno. C. P. 1975. Nickel deposits of the Widgiemooltha Dome—Redross, Wannaway, Widgiemooltha, Dordie. In *Economic Geology of Australia and Papua New Guinea—1. Metals* (ed. C. L. Knight). Australas. Inst. Min. Metall., Melbourne, Monogr. Ser. 5: 82-6.

Gemuts, I. & Theron, A. 1975. The Archaean between Coolgardie and Norseman—stratigraphy and mineralization. In *Economic Geology of Australia and Papua New Guinea—1. Metals* (ed. C. L. Knight). Australas. Inst. Min. Metall., Melbourne, Monogr. Ser. 5: 66-74.

Martin, J. E. & Allchurch, P. D. 1975. Perseverance nickel deposit, Agnew. In *Economic Geology of Australia and Papua New Guinea—1. Metals* (ed. C. L. Knight). Australas. Inst. Min. Metall., Melbourne, Monogr. Ser. 5: 149-55.

Miller, L. J. & Smith, M. E. 1975. Sherlock Bay nickel-copper. In *Economic Geology of Australia and Papua New Guinea—1. Metals* (ed. C. L. Knight). Australas. Inst. Min. Metall., Melbourne, Monogr. Ser. 5: 168-74.

Naldrett, A. J. & Turner, A. R. 1977. The geology and petrogenesis of a greenstone belt and related nickel sulphide mineralization at Yakabindie, Western Australia. *Precam. Res.,* **5,** 43-103.

O'Driscoll, E. S. T. 1971*a*. Deformational concepts in relation to some ultramafic rocks in Western Australia. *Geol. Soc. Aust. Spec. Pub. 3,* 351-66.

——— 1971*b*. The tectonic pattern of the nickel provinces of southwestern Australia. *Soc. Min. Geol. Japan, Special Issue,* 23-33.

Purvis, A. C., Nesbitt, R. W. & Hallberg, J. A. 1972. The geology of part of the Carr Boyd Rocks complex and its associated nickel mineralization, Western Australia. *Econ. Geol.* **67,** 1093-113.

Roberts, J. B. 1975. Windarra nickel deposits. In *Economic Geology of Australia and Papua New Guinea—1. Metals* (ed. C. L. Knight). Australas. Inst. Min. Metall., Melbourne, Monogr. Ser. 5: 129-43.

Ross, J. R. & Hopkins, G. M. F. 1975. Kambalda nickel sulphide deposits. In *Economic Geology of Australia and Papua New Guinea—1. Metals* (ed. C. L. Knight). Australas. Inst. Min. Metall., Melbourne, Monogr. Ser. 5: 100-21.

Schultz, K. 1975. Carr Boyd Rocks nickel-copper deposits. In *Economic Geology of Australia and Papua New Guinea—1. Metals* (ed. C. L. Knight). Australas. Inst. Min. Metall., Melbourne, Monogr. Ser. 5: 125-8.

Sheppy, N. R. & Rowe, J. 1975. Nepean nickel deposit. In *Economic Geology of Australia and Papua New Guinea—1. Metals* (ed. C. L. Knight). Australas. Inst. Min. Metall., Melbourne, Monogr. Ser. 5: 91-8.

Sprigg, R. C. & Rochow, K. 1975. Nickeliferous ochres in the Hinckley (W.A.) and Tomkinson (S.A.) Ranges. In *Economic Geology of Australia and Papua New Guinea—1. Metals* (ed. C. L. Knight). Australas. Inst. Min. Metall., Melbourne, Monogr. Ser. 5: 1008-9.

Turner, A. R. & Ranford, L. C. 1975. Six Mile nickel prospect. In *Economic Geology of Australia and Papua New Guinea—1. Metals* (ed. C. L. Knight). Australas. Inst. Min. Metall., Melbourne, Monogr. Ser. 5: 156-9.

Woodall, R. 1976. Nickel resources of Australia-Summary. *Am. Ass. Pet. Geol. Mem.* **25,** 399-404.

3E

Bauxite

A. M. Murray
Alcoa of Australia Limited

GENERAL AND HISTORICAL INTRODUCTION

Bauxite, the general term for hydrated aluminium oxides (essentially $Al_2O_3.2H_2O$ containing up to 74 per cent alumina), is the main ore-mineral of aluminium. Bauxite deposits of economic significance occur in two regions of Western Australia, the Darling Range in the southwest, and the north Kimberley (Fig. 3.20). Both regions represent remnants of elevated and dissected laterite-capped plateaux, the bauxite forming part of the residual laterite mantle.

Laterite is probably the most abundant surface rock type in Western Australia, extending over large areas in all regions of the state (Prider 1965).

The presence of aluminous laterites has been recognized since 1912, but their commercial potential was not fully examined until 1957, when the Western Mining Corporation initiated as part of a diversification plan, a bauxite-exploration programme, which was sufficiently encouraging for a subsidiary, Western Aluminium N.L., to be formed with North Broken Hill Ltd and Broken Hill South Ltd for evaluation and exploitation of the deposits. By 1963 at least 37 million tonnes had been established, and the participation of the Aluminum Company of America then made it possible to set up a fully integrated bauxite-mining, alumina-refining and aluminium-smelting company: Alcoa of Australia Limited.

Bauxite is currently mined at Jarrahdale, Del Park and Huntly in the Darling Range by Alcoa of Australia Limited, this ore being refined locally at Kwinana and Pinjarra. Other deposits in advanced stages of development occur at Willowdale in the Darling Range (Alcoa), Mount Saddleback immediately east of the Darling Range (Alwest Pty Ltd and Reynolds Metals Company), and the Mitchell Plateau in the north Kimberley (Alumax Bauxite Corporation).

Aluminous laterites are restricted principally to near-coastal environments and all known economic deposits are found within about 100 km of the coast. Tomich (1965) and Geidans (1973) have shown that there is both a spatial and genetic relationship between bauxite formation and rainfall, at least in the Darling Range area. This relationship may date from the Tertiary or Pleistocene.

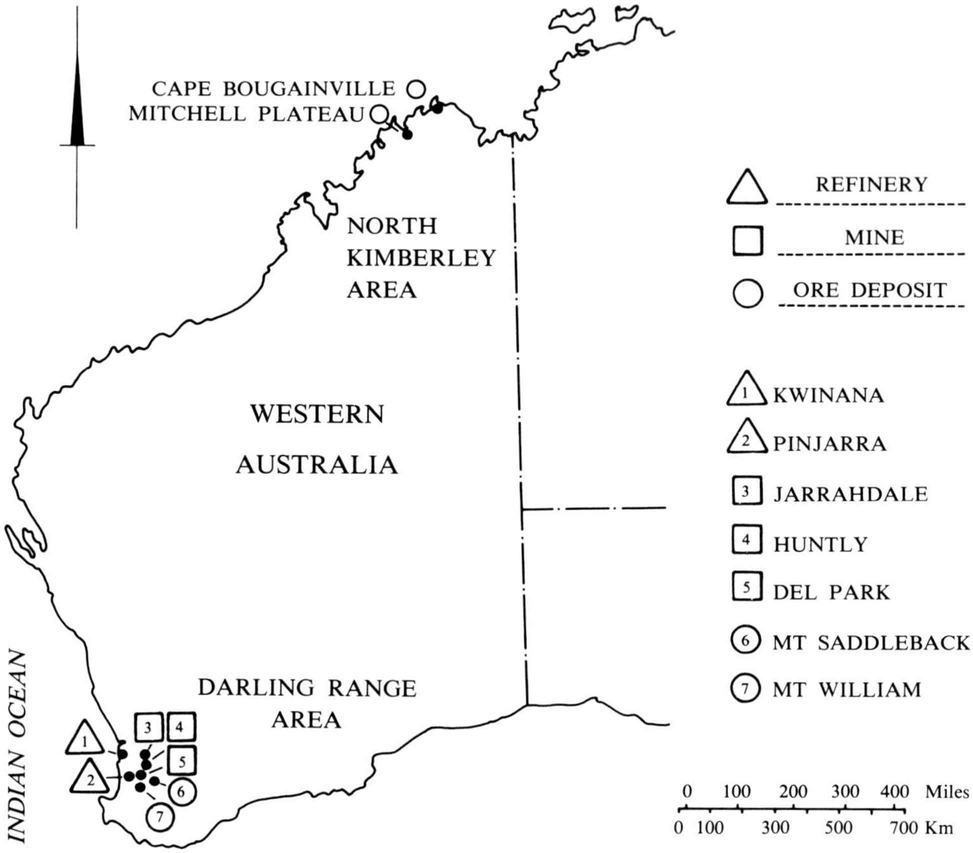

Fig. 3.20 Bauxite deposits of Western Australia

DARLING RANGE BAUXITE DEPOSITS

Location

Bauxite mining is established at two centres in the Darling Range, Jarrahdale and Dwellingup, 45 km and 98 km southeast of Perth respectively. Two mines, Del Park and Huntly, are situated in the Dwellingup area. The mines are located 5 to 10 km east of the Darling escarpment, on heavily forested ridges at altitudes ranging between 230 m and 440 m above sea-level and some 34 km inland from the coast (see Fig. 3.20).

Regional setting

The Darling Range occupies the southwestern extremity of the Precambrian Yilgarn Block, bounded on the west by the Darling Fault. The basement rocks consist of Archaean granites and gneissic granites (dated at over 2.5 billion years), with zones of metamorphic, volcanic and sedimentary rocks northeast of Perth and southeast of Bunbury. These are intruded by Late Proterozoic dolerite dykes, with a principal regional trend of about 330°, which is also reflected in topographic features. The bauxite deposits occur as part of a laterite mantle formed *in situ* by Tertiary and Pleistocene weathering of the basement rocks. This laterite is best developed on the flanks of ridges and is up to 15 m thick. It tends to be thin or absent on crests, where slightly weathered granite or (more commonly) dolerite outcrops. It is also generally absent from valleys, from which it has been eroded away.

Distribution of laterite

The main zone of laterite development extends about 60 km east of the Darling Fault, and about 350 km north to south, from the Mogumber area (31 °S) almost to the south coast near Walpole and Albany (35 °S). However, the principal areas of thick and continuous laterite lie between the Avon and Harris Rivers, a distance of approximately 150 km.

Within the general area of lateritic cover, alumina-rich facies of the laterite (i.e. bauxite) occur as isolated lenses or pockets. Within any area where the laterite has bauxite potential, individual bauxite pockets tend to have sharply outlined margins defined by the change from granite-derived aluminous laterite (bauxite) to dolerite-derived ferruginous laterite (non-bauxitic). Only in a few cases is there a gradual transition from aluminous to ferruginous laterite.

Local geology

In the Jarrahdale and Dwellingup regions, the basement complex consists predominantly of even-grained granitic rock (adamellite), generally with a foliation following the regional trends. Within these are areas of granitic gneisses and migmatitic rocks, commonly containing xenoliths of metabasic rock.

Numerous doleritic dykes cross the area, the majority being 5 to 50 m wide and following the regional south-southeast trend.

Basement rocks are exposed mainly on the steeper valley slopes resulting from river dissection, and on ridge crests, where doleritic dyke rocks are dominant. Elsewhere, the laterite cover is almost continuous, although in most valleys the upper part of the profile has been eroded to expose a thick layer of pallid-zone kaolinitic clay.

Laterite geology

The laterite profile typically comprises an upper ferruginous hardcap 1 to 2 m thick, a middle nodular mottled zone 2 to 10 m thick, and a basal clay zone from a few centimetres to several tens of metres thick (Fig. 3.21,A). The upper hardcap consists of ferruginous and gibbsitic pisolites and angular nodules in a hard hematite-goethite cement. The nodular zone consists of angular gibbsitic or ferru-

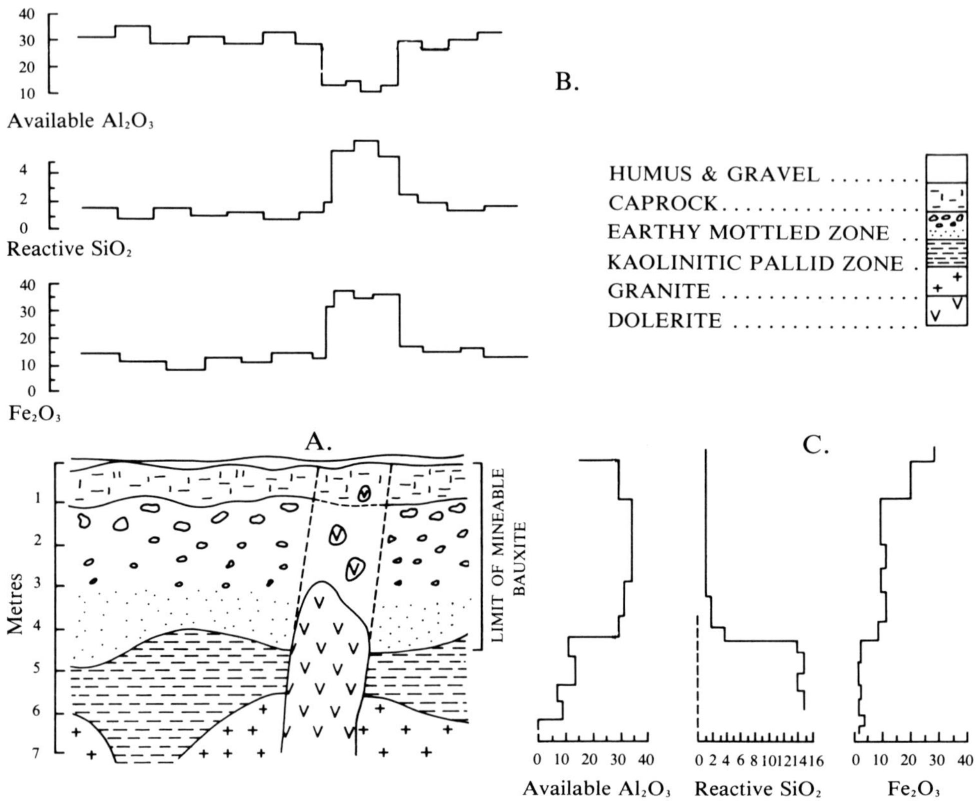

Fig. 3.21 Schematic bauxite profile, Darling Range, Western Australia: (A) geological sketch section showing zones in the laterite profile developed over granite intruded by a dolerite dyke; (B) lateral chemical variation in the bauxite caprock over section A; (C) vertical chemical variation in bauxite profile over granite

ginous concretions with minor pisolites in an earthy friable matrix. The clay zone mainly consists of kaolinitic clay and silica, with minor ferruginous mottling. It overlies the basement granites and dolerites.

Economic grades of alumina occur in the hardcap and earthy nodular zones (Fig. 3.21,A and Plate 1).

In some places banding has been observed in the caprock. Normally this can be attributed to unusual leaching conditions, but in places it could indicate sedimentary zones in the bedrock (Baker 1975).

The laterite is clearly derived from the *in situ* weathering of the basement rocks. Quartz grains, similar in size, shape and abundance to the quartz in the underlying granites occur throughout the lateritic profile, and feldspar crystal forms, pseudomorphed by gibbsite and goethite, are common. Quartz veins can be traced up through the profile, and relatively unweathered cores of bedrock with exfoliation

Plate 1 Bauxite pit at Jarrahdale, showing laterite profile (caprock at surface, friable zone and clay floor), front-end loader and haul truck, and original forest.

skins of progressively more lateritized material are found within the profile. Most significantly, contacts between granite and dolerite exposed in mine faces may be followed upwards through the laterite and can be mapped on the surface.

The laterite/bauxite derived from granitic rocks has common to abundant quartz grains, common feldspar pseudomorphs, and in many places an overall blocky or sandy fabric. In contrast, that derived from dolerite is almost free of quartz, is more ferruginous, and usually has a cellular fabric. The surface caprock also shows these characteristics, but their distinctiveness may be masked by the iron enrichment occurring at this level, particularly where the laterite is derived from granite. Dolerite-derived laterite maintains a more constant iron content throughout the profile.

The iron enrichment takes two forms: (1) cementing of pisolites and nodules by amorphous hematite-goethite; (2) development of highly-ferruginous hematite-maghemite pisolites around nucleii of quartz grains, nodules, or pre-existing gibbsite-goethite pisolites; these secondary ferruginous pisolites tend to be concentrated in zones occupying the planes of weakness in the blocky fabric mentioned above.

Mineralogy

The dominant minerals in the bauxitic laterite are gibbsite, goethite and quartz, with lesser amounts of hematite, maghemite, kaolinite (halloysite) and anatase, with rare muscovite (illite). Boehmite has been reported (Grubb 1966). The proportions of these minerals vary greatly depending on the nature of the laterite as seen by the composition of two specimens shown in Table 3.9 (analyses by x-ray diffraction).

Table 3.9

Mineralogical composition of laterites

High-grade bauxite	%	*Ferruginous pisolitic laterite*	%
Gibbsite	65	Gibbsite	10
Goethite	18	Goethite	20
Quartz	9	Quartz	5
Hematite	7	Hematite	10
Halloysite	1	Maghemite	45
Anatase	0.5	Anatase	trace
Muscovite	0.25		

Gibbsite (Al_2O_3. $3H_2O$ with 65 per cent alumina) contains most of the available alumina in the bauxite, although additional non-available alumina is present in halloysite, illite and boehmite.

The silica content of the bauxite may be mostly ascribed to quartz. However, 1 to 2 per cent is in the form of reactive silica, which varies directly with the amount of kaolinite-halloysite ($Al_2O_3 . 2SiO_2 . 2H_2O$) (Sadlier 1974). These clay minerals occur throughout the profile, but in granitic areas show a very marked concentration in the lowermost pallid zone. This contrasts with doleritic areas, where the clay minerals are more evenly distributed.

Goethite is the dominant iron oxide, occurring with gibbsite in fine or amorphous form throughout the bauxite. Hematite becomes significant in more ferruginous sections, particularly in the caprock, and tends to be concentrated in the nodular, fragmented or pisolitic fractions rather than in the groundmass. As described above, maghemite is found mainly in ferruginous pisolites in the caprock.

Chemistry

In a typical bauxite profile, developed over granite, alumina (both total and available) and iron oxides decrease downwards and reactive silica increases. In all localities, alumina and reactive silica show a marked inverse relationship attributed to the desilication of kaolinite to form gibbsite. The greatest change occurs at the base of the profile where the bauxite zone gives way, often in the space of a few centimetres, to the pallid clay zone (Fig. 3.21,C). Similar changes in chemistry are found laterally across the transition from bauxitic to non-bauxitic laterite (Fig. 3.21,B). Chemical analyses of two specimens are shown in Table 3.10.

Table 3.10

Chemical composition of laterites

	High-grade bauxite derived from granite %	*Ferruginous laterite derived from dolerite* %
Total Al_2O_3	40.5	16.6
Total SiO_2	21.8	9.8
Total Fe_2O_3	10.0	52.4
Total TiO_2	1.2	2.3
Loss on Ignition	23.6	17.1
Balance	2.9	1.8
Available Al_2O_3	37.5	11.5
Reactive SiO_2	0.3	0.4

The balance consists principally of alkalis and trace elements (principally V, Cr, Zr and Ga). In general, the trace elements show higher concentrations in dolerite-derived material, but zirconium and gallium are higher in granite-derived bauxite.

MOUNT SADDLEBACK BAUXITE DEPOSITS

Distribution

The bauxite deposits of the Mount Saddleback region lie immediately east of the Darling Range, about 130 km south-southeast of Perth. They extend from Mount Saddleback itself some 40 km in a north-northwesterly direction beyond the town of Boddington to Mount Wells.

The region has been mapped by the Geological Survey of Western Australia (Wilde 1976) and by geologists of Alwest Pty Ltd and The Broken Hill Pty Co. Ltd (Owen & Hargreaves 1975).

Mount Saddleback is a monadnock, or faulted inlier of basic volcanics, gabbros and metasediments, isolated from and rising above the general level of the Darling Range, wih an extended zone of the same rock formations occupying the lower ground towards the north-northwest. Laterite mantles the ridges of Mount Saddleback, with its thickest development on the slopes, where profiles up to 25 m thick have been recorded.

Bauxite deposits

Owen and Hargreaves (op. cit.) have described the bauxite in detail. It is considered to be derived from the *in situ* lateritization of the basement rocks, which principally consist of basic volcanics and intrusives. Intermediate to acid volcanics, together with metasediments, have been found in the western part of the Saddleback zone (Wilde op. cit.), and the whole belt is enclosed by granitic terrain of the Darling Range type. With the exception of marginal areas developed on granite or metasediments, the laterite and bauxite tends to be quartz-free and ferruginous.

The chief minerals are gibbsite, kaolinite, hematite and goethite, with minor quartz and titanium oxides and rare traces of boehmite. There is less quartz and more iron oxides than in the Darling Range bauxites.

Typical profiles comprise an upper hardcap zone 2 to 6 m thick, containing abundant pisolites and with a high iron content, a middle nodular and vughy zone with minor pisolites and a high gibbsite content, and an irregular lower clay zone consisting mostly of kaolinite with some redeposited cherty silica. Economic grades of alumina are found principally in the middle nodular zone and to a lesser extent in the hardcap zone.

Bedrock structures are occasionally preserved throughout the profile, and contacts between significantly different types of parent rock may be observed in the textures and mineralogy of the hardcap.

Bauxite reserves in the Mount Saddleback area and its immediate vicinity, excluding Boddington-Mount Wells, are reported by Owen and Hargreaves to be in excess of 200 million tonnes averaging 32.2 per cent available alumina.

KIMBERLEY BAUXITE DEPOSITS

Distribution

The bauxite deposits of Mitchell Plateau and Cape Bougainville (Fig. 3.20) border Admiralty Gulf on the northern coast of Western Australia, between lat. 13° 45′S and 15° 30′S and long. 125° 30′E and 126° 15′E. The nearest towns are Wyndham, 250 km to the southeast, and Derby, 350 km to the southwest.

The Admiralty Gulf area of the Kimberley has been mapped and described by the Bureau of Mineral Resources in collaboration with the Geological Survey of Western Australia (Gellatly & Sofoulis 1969, Allen 1971). Geologists of Amax Bauxite Corporation have studied and mapped the bauxite deposits.

Laterite occurs as a residual capping on basic volcanics with interbedded sediments. In the largest area, Mitchell Plateau, some 120 km^2 are covered by laterite, of which 40 km^2 consist of bauxite. At Cape Bougainville there are 85 km^2 of laterite, of which 44 km^2 are bauxite (Joklik *et al.* 1975).

Bauxite deposits

The geology and geochemistry of the bauxite have been described in detail by Joklik *et al.* (op. cit.). It is considered to have formed by *in situ* lateritization of the volcanics and interbedded sediments, and in common with all other Western Australian bauxites, has a dominant mineralogy of gibbsite, hematite, goethite and clay minerals, with minor boehmite, anatase, quartz, and vanadium oxide.

The Kimberley bauxites differ from Darling Range bauxite in having less quartz (traces as against about 20 per cent) and more boehmite. They have a boehmite-to-gibbsite ratio of about 1:10. This largely reflects their derivation from basic rather than from granitic rocks—the source of the Darling Range deposits. In this respect the Kimberley bauxites are more akin to the Mount Saddleback bauxites, which are also derived mainly from basic volcanics and have a similar mineralogy and chemistry.

Joklik *et al.* (op. cit.) report that the bauxite profiles on Mitchell Plateau consist of an upper ferruginous zone, a middle pisolitic zone, and a lower gibbsite zone. All zones may contain economic concentrations of alumina, and bedrock structures (e.g. pillows) may persist through all levels. Profiles on Cape Bougainville are reported to be ferruginous throughout, and tend to have tubular rather than pillowed or pisolitic structures.

Ore reserves in the Kimberley region are reported to be in the order of 1210 million dry tonnes, with grades between 36 per cent and 47 per cent total alumina.

OTHER AREAS OF BAUXITE

Occurrences of gibbsite have been reported from many parts of the state by Simpson (1951), but for the most part they represent only low-grade accumulations of alumina. Excluding the Kimberley area and the southwest region, the highest reported grades were found in lateritized norite at Ora Banda in the Coolgardie Goldfield. This deposit contained 16 per cent total alumina.

However, there are numerous deposits of bauxite in the South-West, to the east and south of the Darling Range proper, extending as far as Walpole and Albany, but they are significantly smaller and more isolated than the known deposits in the Darling Range area.

REFERENCES

Allen, A. D. 1971. Montague Sound 1:250 000 Geol. Series, Explan. Notes. *Bur. Miner. Resour. Geol. Geophys. Aust.* SD/51-12.

Baker, G. F. U. 1975. Darling Range bauxite deposits, W.A. In *Economic Geology of Australia and Papua New Guinea—1. Metals* (ed. C. L. Knight). Australas. Inst. Min. Metall. Monogr. Ser. 5: 980-6.

Geidans, L. 1973. Bauxitic laterites of the south western part of Western Australia. In *Pap. W. Aust. Conf. 1973.* Australas. Inst. Min. Metall.: 173-82.

Gellatly, D. C. & Sofoulis, J. 1969. Drysdale-Londonderry 1:250 000 Geol. Series, Explan. Notes. *Bur. Miner. Resour. Geol. Geophys. Aust.* SD/52-5 and -9.

Grubb, P. L. C. 1966. Some aspects of lateritisation in Western Australia. *J. roy. Soc. W. Aust.* **49**: 117-24.

Joklik, G. F., Jackson, W. D., & Zani. J. A. 1975. Kimberley bauxite deposits, W.A. In *Economic Geology of Australia and Papua New Guinea—1. Metals* (ed. C. L. Knight). Australas. Inst. Min. Metall. Monogr. Ser. 5: 968-80.

Owen, H. B. & Hargreaves, M. R. 1975. Mount Saddleback bauxite area, W.A. In *Economic Geology of Australia and Papua New Guinea—1. Metals* (ed. C. L. Knight). Australas. Inst. Min. Metall. Monogr. Ser. 5: 987-91.

Prider, R. T. 1965. The lateritized surface of Western Australia. *Aust. J. Sci.* **28**: 443-51.

Sadlier, S. B. 1974. Mineralogical aspects of bauxite in the Jarrahdale area, Western Australia. Hons. thesis, Univ. W. Aust.

Simpson, E. S. 1951. *Minerals of Western Australia.* **2**: 366-90. Perth: Govt Printer.

Tomich, S. A. 1965. Bauxite in the Darling Range, Western Australia. *Proc. Australas. Inst. Min. Metall.,* **212**: 125-35.

Wilde, S. A. 1976. Pinjarra 1:250 000 Geol. Series, Explan. notes on the Precambrian of the Pinjarra sheet. *Geol. Surv. W. Aust. Record.*

3 F

Mineral sands

John L. Baxter
Geological Survey of Western Australia

GENERAL

The heavy mineral industry is based on the titanium-bearing oxide minerals ilmenite, rutile and leucoxene, and the refractory mineral zircon. The minerals monazite and xenotime, which contain rare earth elements, are produced as by-products. Table 3.11 indicates the proportions of these minerals in the heavy mineral fraction of some Western Australian deposits. The heavy minerals commonly make up 5 to 40 per cent of the heavy mineral sand. The minerals have been mined from contemporary beaches and dunes along the coast and from buried or 'fossil' shorelines, which are now some distance inland (Fig. 3.22).

Table 3.11

Composition of heavy mineral fractions from Western Australian deposits

Mineral		*Eneabba*	*Capel*	*Yoganup*
Ilmenite	$FeTiO_3$	64.8	73.5	57.0
Altered ilmenite	$(Fe,Ti)O_3$		9.4	4.5
Leucoxene	$TiO_2(Ti,Fe)O_3$	2.2	5.5	6.0
Rutile	TiO_2	10.9	0.4	1.0
Zircon	$ZrSiO_4$	17.1	6.8	12.0
Monazite	$(Fe,La,Di)PO_4$	0.2		0.3
Kyanite	Al_2SiO_5	2.7		
Others		2.3		1.2

Rutile and ilmenite are natural sources of titanium oxide and have been used to produce paint pigment, welding-rod flux and titanium metal. The manufacture of titanium oxide pigment supplies the base for most of the paint used in the world and consequently the demand for the raw material is closely allied with economic conditions. In the period 1974 to 1977, when the economic climate of the world was depressed, so too was the demand for titanium oxide ores.

Zircon is used as foundry sand, in refractory bricks and linings, in ceramics, and

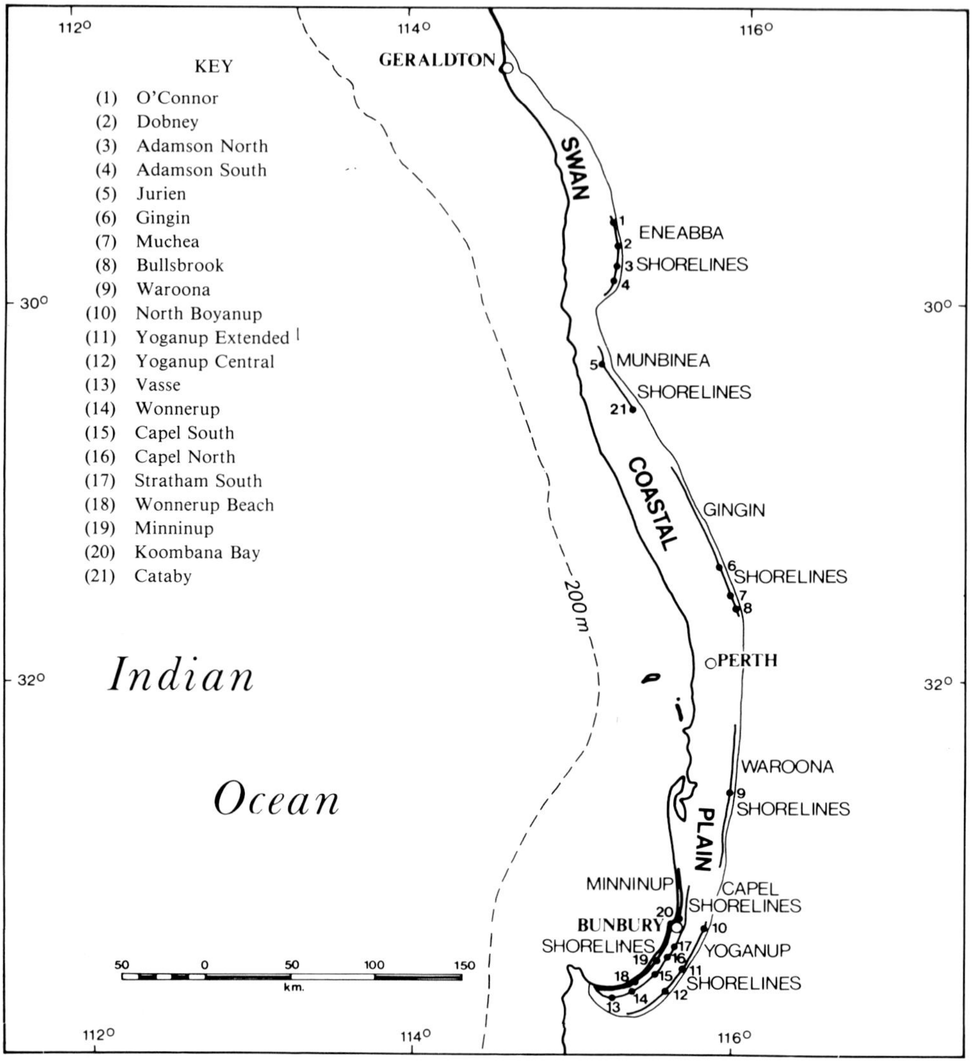

Fig. 3.22 Location of heavy mineral deposits in Western Australia, with the positions of the named shoreline sequences. The continuous line marking the eastern boundary of the Swan Coastal Plain is a scarp.

for the production of zirconium metal. In the past three years the use of zircon in refractory linings has increased dramatically and has led to an increased demand for the mineral. During this time the Australian Government operated a floor price scheme, thus effectively increasing the sale price of Australian zircon from about $30 per tonne in 1973 to more than $200 per tonne in 1975. The reduced demand for zircon because of the slow-down in the world economy, together with the high sale price, led to a slump in the world market. These factors also led to a reduction in the floor price in 1977 to $80 per tonne.

Table 3.12

Production of mineral sands, 1975

Item	*Western Australia*	*Australia*	*World*
Ilmenite concentrate	945 049	991 443	(a) 3 256 000
Leucoxene concentrate	16 871	16 871	
Monazite concentrate	3 505	4 507	12 000
Rutile concentrate	44 087	348 350	366 000
Zircon concentrate	95 958	382 217	510 000

(a) Not available separately
Source: Australian Bureau of Statistics, Perth, W. Aust.

Western Australia produces nearly one-third of the ilmenite used in the world (Table 3.12), and with the establishment of mines at Eneabba can expect to supply a similar proportion of the world's zircon and rutile needs. In March 1973 it was estimated that Western Australia possessed known deposits containing 41.269 million tonnes measured reserves, 32.312 million tonnes of indicated reserves and 34.9 million tonnes of inferred reserves. This information is expanded in Tables 3.13 and 3.14. These reserves should guarantee a continuing supply of heavy minerals to the world market. There are, however, deposits of heavy minerals in South Africa and Sierra Leone, which are currently being developed and which, if they are brought into large-scale production, will compete for a part of the market for all heavy minerals and may lead to a reduction in Western Australia's share of the world supply.

HISTORICAL REVIEW

Black sands containing heavy minerals were first mined in Western Australia in 1949 from the shores of Cheyne Bay on the south coast. Only 158 tonnes of ore were produced before mining ceased. The first commercially important heavy mineral sand deposit on the Swan Coastal Plain (Fig. 3.22) was reported in 1949 at Koombana Bay, near Bunbury. This was followed by the discovery of deposits in the vicinity of Capel and Yoganup. Mines were established at Koombana Bay in 1956 by Cable Sands Pty Ltd, Capel South in 1957 by Western Titanium Ltd, Yoganup Central in 1959 by Westralian Sands Ltd, Wonnerup Beach in 1959 by Ilmenite Minerals Pty Ltd, Capel North in 1964 by Western Mineral Sands Pty Ltd, Wonnerup in 1966 by

Table 3.13

Heavy mineral reserves in Western Australian deposits as reported to the Mines Department

Area	*Demonstrated reserves of heavy minerals*						*Inferred heavy minerals (tonnes)*
	Measured (tonnes)	*Indicated (tonnes)*	*Grade of ore (wt-%)*	*Composition of heavy mineral fraction (wt-%)*			
				İlmenite	*Rutile*	*Zircon*	
Scott Coastal Plain, Bremer Basin, Leeuwin Block		7 563 300	9.9	65.9	3.4	6.1	2 461 000
Southwest (Busselton, Capel etc)	16 420 000	13 120 700	14.4	77.2	1.0	7.3	3 161 000
Midlands (Eneabba, Jurien etc)	24 849 000	9 607 800	9.34	56.8	8.48	21.3	3 158 300
Miscellaneous (excludes Onslow and Lockyer Prospects)		2 021 000					26 100 000
State totals	41 269 000	32 312 800					34 880 300

Table 3.14

Published reserves (including possible ore) of heavy minerals in Western Australia, December 1976

Deposit	Average grade (weight %)	Proved* (tonnes)	Probable* (tonnes)	Possible* (tonnes)	Composition of heavy mineral fraction Ilmenite (weight %)	Rutile (weight %)	Zircon (weight %)
Capel South (a)		6 600 000	1 300 000				
Adamson deposit (a)	9.7	5 226 600	12 615 000			7.8	
O'Connor & Dobney deposit (c)			9 550 000				
Boodanoo (b)				1 000 000	55		
Cheyne Bay (b)	50		304 800		73.5	2.6	19.7
Bremer Bay (b)			371 870		69.0	11.0	20
Minninup (b)	20	116 200			77		5
Wonnerup Beach (d)		130 443			85	2	6
Jurien (a)	9	1 178 000	710 000	1 158 000	55-60	10	10
Jurien (c)	7.4	2 032 000	477 520	609 600			
Gingin (c)	10	1 500 000					
Busselton (c, e)		1 000 000					
Cataby (c)	7.8		700 000		70	7	12
Totals		17 783 243	26 029 190	2 767 600			

* These terms are approximately equivalent to measured, indicated and inferred

Notes: (a) Company report; (b) Geological Survey of Western Australia report; (c) newspaper report; (d) Mineral Resources of Australia Summary Rept 2; (e) includes several deposits on Capel line and on Leeuwin Block.

Table 3.15

*Total production of heavy mineral sand products from individual deposits in Western Australia to 31 December 1977**

Deposit	Company	Ilmenite	Leucoxene	Rutile	Zircon	Monazite	Xenotime	Crude concentrate	Total individual deposit
		Production in tonnes							
Capel South	Western Titanium Ltd	3 331 712(*a*)	14 180	25 578	290 834	30 069	242		3 692 615
Capel North	Western Mineral Sands Pty Ltd	2 181 195			57 687(*c*)	4 938 (*c*)			2 243 820
Stratham South	Cable (1956) Pty Ltd	925 357	30 181		54 709	2 851	6		1 013 104
Wonnerup	Cable (1956) Pty Ltd	260 528	14 189		25 126	949	2		300 794
Yoganup Central and Extended	Westralian Sands Ltd	1 643 254(*b*)	47 530	1 173	222 094(*d*)	8 676			1 922 727
Koombana Bay	Cable (1956) Pty Ltd	227 550			3 481	99(*d*)			231 130
Wonnerup Beach	Cable (1956) Pty Ltd	194 693							194 693
Cheyne Bay	Rare Metals Pty Ltd							158	158
O'Connor	Jennings Mining Ltd (*e*)	413 362	3 025	107 429	58 965				582 781
Adamson	Allied Eneabba Pty Ltd (*e*)	397 231		74 508	98 662	2 365			572 766
	Western Titanium Ltd (*g*)			11 611	12 141				23 752
Jurien	WMC Mineral Sands Ltd (*f*)			16 109	1 488				17 597
State Total		9 574 882	109 105	236 408	825 187	49 947	250	158	10 795 937

* Figures based on those from Department of Mines (Statistics Branch) and supplemented by company information.

(*a*) 146 805 t of upgraded and reduced ilmenite produced from this mine; (*b*) 20 998 t of upgraded ilmenite from this mine; (*c*) up to 1972; (*d*) includes Capel North production 1973-77; (*e*) first production 1974; (*f*) first production 1975; (*g*) first production 1977.

Cable Sands Pty Ltd, and Busselton in 1976 by Westralian Sands Ltd. Ilmenite was the main mineral sought by these companies, but small quantities of zircon, monazite, rutile, and xenotime are produced as by-products (Table 3.15).

In 1964 Laporte Titanium (Australia) Ltd established a factory in the Bunbury region to process ilmenite mined at Capel. This factory produces titanium oxide pigment. In 1968 Western Titanium established at Capel a commercial beneficiation plant to produce synthetic rutile. This was the first commercial plant of its kind in the world. The process effectively increases the percentage of titanium oxide (TiO_2) in ilmenite from 54 per cent to 92 per cent (synthetic rutile) and the product is suitable for producing pigment. These industries have assisted in the development and stability of the heavy mineral sand industry in Western Australia.

In 1970 a syndicate headed by Mr J. Adamson located at Eneabba one of the largest heavy mineral sand deposits in the world. The ensuing prospecting led to discoveries of deposits in similar geological environments at Jurien, Cataby and Gingin (Fig. 3.22). These deposits, unlike those in the Capel-Yoganup area, contain important quantities of rutile and zircon as well as ilmenite. Mines have been established at Eneabba by Jennings Mining Ltd, Allied Minerals N.L. and Western Titanium Ltd since 1974, and W.M.C. Mineral Sands Ltd undertook pilot studies at Jurien in 1975, but the mine site was put on a care-and-maintenance basis in 1977. The Eneabba deposits have introduced Western Australia to the world market for rutile and zircon, and guarantee the continued importance of the state as a future supplier of heavy minerals to the world.

Reviews of the heavy mineral deposits in Western Australia have been published by Lissiman and Oxenford (1975), Welch *et al.* (1975) and Baxter (1977). The reader should consult these works for more details on individual deposits and their regional geological setting.

DEPOSITS ON THE SWAN COASTAL PLAIN

General

Only the deposits on the Swan Coastal Plain will be discussed further, for it is these deposits that have contributed to the development of the heavy mineral sands industry. The Swan Coastal Plain is a strip of generally low-relief terrain 500 km long and up to 30 km wide, which is bounded to the east by series of scarps, shown as a line on Fig. 3.22.

Heavy mineral sands accumulate along the coast as deposits along beaches and in the first line of dunes. On the Swan Coastal Plain there are four sub-parallel dune systems preserved, each of which represents a still-stand of sea-level during the build-up of the Swan Coastal Plain. Three of these shorelines are known to contain heavy mineral deposits (Fig. 3.22); the fourth is slightly inland from the present coast and contains extensive limestone deposits, but is generally poor in heavy minerals. The dune systems are now separated by alluvial floodplains associated with major rivers draining the hinterland. The dune system furthest inland is at the foot of a scarp which extends from Eneabba to Cape Naturaliste. The scarp is known, successively from north to south, as the Gingin, Darling and Whicher Scarps. Correlation of the shoreline sequences is summarized in Table 3.16. The differences in the relative height above sea level of the correlated shorelines is interpreted as being a result of minor crustal warping after deposition of the heavy mineral sands during the Cainozoic Era.

Table 3.16

Correlation of shorelines on the Swan Coastal Plain

	Swan Coastal Plain			
Age	*North*		*South*	
Youngest			Minninup shoreline	(−3 to +3 m)
Intermediate	Munbinea shoreline	(45 m)	Capel shorelines	(3-12 m)
Oldest	Eneabba shoreline	(58-115 m)	Waroona shoreline	(20-60 m)
	Coffee Line shoreline	(60 m)	Yoganup shoreline	(26-76 m)
	Cataby shoreline	(80-90 m)		
	Gingin shoreline	(76 m)		

Figures in parentheses refer to the level above mean sea-level of the deposits

The process of heavy mineral accumulation can be seen on the coast today at Minninup, Wonnerup Beach and Koombana Bay, where concentrations occur on the beach and in the first lines of dunes. The heavy mineral deposit at Koombana Bay contained the wreck of a whaler, the *North America*, which grounded in the 1840s. Periodically the Minninup Beach deposit is partially removed by storm waves, only to be replaced by the ensuing swell-wave pattern. All of these deposits are small in terms of heavy mineral content, but share the characteristics of the larger buried deposits. Typical cross-sections of the heavy mineral deposits at Koombana Bay (Fig. 3.23) and Capel South (Fig. 3.24) indicate the similarity of individual beach sand lenses and their relationships to the interlayered silt, sand and clay deposits. At Capel South repeated beach formation and burial has resulted in a complex distribution of the individual sand units.

ORIGIN OF THE DEPOSITS

The Swan Coastal Plain is composed of beach, dune and river deposits which have built up during the past 65 million years. A modern analogy of the ancient environments responsible for the development of heavy mineral deposits can be seen in the vicinity of the contemporary coastline at Koombana Bay, Minninup and Won-

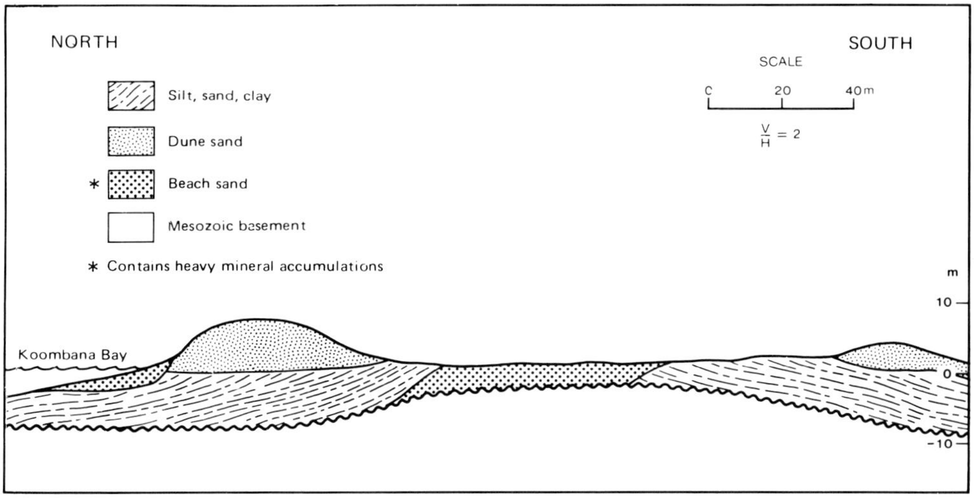

Fig. 3.23 Diagrammatic cross-sections of the Koombana Bay heavy mineral deposit outlining the lithologies present. The inland beach deposit was mined by Cable (1956) Ltd.

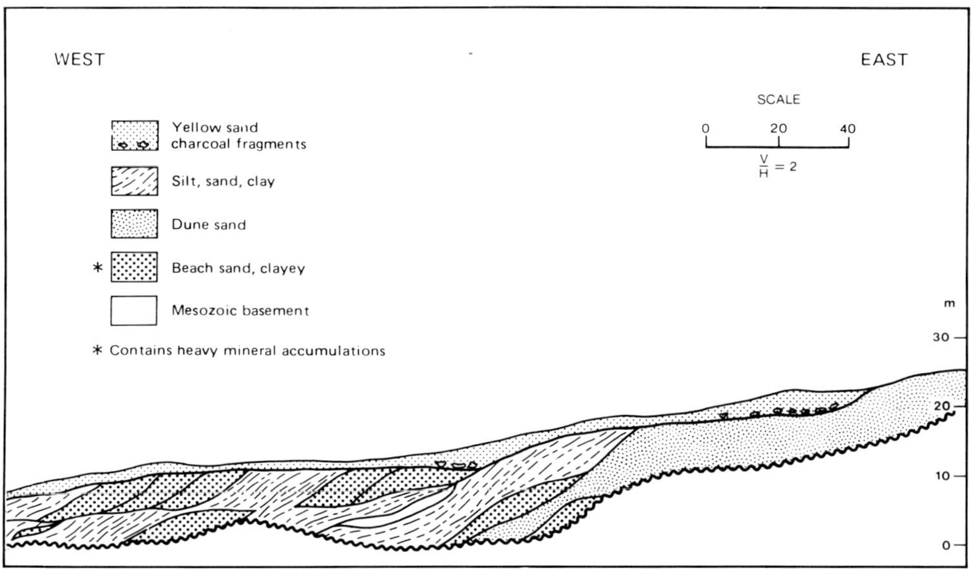

Fig. 3.24 Diagrammatic cross section of the Capel South heavy mineral deposit outlining the lithologies present. The deposit is being mined by Western Titanium Ltd.

nerup Beach. By comparison with the present coastline, and using such information as is available on the position of the Australian coast during the Cainozoic Era, it can be assumed that there was a consistent southwesterly swell along the coast, which built the beaches at the time individual shoreline deposits formed. Erosion of the beach and foredune sand by periodic storms ensured repeated concentration of heavy minerals on the beach. The heavy mineral deposits were then preserved by ultimate seaward migration of the coastline.

REFERENCES

Baxter, J. L. 1977. Heavy mineral deposits of Western Australia. *W. Aust. geol. Surv. Min. Res. Bull.* **10.**

Lissiman, J. C. & Oxenford, R. J. 1975. Eneabba rutile-zircon-ilmenite sand deposit, W.A. In *Economic geology of Australia and Papua New Guinea—1. Metals* (ed. C. L. Knight). Australas. Inst. Min. Metall. Monogr. Ser. 5: 1062-70.

Welch, B. K., Sofoulis, J., & Fitzgerald, A. C. F. 1975. Mineral sand deposits of the Capel area, W.A. In *Economic geology of Australia and Papua New Guinea—1. Metals* (ed. C. L. Knight). Australas. Inst. Min. Metall. Monogr. Ser. 5: 1070-88.

3 G

Minor mineral deposits

REX T. PRIDER
Department of Geology,
University of Western Australia

There are innumerable deposits of minor metal-bearing minerals and industrial rocks and minerals in Western Australia which are of economic significance. Some of these are briefly considered below.

(i) *By-products of the metallurgical treatment of sulphide auriferous ores*—include arsenic, antimony, silver and sulphur.

Arsenic. For some years, at the time of operation of Wiluna Gold Mines which closed down in 1947, arsenic was, next to gold and coal, the most valuable mineral product of Western Australia. The main ore mineral is arsenopyrite (FeAsS containing 46 per cent arsenic). Arsenopyrite was one of the sulphide gangue minerals in the Wiluna gold ore and the arsenic was recovered during the roasting of the sulphide ore. Some arsenic was recovered during the treatment of arsenopyrite-rich gold ore at Meekatharra and Southern Cross. With the closing of these mines the production of arsenic has ceased.

Antimony. The main ore mineral is stibnite (Sb_2S_3 containing 71 per cent antimony) which, like arsenopyrite, occurs in some gold ores such as those of the Blue Spec mine near Nullagine and at Mallina in the West Pilbara. The gold-stibnite deposit at Blue Spec was discovered in 1906 and worked intermittently for gold and antimony between 1935 and 1962. Commercial production recommenced in June 1976, but after a comparatively short time operations are now (1978) being wound down.

Silver. The silver-bearing minerals which occur in the auriferous ores of Kalgoorlie are the various gold-silver tellurides varying from hessite (AgTe) through sylvanite ($Au(Ag,Au)Te_4$) and petzite ($AuAg_3Te_2$) to calaverite ($AuTe_2$ with small amounts of silver). The only silver produced in Western Australia is as a by-product of the refining of gold bullion and nickel.

Sulphur. Western Australia has no deposits of brimstone (native sulphur) which is an important mineral since it is used in the manufacture of sulphuric acid, which in turn is used in making superphosphate. However pyrite (FeS_2), a common gangue sulphide in auriferous and nickeliferous ores, is a potential source of sulphur. Between 1942 and 1970 the demand for sulphur for fertilizer manufacture in Western Australia was partly met from pyrite mined and treated in Western Australia, the main sources being the Iron King pyrite mine at Norseman and the pyrite by-product of gold-ore treatment at Kalgoorlie. However the rescinding in 1970 of the Sulphuric Acid Bounty Act (which provided a bounty for sulphuric acid produced from local materials) and a lower price for imported sulphur resulted in the cessation of local production. Total local production to 1970 amounted to 1 374 984 tonnes.

(ii) *By-products of the metallurgical treatment of sulphide nickel ores*—include copper, colbalt, platinum, palladium and ruthenium.

The production of these metals as by-products of nickel refining in the year ended 31 December 1977 was:

		Value ($A)
Copper	1 830.90 tonnes	1 734 785
Cobalt	200.66 ,,	712 884
Platinum	114.88 kg	527 666
Palladium	298.24 ,,	499 599
Ruthenium	7.32 ,,	13 073

(Figures issued by W.A. Mines Department)

Copper has been dealt with earlier in this chapter [3(B) above]. The only source of the other four metals has been from nickel mining.

These ores, like the auriferous ores, are also a potential source of sulphur.

(iii) *Chromium, vanadium and manganese*—elements which occur mainly in basic iron-rich igneous rocks.

Chromium. The main ore mineral is chromite ($FeCr_2O_4$), which occurs as magmatic segregations in ultrabasic rocks, and the only significant deposit known to date in Western Australia is at Coobina, 1000 km northeast of Perth (Bye 1975). This was first reported in 1925, and 14 500 tons (14 790 tonnes) of ore assaying 43.5 per cent Cr_2O_3 were produced from a number of small lenses in serpentinite between 1952 and 1957. Exploration is continuing in the area.

Vanadium. This is a minor constituent of the mineral magnetite which occurs in basic igneous rocks such as gabbro. There are two important vanadium deposits in Western Australia, but they have not yet been exploited. These are the iron-titanium-vanadium deposit at Barrambie (450 km east-northeast of Geraldton) and the surrounding district (Ward 1975,) and that at Coates Siding near Perth (Blockley 1975*a*), where the better vanadium ore is in laterite overlying magnetite-rich gabbro.

Manganese. This also occurs as a minor constituent of basic igneous rocks, but there are no known economically important primary deposits in Western Australia. However important secondary deposits resulting from the weathering of basic igneous rocks and manganiferous sediments (shales and banded iron-formations) do occur. Of these the most important are the exploited manganese deposits of Peak Hill (Blockley 1975*c*) and various localities—Woodie Woodie, Ripon Hills, Mount Sydney, Skull Springs and Ant Hill—in the East Pilbara (Blockley 1975*b*).

(iv) *Pegmatite minerals*

Pegmatite is the very coarse-grained rock resulting from the crystallization of the end-stages of granitic magma. In these pegmatites we find a concentration of the minor elements of the parent granite magma—elements such as lithium (present in the minerals petalite, lepidolite and spodumene), beryllium (in beryl), molybdenum (in molybdenite), niobium (in columbite), tantalum (in tantalite), tin (in cassiterite), tungsten (in wolframite and scheelite), and many others. Pegmatite is also a source of the industrial minerals feldspar, mica and quartz.

Production of pegmatite minerals has been intermittent and small, from numerous localities which have been described by Hill (1976*a*) and Blockley (1975*a*). The most important Archaean occurrences are: Wodgina, Mount Francisco, Tabba Tabba, Shaw River, Strelley, Moolyella, and Pippingarra in the Pilbara Block; Londonderry (the main producer of feldspar and petalite in the state), Grossmont, Mount Marion, Spargoville, Greenbushes (the main tin producer in the state), and Cattlin Creek in the Yilgarn Block.

There are some less significant pegmatite mineral deposits of which the most important are Gusseys mica deposit in the Kimberley Block, and Yinnietharra (mica and beryl) in the Gascoyne.

The molybdenite and scheelite deposits of Mt Mulgine are, strictly speaking, not pegmatites, but are closely related to them. The molybdenite occurs in quartz veins and disseminations in adjacent altered granite and the scheelite in quartz veins and greisens in metasedimentary rocks close to the intrusive granite of Mt Mulgine.

(v) *Clay minerals*

Clay is used for many different purposes: various ceramics (china clay and earthenware such as fine earthenware, sanitary ware, bricks, pipes, tiles), fillers (for paper), and drilling muds in oil-well drilling. Western Australia is well endowed with clay deposits suitable for most industrial uses. They occur as weathering products in the kaolinized zone of laterite profiles (such deposits are limitless) and as Tertiary and Holocene sediments. The most extensive deposits are those in the laterite profiles where all types of parent rocks such as acid and basic igneous and metamorphic rocks have been completely kaolinized to yield gritty clays (from granitic rocks) and non-gritty clays (from basic rocks). Attempts have been made to produce commercial kaolin from these deposits, but the main problems with the lateritic kaolins are the small proportion of minus 2 micron material, the low whiteness or brightness (even when iron is leached), and too high a viscosity (at standard water-kaolin mixes) for paper coating.

Only a small proportion of clay production is reported to the Mines Department. Much is taken from 'minerals to owner land' or from 'Extractive Industries Licences' under the Land Act. It has been estimated that, while the reported clay production in 1974 was 568 099 tonnes, the actual production amounted to about 1 250 000 tonnes.

The only clays used to date for fine ceramics have been derived from naturally purified clays occurring as sedimentary rocks.

The bentonitic clays (swelling clays) used for oil-drilling purposes are not so abundant. These clays, which consist largely of the clay mineral montmorillonite rather than kaolinite, occur on the floors of Holocene clay pans in the Marchagee district and in some as yet unexploited Lower Cretaceous rocks of the western part of the Carnarvon Basin.

The Western Australian clay deposits have been described by Lipple (1976*b*). The officially reported 1977 production of clays totalled 293 049 tonnes valued at $179 565.

(vi) *Salt deposits (evaporites)*

Salt deposits result from the evaporation of bodies of standing water such as natural salt lakes or man-made sea-water evaporating ponds. These deposits are known as evaporites and some are of economic importance. These are: halite (common salt, NaCl); gypsum ($CaSO_4.2H_2O$), which is used for the manufacture of plaster of Paris; and the potash salts—sylvite (KCl), carnallite ($KCl.MgCl_2.6H_2O$) and alunite ($K_2Al_2(OH)_{12}(SO_4)_4$), which are a source of potash for fertilizers.

These evaporites occur in sedimentary rock sequences of various ages, deposited in suitable environments during arid periods. For example in Western Australia there are thick-bedded salt deposits of Silurian age at depth in the Canning Basin and the Carnarvon Basin, and surface deposits of Recent age in the existing salt lakes in the southern part of the state.

Table 3.17

Salts in sea-water (average)

	Parts per 1000	*Tonnes per cubic kilometre*
NaCl	27.213	28 625 990
$MgCl_2$	3.807	4 004 530
$MgSO_4$	1.654	1 743 880
$CaSO_4$	1.260	1 325 340
K_2SO_4	0.863	907 530
$CaCO_3$	0.123	127 000
Bromides, iodides, and many other compounds	0.076	79 950
Total	35.00	36 814 220

Throughout the southern interior part of the state the topography is very subdued and, with a semi-arid climate, is characterized by the occurrence of salt 'lakes'. These are flat-floored areas of internal drainage which for most of the year are waterless, salt-encrusted flats. Other salt lakes closer to the coast, such as Lake MacLeod north of Carnarvon, are cut-off embayments of the ocean and the salts deposited in them were originally in solution in sea-water. The average amounts of various salts in sea-water are set out in Table 3.17. If sea-water is evaporated, the salts in solution, because of their different solubilities, will be precipitated in an order that is not related to their relative abundance. This order of deposition is calcium carbonate (as limestone), calcium sulphate (as gypsum), sodium chloride (as halite), and finally the 'mother liquor' salts such as potassium and magnesium chlorides, bromides and iodides. The processes of harvesting these salts is dealt with in Chapter 5E below.

(a) *Salt* (common salt)

This is the mineral halite (NaCl) which is one of the most useful mineral substances to man.

Western Australian production for domestic use came initially from salt-lake deposits such as those at Rottnest (the first utilized, from which up to 700 tons (714 tonnes) were obtained annually up to 1952 when production ceased), Hutt Lagoon near Port Gregory (intermittently from the earliest settlement of the area), Lake Spencer near Esperance (formerly the state's chief source of exceptionally pure salt, with a sodium chloride content exceeding 99 per cent on a dry weight basis) and Lake Lefroy near Widgiemooltha (where large-scale production began in 1968). During recent years, *solar salt* has been produced in much larger quantities, for export.

The 1600 km long coastal area between Port Gregory (north of Geraldton) and Broome has a low rainfall and high evaporation rate. These ideal conditions for evaporite deposition, combined with extensive areas of tidal flats which can be cut off from high tides by earthen dykes to form evaporating ponds, make this a suitable environment for the establishment of solar salt plants. Such plants, in response to the demands for salt from Japanese markets, have been established at Shark Bay, Lake MacLeod, Port Hedland and Dampier, resulting in an enormous increase in salt production from an estimated total of 155 000 tonnes for the period 1934-1961 to 3 705 476 tonnes for the single year ending 31 December 1977.

(b) *Gypsum*

There are numerous commercial deposits in salt lakes in the southern half of the state, and the state possesses almost unlimited reserves. There are two main types of deposits: Quaternary barred marine basins such as parts of Shark Bay, Lake MacLeod and Dongara, and inland Quaternary lakes such as Lakes Cowan, Cowcowing, Seabrook, the Yarra Yarra Lakes and many others.

In the former, bedded gypsum has resulted from the evaporation of sea-water and in the latter, gypsum (granules, seed gypsum, or kopai) resulting from the evaporation of surface drainage water occurs on the lake floors and in dunes along the eastern margins of the lakes, resulting from gypsum crystals being blown off the lake flats by the prevailing westerly or northwesterly winds, and accumulating along the east or southeast shores.

Early Western Australian production of gypsum was from the inland salt lakes, the first being from Koorda in 1921. Other inland salt-lake deposits exploited subsequently were those at Lakes Cowan, Seabrook, Cowcowing, and at Hines Hill, Baandee and many other localities. More recently the deposits of Shark Bay have assumed greater importance as producers of gypsum. The 1977 production of gypsum amounted to 105 102 tonnes valued at $282 000.

(c) *Potash*

Potassium salts for use as potassium sulphate fertilizer are the other important evaporite deposits. The first utilized Western Australian deposit was the *alunite* deposit at Lake Campion, where operations commenced in 1944 and continued till 1949, in which period 175 197 tonnes of alunite were treated for the production of 13 145 tonnes of fertilizer. As this final product was of very low grade (27.5 per cent potash) and there was a very low recovery of potash from the alunite which contains 11.4 per cent K_2O) the project was abandoned in 1949.

The present potential source of potash in Western Australia is the 2000 km^2 Lake MacLeod, near Carnarvon. The brines, residual from the evaporation of sea-water and flood waters over a long period of time, contain sodium chloride and the mother liquor potassium and magnesium salts at ten times the concentration of sea-water. Texada Pty Ltd has established a plant to separate the potassium salts from these brines but to date (1978) there has been no economic production, and the operation there is now based entirely on production of common salt.

(vii) *Other industrial minerals and rocks*

Many other economically valuable minerals and rocks occur in Western Australia. The most important of these are described here.

Asbestos

Several different types were exploited in earlier years, the last recorded production being in 1971. The main types are:

(a) *Blue asbestos* (crocidolite) which occurs in thin cross-fibre seams in the weakly-metamorphosed Lower Proterozoic banded iron-formations of the Hamersley Basin. These thin layers are parallel to the bedding in the host banded iron-formations. Mining of these deposits, mainly in the vicinity of Wittenoom and Yampire Gorges, began in 1933 and continued until 1966 with a total production of 154 913 tonnes of fibre.

(b) *White or silky asbestos* (chrysotile) occurs as seams up to 10 cm thick in serpentinites (metamorphosed ultramafic igneous rocks). The main occurrences are in the Archaean rocks of the Pilbara at Nunyerry, Lionel, Soanesville, and Sherlock, with a minor occurrence of excellent quality silky asbestos in Middle Proterozoic dolomitic limestone at the Silver Sheen mine near Meilga, which is in the Ashburton. The total Western Australian production of chrysotile asbestos has amounted to 11 450 tonnes of fibre.

(c) *Brittle asbestos* (tremolite and anthophyllite) is low-grade material occurring in metamorphic rocks at several localities in the Archaean rocks of the Yilgarn Block and one locality in the Pilbara. The total production to date amounts to 518 tonnes. A more detailed account of the asbestos deposits of Western Australia is given by Blockley (1976).

Barite

This mineral is barium sulphate ($BaSO_4$) and is characterized by its high specific gravity (4.45), making it an important additive to drilling mud in oil-well drilling. Barite occurs at many localities in the state, the major deposits being in Archaean rocks at North Pole, Chesterfield and Marble Bar. There are many minor deposits in Archaean, Proterozoic and Cretaceous sedimentary rocks. These are described briefly by Lipple (1976*a*). The 1977 production of barite amounted to 7117 tonnes valued at $21 750.

Construction materials

These include building stone, various aggregates, lime and plaster.

There are many different *building stones* (dimension stone) available in Western Australia but, other than a few building and harbour projects, comparatively little use has been made of them to date. There are excellent granites of various kinds from the uniform medium-grained grey granite of Mahogany Creek (used in the construction of the GPO in Perth), to the porphyritic grey granite of Jerramungup and the porphyritic green granite of Watheroo. Virtually any variety of granitic dimension stone is available in the Darling Range. Sedimentary rocks that have been exploited for local building stone include Donnybrook Sandstone and Coastal Limestone ('Cottesloe stone'), both used in facings for the University of Western Australia buildings. The coastal limestone was used in many early buildings in Fremantle, e.g. the Old Women's Home (now the Maritime Museum). A pink to cream and white spongolite freestone is quarried at Bridgetown and Ravensthorpe for home construction and the comparatively-poorly-consolidated coquinite (composed entirely of small shells) is quarried at Hamelin Pool in Shark Bay for lightweight building stone in that region. A metamorphic rock that has been utilized for building is the flaggy white-to-green quartzite of the Archaean metasediments of the Toodyay area, which has been used for feature walls and pavements.

Aggregate stone for concrete, exposed aggregate panels and for road aggregates and rail ballast is quarried at various places throughout the state, but mainly in the Darling Range. The best all-purpose aggregate stone is dolerite which occurs as dykes in the granitic country rock of the Darling Range. It is quarried, crushed and sized for different aggregates. Granite is also used for aggregate stone. A particularly valuable natural aggregate (not requiring crushing or sizing) is the chert talus deposit at Coomberdale. Materials for exposed aggregate panels include quartz from Gosnells and Mukinbudin.

Vermiculite is a lightweight aggregate which in its expanded form (weighing 80-160 kg/m^3), is used for acoustic and thermal insulating exposed aggregates. It occurs at several localities in the South-West, the main deposit being at Young River.

The main source of *lime* for agricultural lime and lime for cement manufacture is the travertine capstone on the Pleistocene limestone of the coastal belt of the metropolitan area. Formerly the lime required for cement manufacture was obtained by dredging shell from the various shell banks in the Swan River. Gypsum is the mineral used for making *plaster* and its occurrence has been dealt with earlier (see evaporite deposits above).

Western Australian construction materials have been more fully discussed by Connolly (1976*a*).

Gemstones

To date Western Australia has had comparatively little production of precious stones. *Emerald,* a variety of beryl containing a small amount of chromium oxide has been found and exploited to a small extent at Poona, Melville and Menzies. *Diamonds* were found at Nullagine in the Pilbara in 1895 (this area subsequently yielding some 70 stones), and in the Lennard River in the Northern Canning Basin (nine diamonds in 1969). There has been no commercial production, but recent (1978) discoveries in the West Kimberley have lead to a renewed interest in diamond prospecting which has located a number of diamodiferous pipes in which many diamonds have been found. The economic viability of these deposits has not yet been assessed. Although *precious opal* has been recorded from a few localities there has been no official record of production.

On the other hand there are wide-spread occurrences of semi-precious stones. These include many varieties of common opal, chrysoprase, chert, jasper, agate, amethyst, tourmaline and others as described by Connolly (1976*b*).

The production of gemstones for the year ended 31 December 1977 (figures released by the W. Aust. Mines Department) was

		Value ($A)
Emerald	16 525 carats (3505 gm)	17 985
Amethyst	88 kg	370
Chalcedony	6 152 ,,	4 253
Moss opal	16 748 ,,	8 687
Tiger eye	18 850 ,,	9 244
Pilbara serpentinite	455 ,,	1 820
		42 359

Refractories

These materials are used for the manufacture of various heat resistant products from furnace linings to crucibles and spark plugs. The more important refractories are magnesite (magnesium carbonate), corundum (aluminium oxide), and sillimanite, kyanite and andalusite (which are all aluminium silicates of the same composition). The aluminium silicates occur in many minor deposits in metasedimentary Precambrian rocks but the only important refractory that has been exploited on a large scale is magnesite from near Ravensthorpe. All of the known magnesite deposits are of similar nature, and are produced by weathering of basic and ultrabasic igneous rocks. The 1977 production was 23 906 tonnes (including 23 897 from Ravensthorpe), with an estimated value of $637 449.

Talc

This mineral is hydrated magnesium silicate, and in industry the name is also applied to rocks (soapstone) consisting largely of talc. It has a variety of uses based on its properties of extreme softness and smoothness and chemical inertness. The main sources of talc in Western Australia are in Proterozoic rocks at Coodawa near Three Springs (Baker 1976) and at Mount Seabrook near Meekatharra, but there are many other deposits (Wenham 1976) which have been exploited since the first recorded production of this material in 1942, when 270 tonnes were obtained from the Greenbushes-Bridgetown area and 38 tonnes from Mount Monger. The talc production in 1977 amounted to 90 466 tonnes the value of which is not available.

Other minor minerals and rocks exploited in the past, or in small amounts at present, that have not been mentioned above include *diatomite* and *peat* from coastal freshwater lakes and swamps, *garnet* from beach sands in the South-West, *glauconite* (from Gingin), *graphite* (from Young River and Munglinup), *guano* (on the Abrolhos Islands), *ochre* (various iron oxides) from the Murchison, and *silica sands* (for glass making, moulding sands, and fine aggregate for building). Potentially significant deposits of *fluorspar* occur at Speewah (160 km south of Wyndham), Meentheena (80 km southeast of Marble Bar) and at Mt Phillips near Yinnietharra.

There is little doubt that numerous other important minerals and rocks in Western Australia remain to be discovered and exploited.

REFERENCES

Baker, G. F. U. 1976. Talc—Coodawa Deposit at Three Springs, Western Australia. In *Economic Geology of Australia and Papua New Guinea—4. Industrial Minerals and Rocks* (ed. C. L. Knight). Australas. Inst. Min. Metall. Monogr. Ser. 6: 398-407.

Blockley, J. G. 1975*a*. Sundry mineralization in the Archaean of the Western Australian Shield. In *Economic Geology of Australia and Papua New Guinea—1. Metals* (ed. C. L. Knight). Australas. Inst. Min. Metall. Monogr. Ser. 5: 211-14.

——— 1975*b*. Pilbara Manganese Province, W.A. In *Economic Geology of Australia and Papua New Guinea—1. Metals* (ed. C. L. Knight). Australas. Inst. Min. Metall. Monogr. Ser. 5: 1019-20.

——— 1975*c*. Peak Hill manganese deposits. In *Economic Geology of Australia and Papua*

New Guinea—1. Metals (ed. C. L. Knight). Australas. Inst. Min. Metall. Monogr. Ser. 5: 1021.

———. 1976. Asbestos—Western Australia. In *Economic Geology of Australia and Papua New Guinea—4. Industrial Minerals and Rocks* (ed. C. L. Knight). Australas. Inst. Min. Metall. Monogr. Ser. 6: 14-18.

Bye, S. M. 1975. Chromite mineralization within the Coobina Ultramafic. In *Economic Geology of Australia and Papua new Guinea—1. Metals* (ed. C. L. Knight). Australas. Inst. Min. Metall. Monogr. 5: 205-6.

Connolly, R. R. 1976*a*. Construction materials—Western Australia. In *Economic Geology of Australia and Papua New Guinea—4. Industrial Minerals and Rocks* (ed. C. L. Knight). Australas. Inst. Min. Metall. Monogr. Ser. 6: 89-91.

——— 1976*b*. Semi-precious stones—Western Australia. In *Economic Geology of Australia and Papua new Guinea—4. Industrial Minerals and Rocks* (ed. C. L. Knight). Australas. Inst. Min. Metall. Monogr. Ser. 6: 354-5.

Hill, W. B. 1976. Pegmatite minerals (mica, beryl, lithium, feldspar)—Western Australia. In *Economic Geology of Australia and Papua New Guinea—4. Industrial Minerals and Rocks* (ed. C. L. Knight). Australas. Inst. Min. Metall. Monogr. Ser. 6: 235-40.

Lipple, S. L. 1976*a*. Barite—Western Australia. In *Economic Geology of Australia and Papua New Guinea—4. Industrial Minerals and Rocks* (ed. C. L. Knight). Australas. Inst. Min. Metall. Monogr. Ser. 6: 30-2.

——— 1976*b*. Papers on Western Australian clay deposits in *Economic Geology of Australia and Papua New Guinea—4. Industrial Minerals and Rocks* (ed. C. L. Knight). Australas. Inst. Min. Metall. Monogr. Ser. 6: 62-9.

Ward, H. J. 1975. Barrambie iron-titanium-vanadium deposit. In *Economic Geology of Australia and Papua New Guinea—1. Metals* (ed. C. L. Knight). Australas. Inst. Min. Metall. Monogr. Ser. 5: 207-11.

Wenham, M. E. 1976. Talc—Western Australia. In *Economic Geology of Australia and Papua New Guinea—4. Industrial Minerals and Rocks* (ed. C. L. Knight). Australas. Inst. Min. Metall. Monogr. Ser. 6: 397-8.

3 H

MINERAL FUELS

These consist of the fossil fuels coal, oil and natural gas, which have originated by the partial decomposition of organic (plant and animal) material deposited in sediments in past geological times, and radioactive deposits such as those of the element uranium.

3 H(i)

Coal

J. H. Lord
Director, Geological Survey of Western Australia.

HISTORY

When Western Australia was founded in 1829, coal was the major source of fuel other than wood. Consequently it is not surprising that early explorers were interested in coal occurrences. The first discovery was made by A. C. and F. T. Gregory in the bed of the Irwin River on 9 September 1846. In the following year these coal seams were examined and sampled by a geologist, Dr F. von Sommer. Over the years this occurrence has been examined and tested on a number of occasions by various means, but the conclusion has always been the same, that is, that because of its low quality and erratic occurrence it is of no commercial value under existing economic conditions.

Further coal seams were found in drilling on the Greenough River near Eradu in 1906, but commercially these deposits were of no more value than those at the Irwin River.

In the southern portion of the Perth Basin leases for coal were taken out and drilled in 1888 at Fly Brook. Coal was intersected but the seams were thin and the quality poor. Other similar occurrences were found later at Bibilup, Donnybrook, Millbrook and near the Vasse River at Wonnerup.

The find which was to prove the most valuable coal discovery for this state was made in 1883, when coal was located cropping out in a pool on the Collie River near Survey Station T17. As with most mineral finds, there is some debate as to who actually made the discovery. A shepherd, George Marsh, claimed to have located it and shown it to his employer Arthur Perrin. The latter, however, claimed to have located the coal himself but did nothing about the occurrence until 1889, when he interested a Bunbury merchant, David A. Hay, in it. Hay investigated the coal and recovered a bulk sample from the pool.

During the next few years the government initiated prospecting and drilling and drove a tunnel on a coal seam for 305 m to a vertical depth of 53 m. The Collie Coal Mining District was declared open for selection on 23 March 1896, and production commenced in 1898 when the railway reached Collie.

In the early days numerous companies operated on the coalfield until 1920, when Amalgamated Collieries Company of W.A. Ltd was formed and purchased all existing companies. In 1930 the Griffin Coal Mining Co. Ltd was formed and operated as only a small company for many years. In 1949 another company, Western Collieries Ltd, was formed to develop coal deposits to the east and southeast of Collie for use on the Eastern Goldfields.

In 1960 the government awarded its contracts for the supply of coal to Griffin and Western Collieries because of the high price tendered by Amalgamated Collieries, which then ceased operations on the field and surrendered its leases. Since then the two companies have gradually acquired most of the prospective areas of the coalfield and are now sharing the boom in production, which, because of the present world energy crisis, will continue to grow.

GEOLOGY

Despite the existence of large sedimentary basins in Western Australia (see Fig. 3.30, page 146), coal measures* are only poorly developed compared with comparable basins in the eastern states of Australia and other parts of the world. Coal has been located in Permian, Jurassic, Cretaceous and Tertiary rocks in Western Australia, but in all cases it is of poor quality and non-coking.

Canning Basin

The geology of the Canning Basin has been described in several publications (Veevers & Wells 1961, and G.S.W.A. 1975). Onshore the basin covers some 415 000 km^2 and consists mainly of Palaeozoic sediments ranging from Ordovician to Permian, frequently with a thin covering of later Mesozoic and Tertiary deposits.

Coal has been located in the Fitzroy Sub-basin, which lies between the Halls Creek Province and the Fenton Fault in the northern portion of the basin.

While some carbonaceous material has been found in the lower formations of the Permian, the coal measures occur in the Condren Sandstone Member of the Liveringa Formation.

Drilling for coal in the past ten to fifteen years has indicated that the coal measures are widespread but have a variable thickness and shallow dips. The best seam appears to range from 1 m to 2.2 m in thickness. It is a low-grade non-coking coal.

* The term 'coal measures' is applied to an assemblage of strata, consisting generally of sandstones and claystones, containing interbedded coal seams.

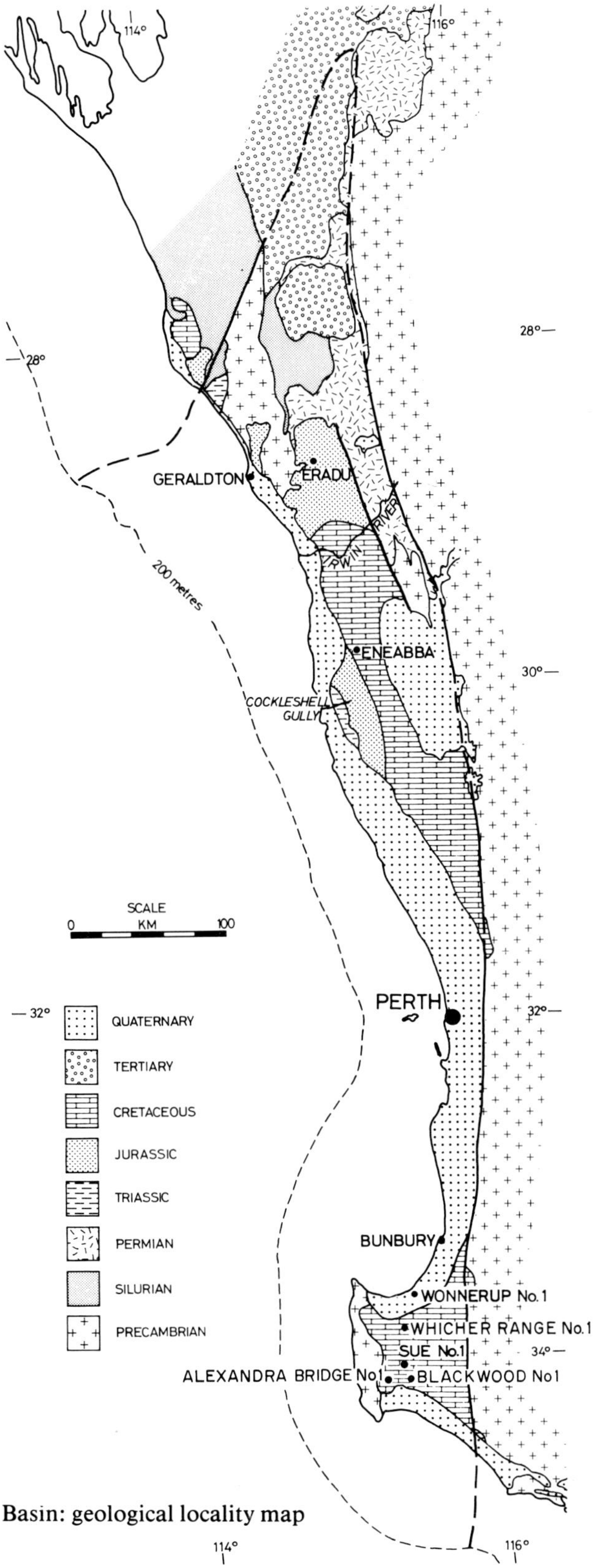

Fig. 3.25 Perth Basin: geological locality map

Perth Basin

The Perth Basin (Fig. 3.25) is elongated along the southern portion of the west coast, extending from the south coast northwards for some 900 km. Generally it is bounded on the west by the outer edge of the continental shelf and on the east by the Darling Fault. This fault is a major tectonic structure with a vertical displacement of over 12 000 m near Perth, and may have been responsible for the extensive faulting which occurs throughout the basin. Two blocks of Precambrian basement crop out as the Northampton and Leeuwin Blocks. As well a subsurface block known as the Beagle Ridge occurs along the west coast.

The detailed geology (Playford *et al.* 1976) shows that the oldest sediments, of Silurian age, crop out in the northern portion of the basin. From Permian onwards, sedimentation, varying from marine to continental, took place during most periods, with coal measures being deposited during the Permian, Jurassic and Cretaceous Periods.

The coal measures occur at reasonable depths for prospecting in the northern portion of the basin and at the southern end in the Bunbury Trough. Exploratory drilling for petroleum, coal and groundwater in recent years has improved our knowledge of the coal measures within the basin (see Fig. 3.25).

The known occurrences of Permian coal measures are:

1. *Irwin River.* Here they have a known maximum thickness of 123 m and are of Lower Permian age. The coal seams are thin and lenticular.
2. *Eradu.* Coal seams occur in the Wagina Sandstone of Upper Permian age and are lenticular.
3. *Bunbury Trough.* The Sue Coal Measures, which occur as a type section in Sue No. 1 oil well (Fig. 3.25) between 1219 m and 3057 m, contain about 85 m of coal in some 70 seams. While they do not crop out, they can be correlated with Collie and have been identified at greater depths in Blackwood No. 1, Whicher Range No. 1 and Wonnerup No. 1 oil wells, but at shallower depths in Alexandra Bridge No. 1 (398 m) and Quindalup No. 2 water bore (112 m) on the coast 11 km west of Busselton.

Coal seams of Jurassic to Cretaceous age occur in two structures of the basin, namely:

1. *Dandaragan Trough.* The Cattamarra Coal Measures of Lower Jurassic age, consisting of 22 coal seams totalling 12 m in thickness, were intersected between 1843 and 1963 m in Eneabba No. 1 oil well.

In 1970 they were located again, covered by 15 to 27 m of Quaternary sands, about 8 km southwest of Eneabba township. The coal could be of use for on-site power generation.

Some minor seams of lignitic coal have been recorded in the upper part of the Yarragadee Formation (Upper Jurassic—Lower Cretaceous).

2. *Bunbury Trough.* Lignitic seams equivalent to the Cattamarra Coal Measures occur in the Blackwood No. 1 and Whicher Range No. 1 oil wells. Also coal has been recorded in the upper part of the Yarragadee Formation.

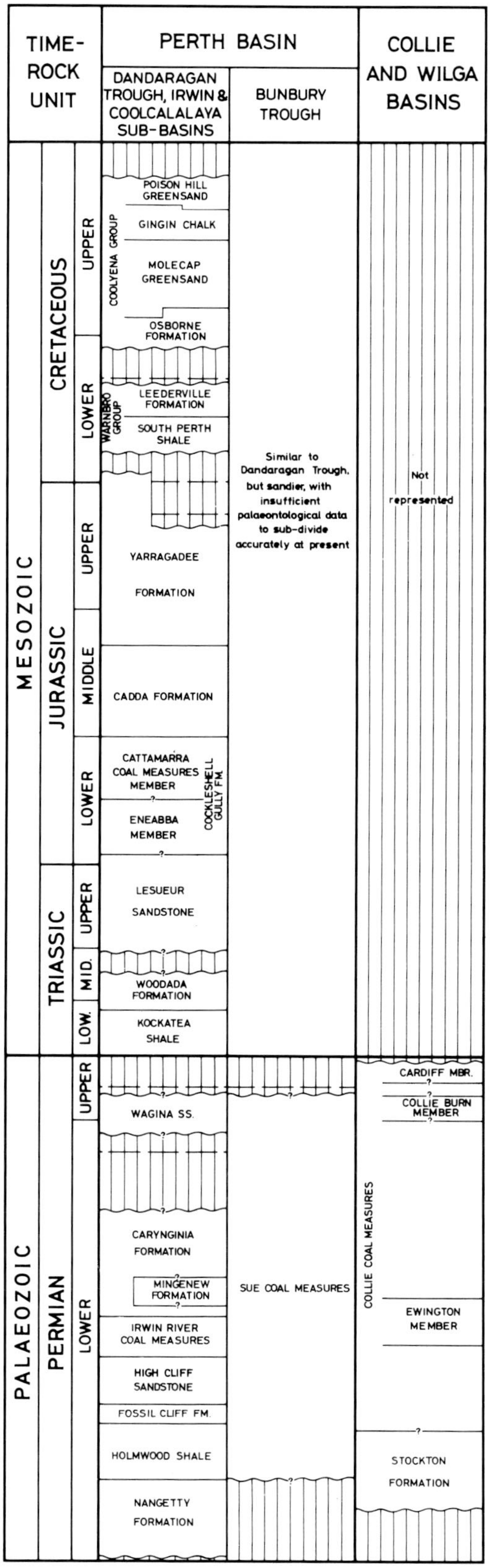

Fig. 3.26 Perth, Collie and Wilga Basins: stratigraphic columns

There are a number of minor occurrences of coal at Fly Brook, Vasse River near Wonnerup, Donnybrook, Bibilup and Millbrook.

Collie and Wilga Basins

The Permian Collie and Wilga Basins have been preserved in depressions on the Precambrian surface of the Yilgarn Block (Fig. 3.27). The first theory suggested was that the down-faulting had preserved these sediments from erosion.

After the first detailed investigation of the Collie Basin (Lord 1952), it was suggested that ice could have formed the depressions, for no evidence of major faulting could be found then or in more recent remapping of the area. Later, drilling showed that the floor of the basin was a cleanly-glaciated Precambrian surface with overlying tillite and glacigene sediments, named the Stockton Formation.

This formation is overlain by the Permian Collie Coal Measures, which in turn are overlain unconformably by the Nakina Formation, probably of Late Tertiary age. The stratigraphic successions of sediments and coal seams in the various sub-basins are shown in Fig. 3.28.

The position and shape of both basins are shown on Fig. 3.27 together with the suggested sub-basins, and on Fig. 3.29 are cross-sections of the Collie Basin. It is thought that the ice which created the depression may have moved from the north-west, but this cannot be proved until it is possible to examine the glaciated floor.

After glaciation the basin and beyond became an area of continental sedimentation and internal drainage, and probably extended to the Perth Basin, where the Sue Coal Measures were deposited. Towards the end of the Permian it probably became an area of numerous smaller swampy areas of sedimentation. Later erosion removed most of the sedimentation and the coal measures have been preserved only in the depressions at Collie and Wilga and in the Perth Basin, where they may have been protected by movement on the Darling Fault.

Initially, sedimentation occurred throughout the Collie Basin and the Ewington Member can be correlated throughout all sub-basins. Later the Cardiff Sub-basin separated from the other two sub-basins, and although sedimentation continued in the three basins, only the Muja and Shotts Sub-basins can be correlated. Still later sedimentation in the Upper Permian was confined to the Cardiff and Muja Sub-basins. The absence of sedimentation in the Shotts Sub-basin suggests that it has either been removed by erosion, or Shotts may not be a true sub-basin but merely a slight depression or shelf on the basement.

The sedimentation in the Wilga Basin may have occurred in the Upper Permian contemporaneously with the Cardiff and Muja Members. The vegetable matter which forms the coal seams is thought to have been mainly washed or floated into the area of deposition. The evidence for this is the scarcity of *in situ* tree stumps and the nature of the floor and roof of the seams.

Other basins

Bremer Basin is the only other basin where any significant carbonaceous sediments occur. Along the south coast small isolated occurrences of lignite or peat have

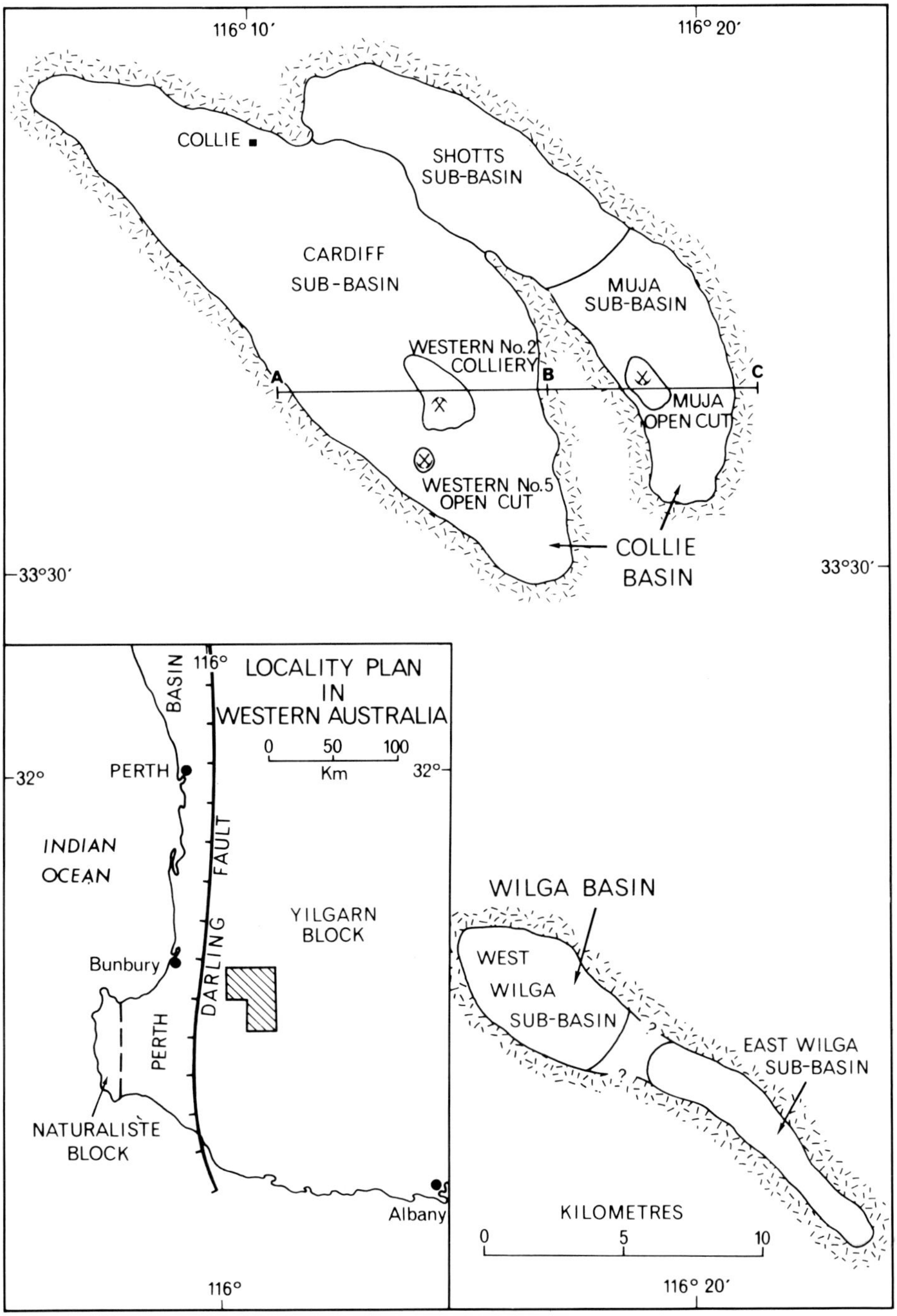

Fig. 3.27 Collie and Wilga Basins: locality plan

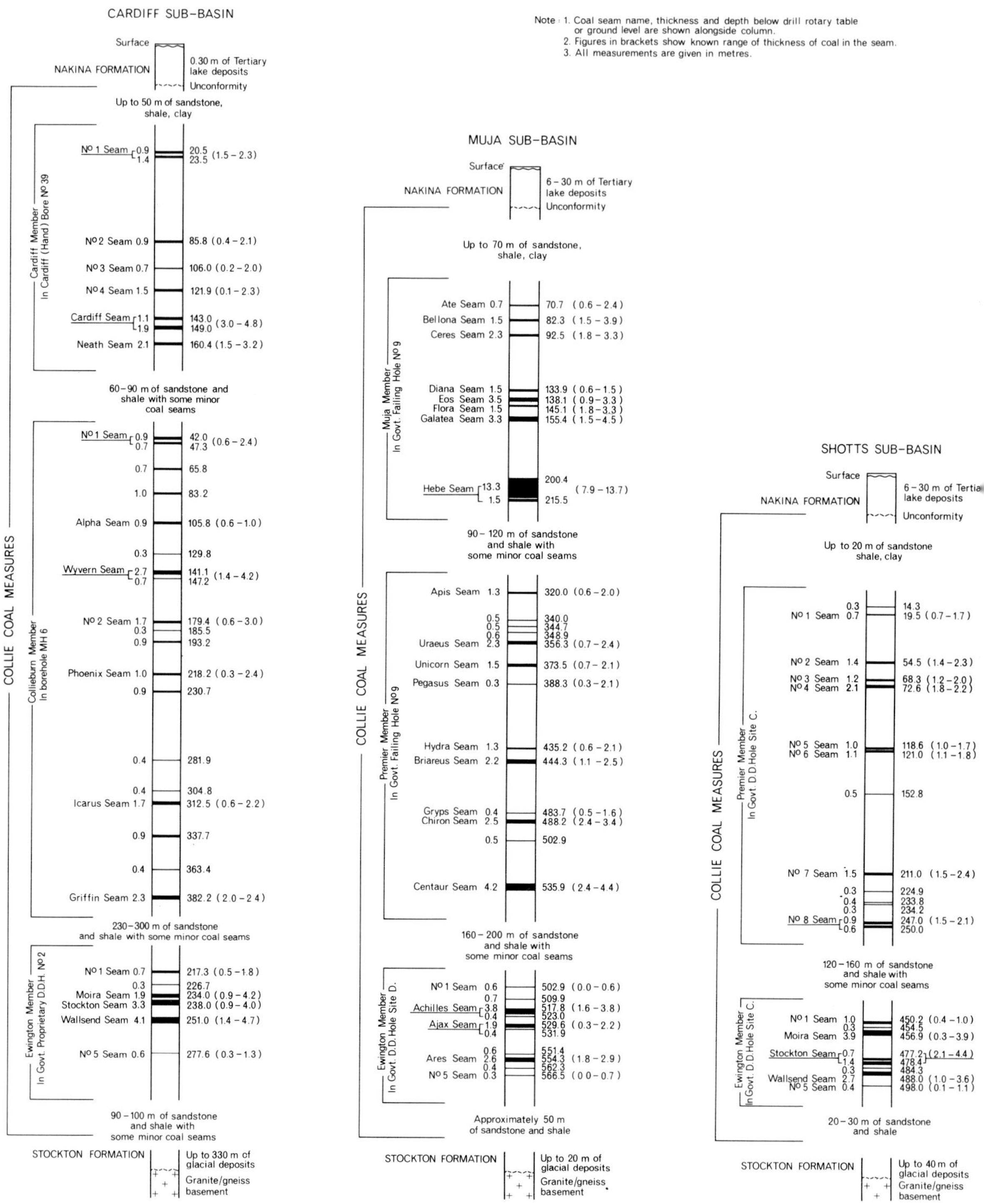

Fig. 3.28 Collie Basin: stratigraphic succession of coal seams

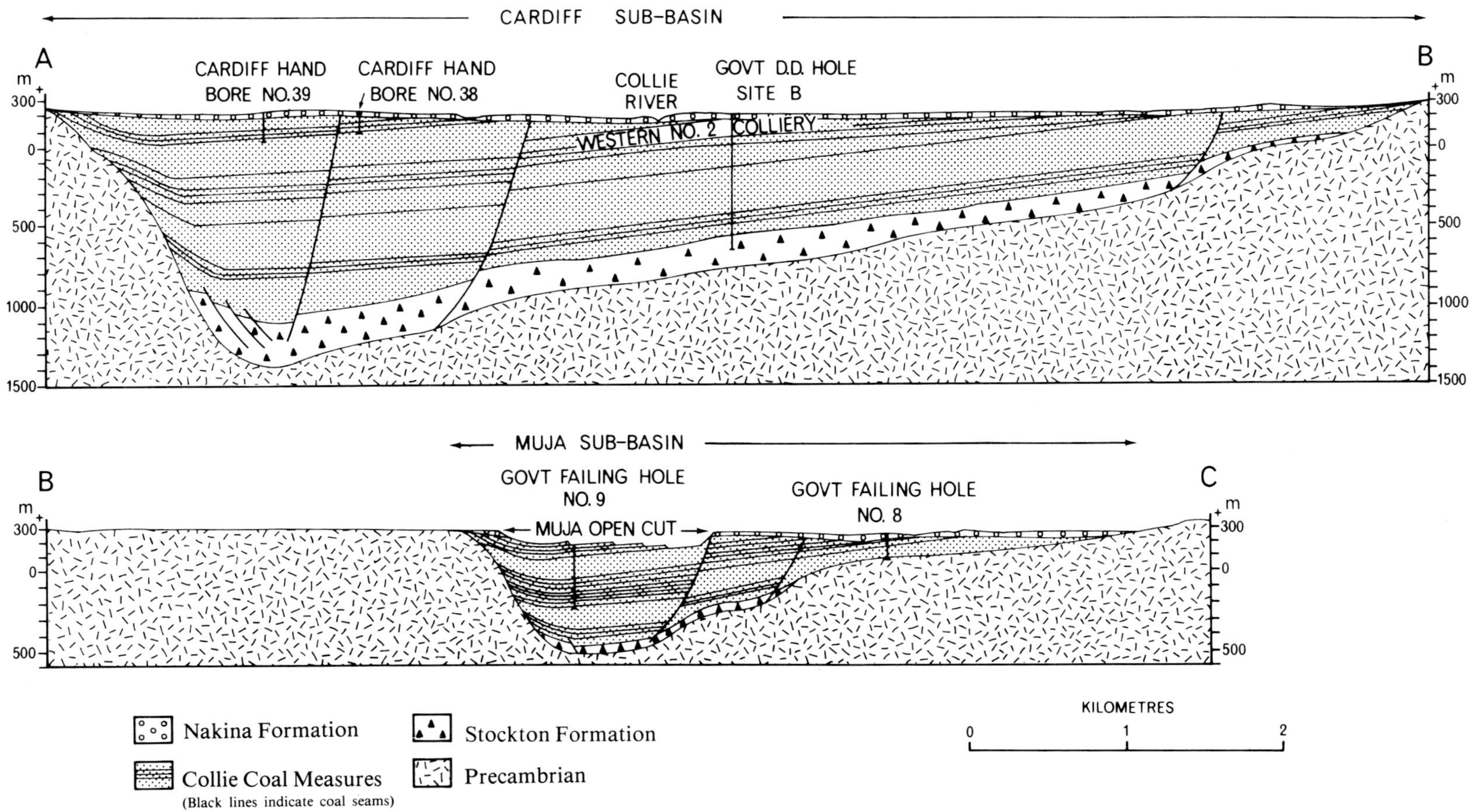

Fig. 3.29 Collie Basin: typical cross-sections

Table 3.18

Analyses of Western Australian coals

Locality	*Moisture* %	*Ash* %	*Volatile matter* %	*Fixed carbon* %	*Specific energy* MJ/kg	*Specific energy (d.a.f.)* MJ/kg
CANNING BASIN						
Lower Liveringa Stn—						
well seam—3.7 m thick	8.9	27.4	29.7	34.0	18.0	—
TR3518H—Myroodah No. 7 Bore—						
0.7 m at 47.55 m	20.0*	21.1	19.1	39.8	19.7	33.4
TR3518H—Myroodah No. M11—						
0.76 m at 57.91 m	20.0*	31.6	20.8	27.6	14.9	30.8
TR3431H—Bore A1—near Audreys						
Bore—1.55 m at 21.94 m	20.0*	20.6	27.8	31.5	18.9	31.8
TR3431H—Bore A4—near Audreys						
Bore—1.61 m at 37.49 m	20.0*	16.5	27.9	35.6	20.3	32.0
PERTH BASIN—NORTHERN						
Eradu—shaft—4.9 m seam	30.0	17.7	20.7	31.6	14.1	—
Irwin River—new tunnel (1946)—1.8 m	29.7	17.9	31.7	21.0	13.9	—
Eneabba—						
No. 1 oil well	3.7*	14.6	34.9	46.8	27.9	34.2
Taylor-Woodrow area	22.1	17.4	27.7	32.8	18.3	—
Taylor-Woodrow area (washed)	24.6*	11.2	29.8	34.4	19.3	
Hill River No. 2A Bore	20.0*	9.8	31.9	38.3	20.7	29.5
PERTH BASIN—BUNBURY TROUGH						
Whicher Range No. 1 well at 4120 m	2.3*	10.7	31.8	55.2	29.9	34.4
Sue No. 1 well at 1409 m	10.2*	15.1	34.4	40.3	21.7	29.1
Sue No. 1 well at 2802 m	2.2*	11.7	23.2	62.9	28.9	33.6
Blackwood No. 1 well at 2483 m	5.7*	15.5	30.6	48.2	23.8	30.2
Alexandra Bridge No. 1 well at 612 m	20.8	8.5	26.7	44.0	20.8	29.5
Quindalup No. 2 water bore at 207 m	6.4*	15.5	26.2	51.9	24.8	31.8
Donnybrook	36.3	13.5	21.7	22.6	14.1	—
Bibilup	36.5	4.6	32.1	26.8	16.4	—
Fly Brook	44.4	3.8	23.1	28.7	14.3	28.2
COLLIE BASIN						
Western No. 2 Colliery—run-of-mine	25.7	4.0	26.6	43.7	20.8	29.6
Western No. 5—						
Open Cut	25.5	8.3	26.4	39.8	19.2	29.0
Run-of-mine	24.7	8.0	27.2	40.0	19.7	29.3
Muja Open Cut—run-of-mine	26.6	2.5	27.8	43.1	20.9	29.4
WILGA BASIN						
TR3725H—Bore G2—						
2.7 m at 82 m	19.3	7.3	32.4	41.0	20.6	28.1
2.0 m at 109 m	27.2	8.9	27.8	36.1	18.2	28.5
BREMER BASIN						
Fitzgerald River—GSWA Bore 1B	52.4	15.1	19.1	13.4	9.3	28.5
Fitzgerald River—GSWA Bore M	56.8	6.5	21.6	15.1	10.6	28.7

* Reduced to a standard 20% or air dried before analysis.

been located, and these extend sporadically northwards to Norseman. These deposits are of Tertiary age and of no economic significance at present.

QUALITY

The known coal occurrences in this state are of poor quality, being sub-bituminous steaming coals with a high moisture content, relatively low specific energy (calorific value) and in many cases with a high ash content. The only coals which show any possible coking properties have been found in deep oil wells. Table 3.18 shows typical analyses of our known coals.

The Collie Basin produces the best coal which can be economically extracted. This coal has a particularly low ash and sulphur content, high moisture content and a specific heat of about 20 MJ/kg as mined. It is an adequate coal for local power generation and industry and makes a very good char.

Collie coal decomposes fairly rapidly when exposed to the atmosphere, and this defect, combined with its low quality, has made it unattractive for export up to now.

If a power station was required to the north of Perth, it might be possible to develop the prospect at Eneabba, which has a high ash content and a lower specific energy than Collie coal.

RESERVES

There are only two areas where coal has been defined sufficiently by drilling to establish reserves. These areas are at Collie and Eneabba.

The first comprehensive estimate of resources at Collie was in 1956 (Low 1958) and resulted from ten years investigation and 14 124 m of drilling by the Geological Survey. Since then these resources have been modified as a result of further company exploration and mining.

The total resources of the Collie Basin are now (1978) considered to be 1915 million tonnes. Only seams with 1.85 m (6 feet) or more of clean coal have been considered in these resources. It must be understood that not all of this coal will be recoverable.

Table 3.19

Collie Basin: amount of total coal reserves shown to be extractable to 1978

	Measured (megatonnes)	*Indicated (megatonnes)*	*Inferred (megatonnes)*	*Total of all categories (megatonnes)*
Open Cut	147	120	19	286
Colliery	14	81	24	119
Total	161	201	43	405

As more detailed drilling is done by companies, reserves of extractable coal under present economic conditions, are being established. These reserves will increase as more testing is done and economic conditions alter particularly due to the energy crisis. Table 3.19 shows reserves of coal at Collie.

At Eneabba the total reserves are estimated to be 38 million tonnes, allowing a maximum depth of overburden of 91 m, or 28 million tonnes with a maximum overburden of 61 m. Further prospecting may establish additional reserves.

PRODUCTION

Commercial production of coal in this state has only been achieved from the Collie Basin, where it has been continuously mined since 1898. At Irwin River, Eradu and Wilga, bulk samples have been mined from time to time to test the viability of the deposits, all without commercial success. At other localities testing has been confined to drilling to obtain samples.

Table 3.20

Collie Basin: annual production statistics

Year	*Tonnes mined*	*Av. price per tonne ($ Aus.)*	*Men employed*	*Tonnes per man*
1899	3 564	0.98	70	51
1900	120 305	1.14	400	301
1905	129 403	0.86	351	369
1910	266 361	0.86	521	511
1915	291 253	0.94	498	585
1920	469 413	1.50	830	566
1925	444 460	1.63	677	656
1930	509 448	1.55	896	569
1935	545 783	1.16	689	792
1940	548 058	1.33	713	769
1945	552 057	2.08	860	642
1950	827 382	3.11	1 099	753
1955	918 253	6.73	1 386	663
1960	937 151	5.20	984	952
1965	1 009 641	4.37	760	1 329
1970	1 216 898	4.81	635	1 916
1972	1 167 486	5.06	617	1 892
1974	1 446 048	6.32	685	2 111
1976	2 268 728	9.02	860	2 638
1978	2 403 503	11.92	870	2 763

Production and prosperity at Collie has fluctuated over the years owing to wars, economic conditions and the fuel policy of the state government. Initially the major consumer was the railways, with more and more being required for industrial and domestic power generation. During the 1960s there was a move towards the greater use of imported fuel oil for power, and the use of coal by the railways sharply decreased. The later rapid increase in oil prices brought about an energy crisis in 1973, and made Collie coal a more economical and secure source of energy. Since that time power stations and industry, but not the railways, have been gradually returning to the use of coal.

As shown in Table 3.20 production is increasing rapidly and could reach 5 million tonnes per year during the 1980s and, depending on industrial development, 10 to 15 million tonnes per year before the end of the century.

The price of coal rose to $6.94 per tonne in 1954, but through the re-organization of the industry the price gradually fell to $4.23 in 1966 and only exceeded the 1954 price in 1975.

All attempts to develop an export market have failed (1977) because of the quality of the coal. In the future, owing to the energy crisis, there may be an overseas market for such coal, but the long-term economy of the state currently demands that it should be reserved for local consumption, unless large additional deposits are found elsewhere in Western Australia.

CONCLUSION

Although Western Australia has some very large mineral resources such as iron-ore, bauxite, nickel and natural gas, it has only modest coal resources. The possibility of finding large new coal deposits (and particularly coking coal) is remote.

At Collie the state has a valuable resource of low-grade coal suitable for power generation and sufficient to last well into the next century. Also there is a limited resource at Eneabba which might be increased with further exploration.

Unless new, large coal deposits are located, the known coal should be reserved as a relatively cheap source of power for Western Australia until some cheaper alternative source is available.

REFERENCES

Cockbain, A. E. & van de Graaff, W. J. E. 1973. The Geology of the Fitzgerald River lignite. *W. Aust. geol. Surv. Ann. Rept 1972*: 81-92.

Geological Survey of Western Australia, 1975. Geology of Western Australia. *W. Aust. geol. Surv. Mem.* **2**, 514 pp.

Johnstone, M. H. 1964. A preliminary investigation of the Jurassic coal in the Perth Basin. *Proc. Australas. Inst. Min. Metall.* **211**: 61-73.

Johnson, W., de la Hunty, L. E. & Gleeson, J. S. 1954. The Geology of the Irwin River and Eradu Districts and surrounding country. *W. Aust. geol. Surv. Bull.* **108**.

Lord, J. H. 1952. Collie Mineral Field. *W. Aust. geol. Surv. Bull.* **105**(1).
——— 1975. Papers on Western Australian coal basins in *Economic Geology of Australia and Papua New Guinea—Coal* (ed. C. L. Knight). Australas. Inst. Min. Metall. Monogr. Ser. 6.
Low, G. H. 1958. Collie Mineral Field. *W. Aust. geol. Surv. Bull.* **105**(2).
Playford, P. E., Cockbain, A. E. & Low, G. H. 1976. Geology of the Perth Basin, Western Australia. *W. Aust. geol. Surv. Bull.* **124**.
Veevers, J. J. & Wells, A. T. 1961. The Geology of the Canning Basin, Western Australia. *Bur. Miner. Resour. Geol. Geophys. Aust. Bull.* **36**.

3 H(ii)

Oil and gas

PHILLIP E. PLAYFORD
Geological Survey of Western Australia

HISTORICAL REVIEW

Interest in the prospects of finding oil in Western Australia was first aroused soon after the turn of the century, when supposed signs of oil were reported in the Warren River area near the south coast. This resulted in a short-lived oil boom and the drilling of four shallow holes in the area from 1902 to 1904, but without success.

The first confirmed oil-showings in the state were obtained in 1918, when a water driller, Harry Price, reported traces of oil in a bore east of Fitzroy Crossing in the Kimberley District (Canning Basin). The oil occurrence was confirmed by the Geological Survey, and the Freney Kimberley Oil Company Ltd was formed soon afterwards to explore the area. This company, under the direction of Arthur Wade, was the pioneer of scientific oil exploration in this state. Its first exploratory well, on the Mt Wynne Anticline, was spudded in June 1922. The company also began drilling that year at Price's Creek, close to the water bore where oil traces had been reported in 1918. Although further showings were reported from three holes in the Price's Creek area, no discoveries were made.

Also in 1922 bitumen was found by Walter Okes in vesicles of a basalt flow in the Ord River area of the East Kimberley District (Ord Basin). As a result the Okes Durack Kimberley Oil Company N.L. was formed to explore in the vicinity of the bitumen occurrence. It drilled a single dry exploratory hole in 1923-24.

A well-known American oil geologist, F. G. Clapp, visited Western Australia in 1924 and reported very unfavourably on its oil prospects. His report was an important factor in discouraging exploration in the state. However, the Freney Kimberley Oil Company still had faith in the potential of the Canning Basin, and despite a continuing lack of success, the company continued drilling sporadically up to 1942, when operations ceased because of the war. Other companies and individuals carried out preliminary geological exploration during this period in various basins, but no drilling resulted from this work.

The first large-scale exploration with modern techniques began in the state in 1952, when West Australian Petroleum Pty Ltd (Wapet) began operations in the Carnarvon, Canning and Perth basins (Fig. 3.30). Its concessions had originally been taken up by Australian Motorists Petroleum Co. Ltd (Ampol), which company became a member of the Wapet consortium. The following year Wapet had spectacular success with its first hole, striking oil at Rough Range in the Carnarvon

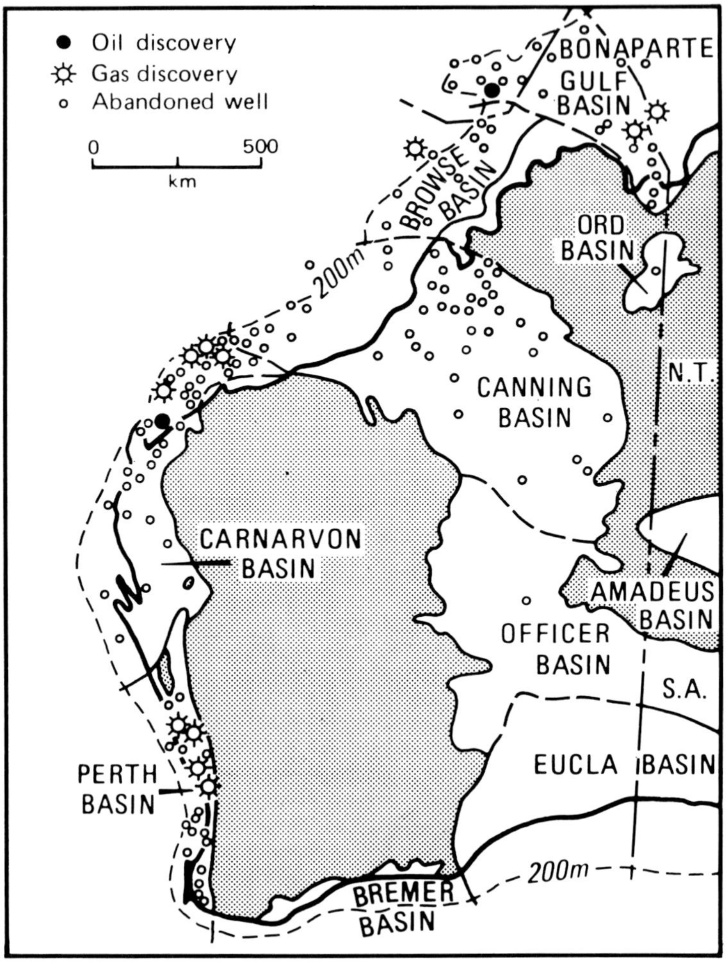

Fig. 3.30 Phanerozoic sedimentary basins of Western Australia, showing locations of principal exploratory wells and oil and gas discoveries

Basin. This discovery was a major turning-point for oil exploration in this state, and indeed in Australia as a whole, for it convinced most people for the first time that oil could be found in this country. Even though subsequent drilling showed the Rough Range accumulation to be very small and uneconomic it served the purpose of stimulating exploration throughout Australia (Playford & Johnstone 1959).

After the Rough Range discovery there was an immediate expansion in geophysical exploration and drilling, but activity declined after 1956 because of a lack of further discoveries, and it reached a low point in 1961. However, in that year Australia's first commercial oilfield was found at Moonie in Queensland, and this resulted in a considerable expansion in exploration throughout Australia. The increased activity in this state led to the discovery of the Barrow Island oilfield in 1964. It was developed as the state's first commercial field, production beginning in 1967. A commercial gasfield was found at Dongara in the Perth Basin during 1966, and this field began production in 1971.

There has been little success with onshore exploration since 1966, only minor discoveries of gas and no discoveries of oil having been made since then. Consequently there has been a steady decline in onshore exploration.

Exploration in the offshore areas of the state began in 1961, when marine seismic surveys were conducted by Wapet in the Exmouth Gulf-Carnarvon area. In 1968 B.O.C. of Australia Ltd (now Woodside Petroleum) drilled the first offshore well in Western Australian waters, at Legendre No. 1 in the northern Carnarvon Basin. It made a small (non-commercial) oil discovery, giving a big impetus to offshore exploration. This eventually led to the discovery in 1971-72 of Woodside's major gas-condensate fields in the offshore northern Carnarvon Basin, at North Rankin, Goodwyn and Angel. Another major gas discovery was made in 1973 by Wapet at West Tryal Rocks. In addition there have been several smaller (currently non-commercial) discoveries of gas and oil in various offshore areas.

Government approval for development of the giant North Rankin field was announced in August 1977. Part of the gas is to be exported (as liquefied natural gas), and part will be consumed in Western Australia. Development of the Goodwyn and Angel fields is expected to follow later.

SEDIMENTARY BASINS

About 45 per cent of the land area of Western Australia and 90 per cent of the continental shelf are underlain by Phanerozoic sedimentary basins (Fig. 3.30). This is a total area of some 1.7 million km^2. Every system of the Phanerozoic is represented in these basins (Playford *et al.* 1975). The stratigraphy of these basins is shown in Table 2.1, pages 34-5.

Explorations carried out in each of the sedimentary basins indicate that the most prospective are the Perth, Carnarvon, Canning, Browse and Bonaparte Gulf Basins. The others are believed to have prospects ranging from very low to nil.

Perth Basin

The Perth Basin is a long, narrow trough of sediments extending for nearly 1000 km along the southwest coast (Fig. 3.31). It is bounded on the east by the Darling Fault, which is believed to have a maximum vertical displacement exceeding 10 km. The age of the sedimentary sequence ranges from Silurian to Tertiary, and the most prospective section is of Permian, Triassic, Jurassic, and Cretaceous age. The basin is strongly faulted, and as seismic records are commonly poor (especially in the coastal strip underlain by Quaternary dune limestone), it is difficult to define drillable structures adequately. Most exploration has been carried out in the Dandaragan Trough, which contains the thickest section in the basin (Jones 1976).

Wapet has found gas in the Perth Basin at Dongara, Mondarra, and Yardarino in

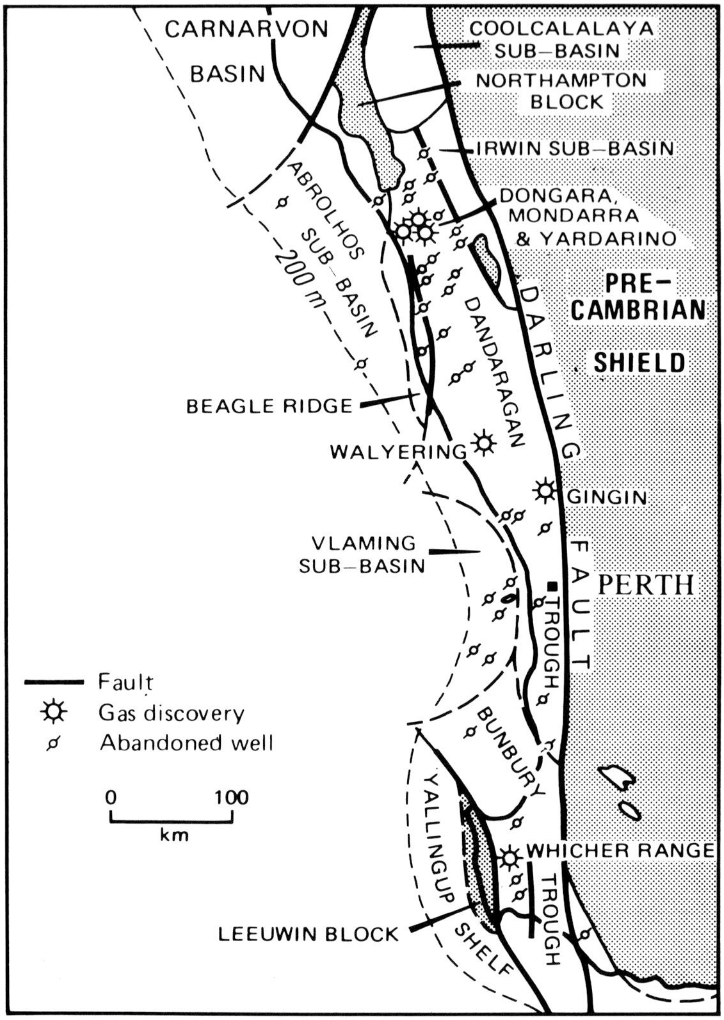

Fig. 3.31 The Perth Basin: structural subdivisions and location of principal exploratory wells and discoveries

Lower Triassic and Upper Permian sandstones, and at Gingin and Walyering in Lower Jurassic coal measures. Another gas discovery (non-commercial) was made by Union Oil Development at Whicher Range, in Lower to Upper Permian coal measures.

A thin oil column is present below the gas at Dongara and Yardarino, and minor amounts of oil have been recovered by Wapet from Lower Cretaceous sandstone in Gage Roads No. 1 well (offshore) and from Lower Triassic sandstone in Mt Horner No. 1 well.

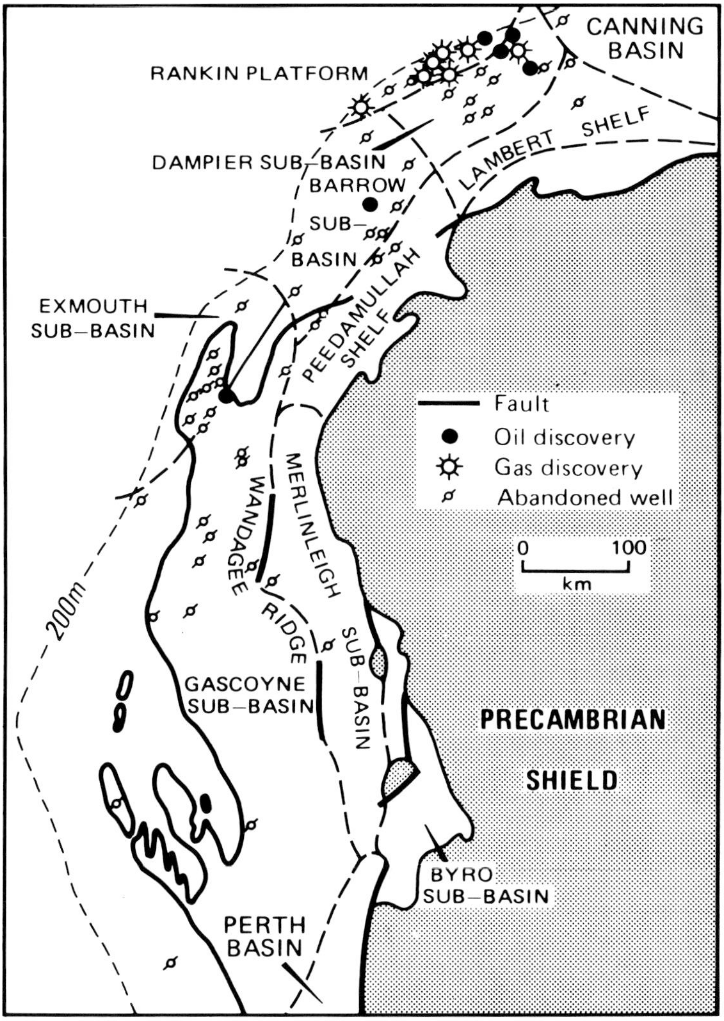

Fig. 3.32 The Carnarvon Basin: structural subdivisions and location of principal exploratory wells and discoveries

Production from Dongara began in 1971, and from Mondarra in 1972. These fields arc supplying gas to Perth and its adjoining industrial areas. Production in August 1977 amounted to about 73.93 million m^3, of which 90 per cent is being supplied by the Dongara field. Small amounts of gas were also produced into the Dongara-Perth pipeline from the Gingin and Walyering fields, but both have now been abandoned.

The Dongara gasfield is the largest known in the Perth Basin. It occurs in a faulted anticlinal structure on the west side of the Mountain Bridge Fault. Production is from the Lower Triassic Dongara Sandstone Member of the Kockatea Shale and from the Permian Wagina Sandstone, Carynginia Formation, and Irwin Coal Measures at depths of around 1600 m. Original reserves amounted to about 10 600 million m^3 of gas, of which 4150 million m^3 had been produced to August 1977, and at least 50 000 m^3 of oil, of which 25 000 m^3 had been produced to the same date. In addition 21 000 m^3 of condensate had been produced from the field.

The Mondarra field also occurs in a faulted anticline, and production is from the same Triassic and Permian units as at Dongara. Original gas reserves were estimated to be 530 million m^3, and 330 million m^3 had been produced by August 1977.

Yardarino is a much smaller field, but it will probably be developed shortly.

The Gingin and Walyering fields produced from low-permeability sandstone in the Lower Jurassic Cattamarra Coal Measures Member of the Cockleshell Gully Formation, between about 3370 m and 3390 m at Walyering and 3870 m and 4160 m at Gingin.

Gas occurred in Whicher Range No. 1 in sandstones of the Permian Sue Coal Measures, below 4000 m. Although the gas column is thick (over 600 m), the reservoir sandstones have very low permeabilities, and the field was relinquished as being uneconomic. The oil occurrences at Gage Roads (in the Lower Cretaceous Gage Sandstone Member) and Mt Horner (in the Lower Triassic Kockatea Shale) were also judged to be uneconomic.

The best prospects for future discoveries in the Perth Basin appear to be offshore, especially in the Abrolhos Sub-basin. Onshore there are still prospects for the discovery of further gasfields, but it does not now seem likely that large fields will be found.

Carnarvon Basin

The Carnarvon Basin contains up to 20 000 m of sediments ranging from Silurian to Tertiary in age, most of which are marine. The Triassic, Jurassic, and Cretaceous sequences are believed to be the most prospective (Thomas & Smith 1976).

North and northeast-trending faulting is widespread in the basin, with normal movement predominating. A belt of large anticlines occurs in Tertiary and Cretaceous rocks in the coastal part of the basin extending from Shark Bay to Barrow Island (Fig. 3.32). These conspicuous folds attracted attention in the early stages of exploration in the basin, and the first exploratory well was drilled on one of them, Rough Range Anticline, in 1953-54.

The Rough Range discovery well flowed oil from the Lower Cretaceous Birdrong Sandstone at a depth of about 1100 m at rates of about 80 m^3 per day. This discovery in 1953 aroused great hopes that major oilfields would be found in the belt of coastal folding, especially around Exmouth Gulf. However, subsequent drilling showed that the oil accumulation below Rough Range is very small (original reserves about 12 700 m^3, of which 2700 m^3 were produced in testing), covering an area of less than 10 ha. Moreover, the other anticlines drilled between Exmouth Gulf and Shark Bay all proved to be dry.

The last of the Tertiary anticlines to be tested was Barrow Island, at the northeast

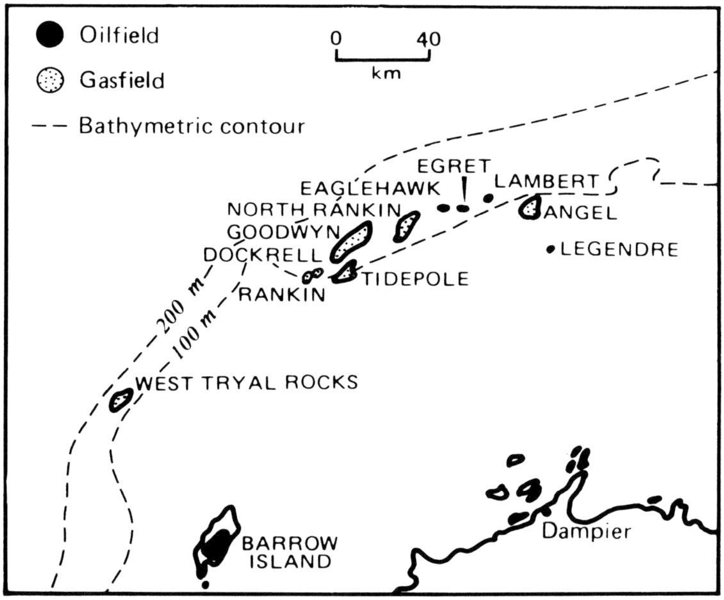

Fig. 3.33 The North-West Shelf: oil and gas discoveries and bathymetry

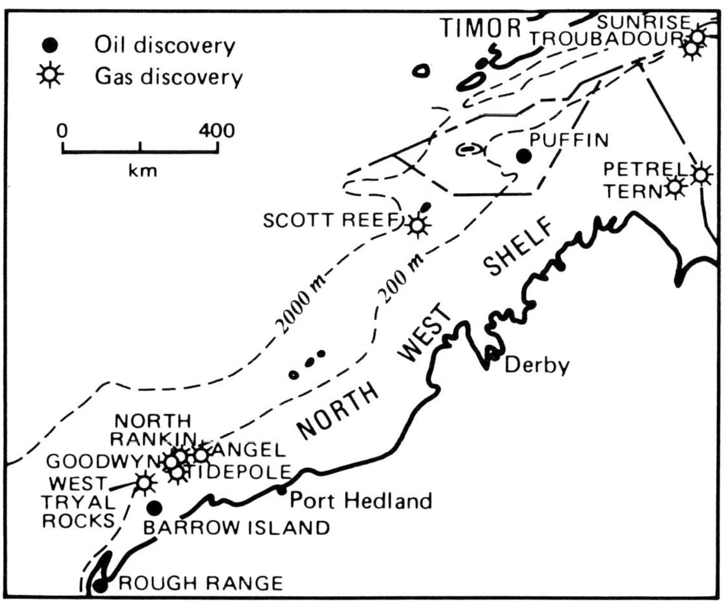

Fig. 3.34 Northern Carnarvon Basin: oil and gas discoveries and bathymetry

end of the fold belt. Barrow Island No. 1 was drilled in 1964, near the surface culmination of the structure (Crank 1973). It was an oil discovery, flowing at rates of up to 157 m^3 per day from the interval 2059-2069 m in Jurassic rocks. Subsequent drilling showed that the Jurassic reservoir sandstones are lenticular and are mainly gas bearing, but it was realized during the evaluation drilling programme that a low-permeability sandstone at depths of around 650 m in the Lower Cretaceous section contained oil and gas. This unit, subsequently referred to as the Windalia Sand Member of the Muderong Shale, had not previously been considered prospective. Drilling showed that it constituted a commercial oilfield (with a small gas cap), and production from the field began in 1967. Production is stimulated by fracturing and water injection. It is estimated that about 38.9 million m^3 of the 120.6 million m^3 of oil originally in place in this unit will be recovered. Of this amount, 21.2 million m^3 had been produced to 31 July 1977. The monthly production for July 1977 amounted to 150 000 m^3.

Three deep test wells were drilled from 1973 to 1976 on Barrow Island so that the Middle and Upper Jurassic sequence (Dingo Claystone) could be more fully evaluated. High-pressure gas and condensate were encountered in several wells, but the accumulations are believed to be too small for economic development.

By the late 1960s exploration in the Carnarvon Basin had indicated that the best prospects were in the northern offshore part of the basin. This area is at the southern end of the North-West Shelf (Fig. 3.33), a broad area of continental shelf extending for some 2000 km from Exmouth Gulf in the Carnarvon Basin to the Timor Sea in the Bonaparte Gulf Basin. The non-commercial oil discovery at Legendre in 1968 was followed by major gas-condensate discoveries at North Rankin (1971), Goodwyn (1971), and Angel (1972), and West Tryal Rocks (1973), all in the offshore northern Carnarvon Basin (Fig. 3.34). The Angel discovery was made in the Dampier Sub-basin in Upper Jurassic to Lower Cretaceous sandstone equivalent to the Barrow Formation. The other fields occur in the Rankin Platform, a major structural uplift near the western margin of the continental shelf. The reservoir sandstones there form part of the Middle Triassic to Lower Jurassic Mungaroo Formation, and these are capped by Lower Cretaceous shales, which provide the seal and give drape closure. The sandstones have excellent reservoir characteristics. The source of the gas and condensate in all of these fields is believed to be Jurassic shales in the Dampier Sub-basin. Total proved and probable recoverable reserves of the four potentially economic fields are estimated to exceed 410 000 million m^3 of gas and 57 million m^3 of condensate.

North Rankin is the largest gasfield in Australia, and it is classed as a giant by world standards, ranking about thirty-fifth in the world list. It occurs in a large Triassic-Lower Jurassic horst block on the south-east margin of the Rankin Platform. The discovery well encountered some 310 m of reservoir sandstone over a gross interval of 564 m, between depths of 2687 m and 3251 m. The field is apparently filled to spill point. Estimated reserves amount to 242 700 million m^3 of gas and 29.2 million m^3 of condensate.

North Rankin is planned as the first field for development by the Woodside consortium. Two production platforms are expected to be erected over the field, and these will be linked by pipeline to the coast and to Perth. A plant for liquefying natural gas will be constructed on the Pilbara coast near Dampier and the liquefied gas will be exported, possibly to the United States and Japan.

The Goodwyn field also occurs in a horst of Mungaroo Formation, with drape closure provided by the overlying Cretaceous shale. It covers a larger area than the North Rankin field, but the productive column is thinner. Moreover, whereas North Rankin has a common gas-water contact in the various sandstone reservoirs, Goodwyn has a series of separate gas-water interfaces stepping upwards to the northwest. Four gas wells have been drilled on the field. Goodwyn No. 3 occurs in a separate fault compartment which also has a separate pressure system, and (unlike the others) contains an oil leg. Reserves of the Goodwyn field are estimated to be 72.78 million m^3 of gas and 14.9 million m^3 of condensate.

The Angel field is less well defined than North Rankin and Goodwyn, for seismic records over the structure are poor, and only three wells have been drilled. The field occurs in a faulted anticlinal structure. Reserves are estimated to be 40 350 million m^3 of gas and 9.5 million m^3 of condensate.

The West Tryal Rocks field is situated near the southwestern end of the Rankin Platform. It is similar to the other two fields on the platform (North Rankin and Goodwyn), occurring in a horst of Mungaroo Formation. Only two wells have been drilled, and it is not yet possible to make a reliable estimate of reserves. However, it seems clear that they will exceed 60 000 million m^3. However, unlike the other fields in the northern Carnarvon Basin, gas from West Tryal Rocks has a high content of inert gases (nitrogen 16.5 per cent, carbon dioxide 10.5 per cent).

The reserves of the other fields that have been found in the offshore Carnarvon Basin (shown on Fig. 3.34) are generally very small, and all are believed to be uneconomic at present. The Tidepole gasfield is the largest of these fields, and it may eventually be developed.

Major interest in the Carnarvon Basin is now centred on the Exmouth Plateau area, west of the Rankin Platform (Fig. 3.33). This is a major faulted uplift in water 800 to 2000 m deep, and the stratigraphy is believed to resemble that of the Rankin Platform.

Canning Basin

The Canning Basin (Fig. 3.35) is the largest of the state's sedimentary basins, covering an area of some 530 000 km^2. Most of the section in the basin is marine, and it ranges from Ordovician to Tertiary in age (Horstman *et al.* 1976, Warris 1976). The aggregate thickness of these rocks exceeds 13 000 m.

The results of exploration in the Canning Basin, both onshore and offshore, have been disappointing. The initial exploration, by Freney, was regarded as encouraging, for signs of oil were reported from Ordovician rocks at Prices Creek (on the Lennard Shelf) and from Permian rocks in the Fitzroy Trough. Wapet's early exploration also gave some encouragement, small quantities of oil and gas being obtained from Carboniferous and Devonian rocks in Meda No. 1 (drilled 1958), and

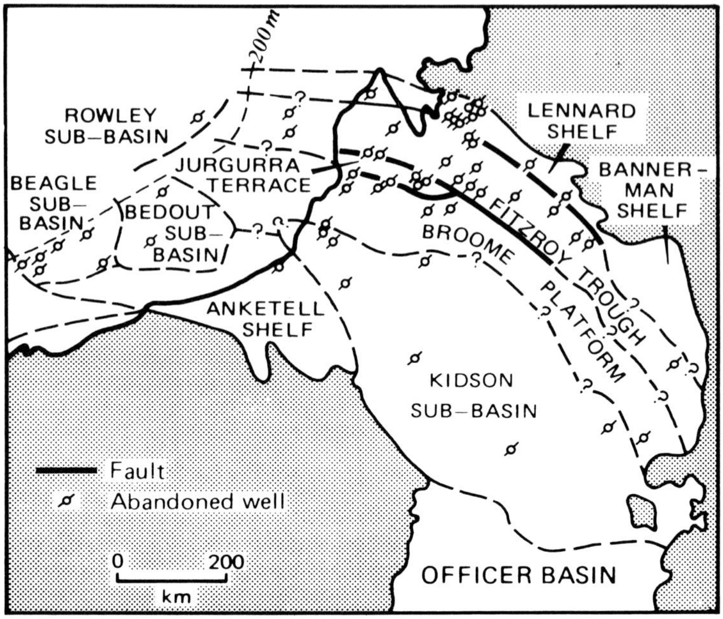

Fig. 3.35 Canning Basin: structural subdivisions and principal exploratory wells

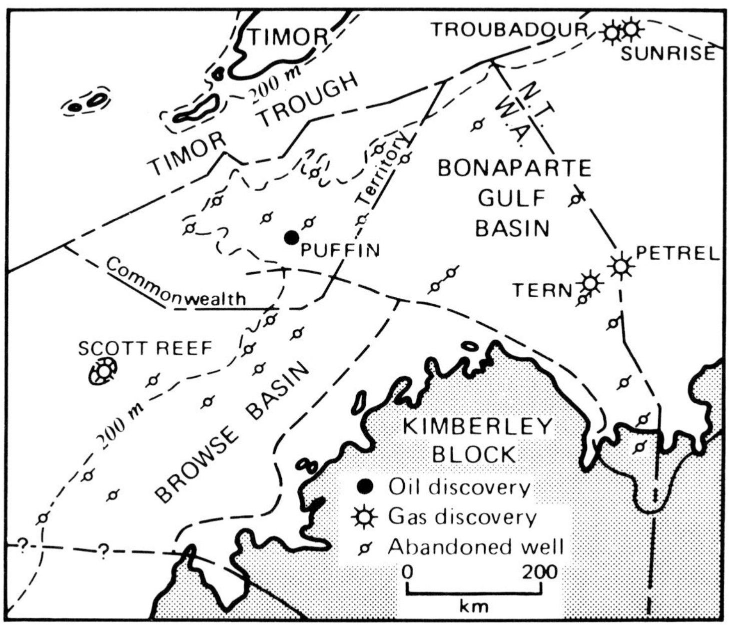

Fig. 3.36 Browse and Bonaparte Gulf Basins: principal exploratory wells and discoveries

small but significant oil showings being observed in Ordovician rocks in several wells. However, no commercial discoveries were made, either onshore or offshore. Moreover, the Mesozoic section, which is so prospective in the adjoining Carnarvon Basin, has proved dry in wells drilled to date in the Canning Basin.

Exploration in the Canning Basin was initially concentrated on a series of large anticlinal structures in the Fitzroy Trough. Most of these have now been drilled, though without success. There has also been considerable interest in the prospects of the Devonian reef complexes known from outcrop along the northern margin of the basin and from the subsurface on the Lennard Shelf and Jurgurra Terrace. However, only one well, Meda No. 1, has recovered significant hydrocarbons from these rocks. A small gas flow (up to 2880 m^3 per day) was obtained from the Devonian reef complex encountered in this well, over the interval 2010-2041 m, and salt water with 7 per cent of oil was recovered from the interval 1558-1565 m in the overlying Lower Carboniferous part of the Fairfield Group. Other wells intended to test Devonian reefs on the Lennard Shelf have either failed to penetrate reefal carbonates, or those carbonates have proved to be flushed with fresh water. However, it is believed that the Devonian reef complexes have not been adequately evaluated in the subsurface and that they are still to be regarded as prospective.

Future exploration in the Canning Basin is likely to be concentrated in the northern onshore part (for Devonian and Ordovician objectives) and offshore (for Mesozoic objectives). There are believed to be moderately good prospects for commercial discoveries.

Browse Basin

The Browse Basin contains a sedimentary sequence ranging from Carboniferous to Tertiary in age, and which is at least 12 000 m thick (Crostella 1976). The basin is wholly offshore (Fig. 3.36), much of it in deep water (greater than 200 m). The best prospects are believed to be in the Mesozoic section.

The first well to be drilled in the Browse Basin was Woodside's Scott Reef No. 1 (1971). It discovered gas and condensate in Lower to Middle Jurassic sandstone and Upper Triassic dolomite and sandstone at depths of 4296-4389 m. The Jurassic and Triassic sequences are separated by an angular unconformity, and the reservoir is capped by Jurassic claystone and volcanics. Net pay in Scott Reef No. 1 was about 28 m, and no hydrocarbon-water contact was detected, the well being in impermeable rocks at its total depth (4734 m).

The Scott Reef discovery is probably unique in the world, for it occurs beneath the lagoon of an atoll on the continental slope, the atoll being separated from the continental shelf by water up to 550 m deep. The foundation of the atoll is a Mesozoic anticline, and reef growth has occurred since the Miocene or possibly earlier, keeping pace with subsidence of the continental margin.

The success of the initial well on the Scott Reef structure gave grounds for optimism that a major field would be developed in the area, especially as the well had not penetrated the hydrocarbon-water contact. However, Scott Reef No. 2A encountered only poor reservoirs in the Jurassic, and was abandoned before reaching the Triassic at its total depth of 4820 m. This is the deepest well to have been drilled in Western Australia. The results of this well have significantly downgraded the potential of the Scott Reef field. Although further drilling would be required for more conclusive evaluation of the field, its prospects for economic development in the near future do not seem bright.

Wells drilled on the continental shelf in the Browse Basin proved dry, although some minor hydrocarbon shows have been recorded. This part of the basin generally lacks attractive structures. However, several large anticlines have been defined in deep water on the continental slope, and these will be drilled during the next phase of exploration. The first of such test wells was Caswell No. 1.
1977).

Bonaparte Gulf Basin

Most of the Bonaparte Gulf Basin is situated offshore, in areas administered by Western Australia and the Northern Territory (including the Territory of Ashmore and Cartier Islands). The basin contains a sedimentary sequence having an aggregate thickness of about 20 000 m, and ranging from Cambrian to Tertiary in age (Laws & Brown 1976). Most of the section is marine, and the best prospects are in Permian and Mesozoic rocks.

The first significant flow of hydrocarbons to be obtained in the basin was from the onshore Bonaparte No. 2A well, drilled by Alliance Oil Development Australia N.L. in 1964. The discovery was non-commercial, but flows of up to 43 500 m^3 per day of gas were recorded from the Lower Carboniferous Milligans Beds over the interval 1437-1470 m. Subsequent onshore exploration proved to be discouraging, and since the late 1960s, exploration activity has been almost wholly offshore.

Several gas discoveries and a single oil discovery have been made in the offshore Bonaparte Gulf Basin, but none is believed to be commercially viable at present. The gas discoveries have been at Tern in Western Australia, and at Troubadour, Sunrise, and Petrel in the Northern Territory. Arco's Tern field may contain 30 billion cubic metres or more of gas, and it could become economic in the future. The small oil-field that has been found in the basin is at Puffin in the Territory of Ashmore and Cartier Islands.

Although no large discoveries have been made to date, exploration results have been sufficiently encouraging to suggest that the basin has reasonable prospects for commercial discoveries in the future. However, most of the well-defined structures have now been drilled (an exception being the large Kelp anticline in the Northern Territory), and future exploration is expected to pay more attention to potential stratigraphic traps.

Ord Basin

The Ord Basin straddles the border between Western Australia and the Northern Territory. It contains some 760 m of Cambrian and possibly Devonian sediments resting on basalts of probable Early Cambrian age.

The Okes Durack bore was drilled on an anticlinal structure in the basin in

1923-24. It bottomed in basalt, and was dry. Since then there has been no significant exploration in the basin, and its hydrocarbon prospects are regarded as nil.

Amadeus Basin

Very little of the Phanerozoic part of the Amadeus Basin extends into Western Australia. The sediments are of Ordovician, Silurian or Devonian, and of Permian age, and they are believed to have no significant prospects in this area.

Officer Basin

The Officer Basin covers an extensive area in Western Australia (about 300 000 km^2) and South Australia (Krieg *et al.* 1976). Most of the section in the basin is of Proterozoic age, with only a thin cover of Phanerozoic sediments (probably less than 1000 m through the Western Australian part of the basin). The oldest Phanerozoic deposits are not adequately dated, but they are thought to be early Palaeozoic in age. They are overlain by thin Permian (glacigene) and Cretaceous deposits.

Hunt Oil Co. carried out geophysical exploration in the Officer Basin in this state from 1961 to 1966. The company also drilled four stratigraphic wells and a single test well, Yowalga No. 2, in 1966. This well entered presumed Proterozoic rocks at shallow depth and was abandoned. Since then there has been no significant exploration in the Western Australian part of the basin, and its prospects are believed to be very low.

Eucla Basin

The Eucla Basin is occupied by a thin flat-lying sheet of Cretaceous and Tertiary sedimentary rocks, with minor developments of Permian glacigene deposits in the northern part (Lowry 1976). The basin also extends into South Australia. The maximum thickness of Phanerozoic sediments probably does not exceed 750 m onshore, but it may thicken to as much as 1500 m on the continental shelf.

Three onshore stratigraphic wells were drilled in the Western Australian part of the basin in 1960 and 1964, all of which entered basement rocks at shallow depths. There has also been a small amount of onshore geophysical exploration, and the conclusion reached was that the section is everywhere too thin and lacking in structure to have any hydrocarbon potential. Attention then turned offshore, where geophysical exploration showed that the section on the continental shelf is also thin and flat-lying and the only possible petroleum prospects appear to be in the thicker sedimentary sequence developed below the continental shelf.

Bremer Basin

The Bremer Basin contains a thin Tertiary sequence onlapping an irregular topography developed in the Precambrian basement. The maximum thickness of sediments probably does not exceed 200 m onshore and 1200 m on the continental shelf. However, the section apparently thickens considerably below parts of the continental slope in very deep water. Three stratigraphic wells have been drilled onshore, and all encountered basement rocks at shallow depth. A reconnaissance seismic survey has been carried out offshore, and it is clear that the hydrocarbon potential is poor there also. There may be some prospects on the continental slope, but the depth of water and restricted lateral extent of the sediments make this area unattractive.

FUTURE OUTLOOK

The most important petroleum development in Western Australia during the next few years is expected to be the exploitation of the large gasfields in the northern Carnarvon Basin. This will be the largest single industrial undertaking in the state's history, and it will be of vital importance to our future energy supplies. Development of these fields should also stimulate exploration in other areas of the state.

The strong hopes of a few years ago that major oilfields would be found in Western Australia have undoubtedly dimmed, as there have been no important oil discoveries since Barrow Island in 1964, and exploration data indicate that generative sediments in many areas are gas prone. However, the offshore potential for further gas discoveries seems good, and there is a reasonable chance that significant oilfields will eventually be found. Moreover, there are still prospects for economic discoveries onshore in several areas, especially in the Perth and Canning Basins.

The main frontier areas for exploration in this state are in deep water on the continental slope, and increased attention will be paid to these areas during the next few years.

REFERENCES

Crank, K. 1973. Geology of Barrow Island oil field. *APEA J.* **13,** pt 1: 49-57.

Crostella, A. 1976. Browse Basin. In *Economic Geology of Australia and Papua New Guinea—3. Petroleum* (ed. C. L. Knight). Australas. Inst. Min. Metall. Monogr. Ser. 7: 194-9.

Horstman, E. L., Lyons, D. A., Nott, S. A. & Broad, D. S. 1976. Canning Basin onshore. In *Economic Geology of Australia and Papua New Guinea—3. Petroleum* (ed. C. L. Knight). Australas. Inst. Min. Metall. Monogr. Ser. 7: 170-84.

Jones, D. K. 1976. Perth Basin. In *Economic Geology of Australia and Papua New Guinea—3. Petroleum* (ed. C. L. Knight). Australas. Inst. Min. Metall. Monogr. Ser. 7: 108-26.

Krieg, G. W., Jackson, M. J. & van de Graaff, W. J. E. 1976. Officer Basin. In *Economic Geology of Australia and Papua New Guinea—3. Petroleum* (ed. C. L. Knight). Australas. Inst. Min. Metall. Monogr. Ser. 7: 247-53.

Laws, R. A. & Brown, R. S. 1976. Bonaparte Gulf Basin—south eastern part. In *Economic Geology of Australia and Papua New Guinea—3. Petroleum* (ed. C. L. Knight). Australas. Inst. Min. Metall. Monogr. Ser. 7: 200-12.

Lowry, D. C. 1976. Eucla Basin. In *Economic Geology of Australia and Papua New Guinea—3. Petroleum* (ed. C. L. Knight). Australas. Inst. Min. Metall. Monogr. Ser. 7: 96-8.

Playford, P. E., Cope, R. N., Cockbain, A. E., Low, G. H. & Lowry, D. C. 1975. Phanerozoic. In *The Geology of Western Australia*. Geol. Surv. W. Aust. Mem. **2**: 223-8.

Playford, P. E. & Johnstone, M. H. 1959. Oil exploration in Australia. *Bull. Am. Assoc. Pet. Geol.* **43**: 397-433.

Thomas, B. M. & Smith, D. N. 1976. Carnarvon Basin. In *Economic Geology of Australia and Papua New Guinea—3. Petroleum* (ed. C. L. Knight). Australas. Inst. Min. Metall. Monogr. Ser. 7: 126-65.

Warris, B. J. 1976. Canning Basin, off-shore. In *Economic Geology of Australia and Papua New Guinea—3. Petroleum* (ed. C. L. Knight). Australas. Inst. Metall. Monogr. Ser. 7: 185-93.

3 H(iii)

Uranium

J. D. CARTER
Geological Survey of Western Australia

INTRODUCTION

The radioactive element uranium was first recognized in pitchblende in Germany in 1789, the name being derived from the planet Uranus. Early interest in the metal centred on the therapeutic properties of one of its many disintegration products, radium, and it was not until the development of nuclear weapons that uranium achieved prominence. A large proportion of the world's population may soon depend upon energy derived from the metal. Despite the prevailing controversy surrounding its use, some authorities claim that by the end of this century as many as 40 countries could be producing up to 85 per cent of their electricity from nuclear energy plants.

Uranium was first discovered in Western Australia at Wodgina in 1910. So far in the state the only major uranium orebody known is that at Yeelirrie, a pastoral property about 700 km northeast of Perth. Yeelirrie is the third largest deposit in Australia, being surpassed only by the Jabiluka and Ranger orebodies in the Northern Territory.

URANIUM GEOLOGY AND RADIOACTIVITY

Uranium minerals

Uranium is one of the less common elements in the earth's crust. It is more abundant than silver and gold but slightly less abundant than such metals as nickel, copper and lead. Uranium is never found in its elemental state but always in combination with other elements, with which it forms more than 100 minerals. The most important, and the one which to date has yielded most commercial uranium, is uraninite (UO_2), which includes the amorphous variety known as pitchblende. No commercial occurrence of uraninite or pitchblende is known in Western Australia, although these minerals have been identified at several localities.

A characteristic of uraninite is the ease with which it oxidizes to brightly-coloured yellow, orange, red, or green uranium minerals. The most important as a source of uranium is carnotite (approximately $K_2O.2UO_3.V_2O_5.2H_2O$), the only uranium mineral identified so far at Yeelirrie.

The chief uranium mineral at Wodgina is kasolite, a hydrous silicate of uranium. A number of other uranium-bearing minerals have been recorded, but many examples, although containing uranium, are not recognized ores of the element. These include (from the Pilbara Goldfield) uraniferous niobates and tantalates such as fergusonite and euxenite.

Uranium deposits

Classifications of uranium deposits depend both upon the types of rocks containing uranium ores, and the mode of occurrence of mineralization. Among host types of rocks are sandstone, quartz conglomerate, pegmatite and calcrete. Classifications defined by mode of occurrence include mineralized veins and lodes, and irregular replacement bodies, the former cutting one or more rock types, and the latter replacing them.

Although most uranium deposits in Australia are the vein and lode type found in the Northern Territory, only minor uranium vein mineralization has been found in Western Australia. An example occurs at Mundong Well in the Ashburton Goldfield. No replacement uranium body is known in Western Australia.

Sandstone-type deposits from which, in the United States, most of the Western world's uranium is now obtained, have not been positively identified in the state. Quartz conglomerates constitute the second most important uranium source of the Western world, with production centred in Canada and South Africa. Again Western Australia is not known to have an important representative of this type, though low-grade uranium accumulations have been detected in conglomerates in the Pilbara Goldfield. Uranium-bearing pegmatites, typically cutting granitic rocks, are known from the Pilbara Goldfield and elsewhere, but no uranium production of any note has come from such rocks.

Perhaps the most striking feature of uranium in Western Australia is that, although a wide variety of rocks are found which are known to be favourable for uranium accumulation in other parts of the world, none has yielded a commercial deposit; yet calcrete which is not known to carry commercial uranium elsewhere, is widely mineralized. Calcrete, a surface and near-surface secondary limestone, is typically associated with valley-fill sediments in drainage systems in inland parts of the state, north of latitude 30°. It is the host rock for uranium at Yeelirrie.

Radioactivity

All uranium minerals are radioactive and emit radiations which can be detected. This is the property by means of which the Yeelirrie deposit was discovered. Uranium undergoes spontaneous decay, resulting after a number of steps in the formation of the more stable element lead. At each stage of decay, particles and rays are given off which are detectable by instruments. These emissions, alpha and beta particles, and gamma rays, do not travel far, and even gamma rays, which have the greatest penetration, are stopped by a metre or so of soil. Most methods of prospecting for uranium use the property of greater penetration by gamma rays.

Instruments measure the strength of gamma emissions in electronvolts. The well-known type of detector, the Geiger counter, is now superseded by the more efficient scintillation counters and spectrometers. Radiometric surveys using detectors may be made on foot, from vehicles or low-flying aircraft, or below ground by probing boreholes. A comprehensive range of radiometric surveys has been put to use in uranium search in Western Australia, and the discovery of the Yeelirrie deposit was the direct result of an airborne radiometric survey.

SEARCH FOR URANIUM IN WESTERN AUSTRALIA

Since the first discovery of uranium-bearing minerals at Wodgina in 1910, exploration for the element has passed through three phases of increasing intensity in response to prevailing demands. The earliest, and the least important, commenced with the Wodgina discovery when the mineral's radium content was sought for therapeutical purposes. It was realized early that the Pilbara pegmatites were unlikely to yield radium in useful quantities, and interest was not sustained. Immediately after the war, in 1947, exploration was prompted by the demand for uranium for military purposes, but this phase passed its peak by about 1956 without a worthwhile deposit being identified. The current phase began in the late 1960s with the realization that anticipated nuclear power generation would create a market for uranium larger than the existing known reserves. Although the amount of exploration carried out is considerable, the only large reward so far is Yeelirrie.

Early exploration

During 1905 the first geological reconnaissance of the Pilbara Goldfield was completed, and the then government geologist observed that tantalite-bearing pegmatites such as those which occur at Wodgina sometimes carry radioactive minerals elsewhere, and that careful search may result in their discovery (Gibb Maitland 1906). In 1910 this prediction was realized when a Mr H. Marshall sent to the Geological Survey a specimen of yellow ochre obtained from pegmatite on mineral lease No. 46 at Wodgina. Gibb Maitland, with characteristic astuteness, arranged for the specimen to be tested for uranium, which turned out to be a major constituent. Gibb Maitland showed more than ordinary prescience when mentioning that the importance of this find 'lies in the enormous stores of energy which radio-active minerals possess'. Formerly the name 'pilbarite' was given to the ochre, but this has now been shown to be a mixture of kasolite and thorogummite (Honea 1957).

Pegmatites containing very small amounts of uranium-bearing accessory minerals were identified elsewhere in the state, and until the 1950s were the only known occurrences of the element. None have commercial potential and none were mined for uranium in any acceptable sense, including the pegmatite of Lookout Hill (sometimes incorrectly called Radium Hill) 5 km south-east of Holleton about which there have been some misleading statements published, suggesting that it was worked for uranium.

Post-war exploration

Owing to its demand for military purposes, exploration for uranium was intensified throughout Australia immediately following the war. Participation by the state consisted of surveys of beach sands and river sands for radioactive minerals. Between 1947 and 1949 beach sands from Geraldton to the Great Australian Bight and along parts of the Pilbara coastline were sampled but without success. Surveys of river sands in the southwest of the state were similarly disappointing. There were further geological examinations of uranium-bearing pegmatites of the Pilbara in 1947 and 1948, but no significant discoveries were made.

However, the publicity accompanying these surveys brought about more positive results, and in 1952 the first non-pegmatite uranium find was made at the Wilgie Mia ochre cave in the Weld Range. R. H. Stowe, who was on holiday near Cue, detected radioactivity with a Geiger counter. The small amounts of minerals subsequently recovered were considered to be soddyite, but have since been shown to be boltwoodite associated with a little carnotite (D. Burns pers. comm.). The uranium minerals were probably deposited in minute quantities by waters circulating in ochreous ironstone material.

In the following year, 1953, yellow uranium-vanadium ochre was found associated with weathered granite and superficial deposits a few kilometres south of Norseman, near Lake Dundas. This second discovery of non-pegmatite uranium is noteworthy, since it gave the first indication of the physico-chemical regime under which secondary uranium is precipitated in many parts of the Eastern Goldfields, including Yeelirrie.

Without doubt the most important innovations of the early post-war phase of prospecting were the rationalization of exploration by defining geological situations in the state known to be favourable elsewhere for uranium mineralization, and the introduction of airborne radiometric surveys.

A type of uranium deposit other than pegmatite, and one that was early sought after, was mineralization in quartz conglomerate, and for this, Lower Proterozoic conglomerates in the Pilbara Goldfield were examined in 1955 by Enterprise Exploration Company Pty Ltd. It was not until 1970, however, that exploration by Cominco Exploration Pty Ltd confirmed the validity of the analogy. Cominco's boreholes intersected low-grade uranium values associated with thucholite pellets and heavy minerals similar to those of the Blind River deposits of Canada, in conglomerates of the Hardey Sandstone outcropping near Nullagine. No commercial deposit has been proved there but exploration is continuing.

The upsurge of uranium exploration in the Kimberley Goldfield in 1954 was encouraged by analogies drawn between the geology of the Halls Creek area and uranium fields in Northern Territory such as Rum Jungle, but searches for similar mineralization have been disappointing.

Airborne radiometric surveys employing scintillometers, introduced into the state by the Commonwealth Bureau of Mineral Resources in 1954, in country near Halls Creek, have proved to be the most effective means of searching for uranium, and led to the discovery of Yeelirrie. The 1954 Kimberley survey was followed in 1955 by flights over parts of the Pilbara Goldfield, Wilgie Mia and Norseman, and these surveys became a regular and statewide feature of geological investigations. Their effectiveness has been proved, and a success worthy of mention was the first discovery in 1961 of carnotite in the state, found in calcrete approximately 20 km southeast of Mundong Well (Gardener & Jones 1967).

Recent exploration

Interest in uranium revived in the mid-1960s in response to the need for the generation of electricity by nuclear power. An increase in tempo of exploration brought about discoveries of quartz conglomerate, calcrete and vein type deposits.

Initially many exploration companies paid attention to radioactive Proterozoic conglomerates found in the Kimberley Goldfield. Radioactivity in these conglomerates, principally those in the King Leopold Sandstone, was eventually found to be due to thorium minerals rather than to uranium. However, as mentioned earlier, the Hardey Sandstone in the Pilbara Goldfield was shown in 1970 to carry quartz conglomerate-type mineralization.

In 1970 geologists of Western Mining Corporation discovered uranium at Yeelirrie. This success was the result of airborne radiometric surveys carried out by the company and the Commonwealth Bureau of Mineral Resources. Western Mining Corporation surveyed parts of the Meekatharra district in 1969 and correlated airborne radiometric anomalies there with concentrations of uranium in calcretes filling drainage channels. As soon as the results of the airborne survey of the Sandstone area by the Commonwealth Bureau were published in 1970, the presence of radiometric anomalies associated with calcrete-filled channels near Yeelirrie at once attracted the attention of the company. Since then calcrete-filled drainage channels have become a prime target for exploration.

In 1972 small veins carrying copper ores near Mundong Well in the Gascoyne Goldfield were shown also to contain small amounts of pitchblende and kasolite. These discoveries created a surge of exploration activity in the Ashburton and Gascoyne Goldfields and resulted in discoveries of small occurrences of uranium mineralization of various types such as those of the conglomerate and calcrete types, but so far none has been shown to be of commercial significance.

In recent years, exploration of calcrete-filled drainages has led to investigations of saline lakes into which the channels discharge. Discoveries of small accumulations of uranium minerals within lacustrine sediments may represent mineralization of the sandstone type, but exploration is still too early to allow definite statements to be made on their genesis.

YEELIRRIE

In 1972 Western Mining Corporation announced the discovery of carnotite in calcrete over a wide area at Yeelirrie, and eventually estimated ore reserves as 46 000 tonnes of U_3O_8 at an average grade of 0.15 per cent. Of this 24 000 tonnes are above 0.36 per cent.

Near Yeelirrie there are some of the oldest rocks of the state. These consist of greenstone belts of metamorphosed layered rocks, including volcanic and sedimen-

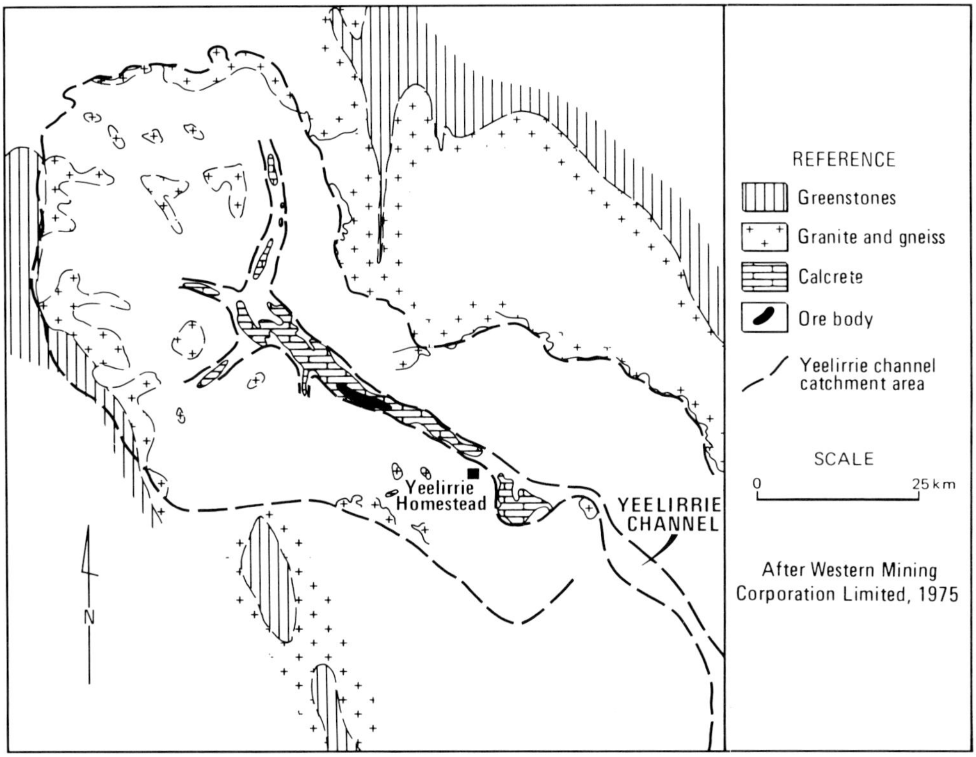

Fig. 3.37 Yeelirrie area: geological sketch map

tary representatives, intruded and enveloped by granites and gneisses. Their distribution is shown on Fig. 3.37. A profile illustrating the landscape features and the distribution of the rocks is represented on Fig. 3.38.

Calcrete, where most of the carnotite is found, has formed over extensive areas and may be up to 6.5 km in width and 20 m in thickness. It has many forms, but one of the more common is a hard white porcellaneous and cavernous variety. It may also be friable, or completely powdery. Carnotite has been deposited as thin films on cavity walls in the calcrete, or dispersed through earthy calcrete, and associated clays with quartz grains.

Processes responsible for precipitating carnotite in the drainage channels are not thoroughly understood. Deposition of carnotite ($K_2O.2UO_3.V_2O_5.2H_2O$) requires concentrations of potassium, uranium and vanadium to collect in drainage channels. Potassium and uranium can be derived by the weathering of granitic rocks such as those shown in Fig. 3.38, and vanadium is available from certain rocks in greenstone belts. Drainages carry these elements into calcrete-filled channels where, as a consequence of evaporation, high concentrations are achieved. Solutions so charged come into contact with calcrete within oxygenated near-surface zones with waters of

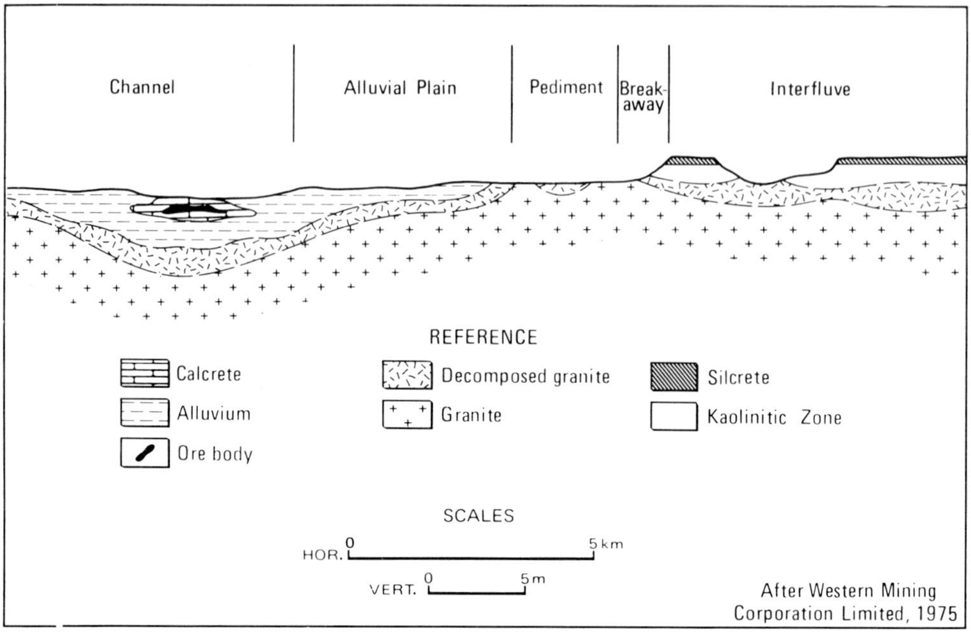

Fig. 3.38 Yeelirrie area: sketch section to illustrate geology and geomorphology

slightly higher pH. In this environment the solubility product of carnotite may be exceeded and precipitation of the mineral takes place (Butt *et al.* 1977, p. 16).

So far, Yeelirrie represents the only large deposit of uranium-mineralized calcrete yet discovered, though there are many other small examples, both in Western Australia and other parts of the world.

REFERENCES

Butt, C. R. M., Horwitz, R. C. & Mann, A. W. 1977. Uranium occurrences in calcrete and associated sediments in Western Australia. *CSIRO Miner. Res. Lab. Aust. Rept FP16.*

Gardener, J. E. F. & Jones, W. R. 1967. Ground inspection of airborne radiometric anomalies, Carnarvon Basin, Western Australia 1961. *Aust. Bur. Miner. Resour. Rec. 1967/14* (unpubl.).

Gibb Maitland, A. 1906. Third report on the geological features and mineral resources of the Pilbara Goldfield. *Geol. Surv. W. Aust. Bull.* **23**.

Honea, R. M. 1957. Identity of pilbarite with thorogummite and kasolite. *Amer. Min.* **42**: 908-10.

Western Mining Corporation Limited (Exploration Division), 1975. A general account of the Yeelirrie uranium deposit (unpubl.).

31

Ground-water and its importance to the mineral industry

E. P. D. O'Driscoll
Geological Survey of Western Australia

The climate of Western Australia is either arid or semi-arid over four-fifths of the state, and impressively large tracts are occupied by poorly-watered areas such as the Gibson Desert, Great Sandy Desert, Victoria Desert and the Nullarbor Plain.

Water is the only mineral on which man is completely dependent for life, hence it is fortunate that in much of the state ground-water reserves are much larger and more readily available than those of surface water.

Ground-water originates as rain, which penetrates downward through the soil and into the underlying soil or rock, through which it may continue to move slowly under gravity. Unlike other minerals therefore it is renewable, although the natural rate of movement is slow and replenishment may take a long time. Ground-water occurrence is controlled by factors such as rainfall, topography, rock type and structure, and evapotranspiration losses. The relative importance of these factors varies with locality. Evapotranspiration losses are usually a dominant factor however, because the naturally established vegetation is usually of a type best suited to using up all the available soil moisture in an average year, which means that an excess of downward-percolating rainwater in wet years or along watercourses and swamps is the only replenishment to a permanent water-table existing below the trees. Because the vegetation extracts pure water from the soil moisture, leaving the salt still in solution in the remainder, the ground-water salinity is much higher than that of the rainfall, although the proportions of the various ions remain constant. In low-rainfall areas such as the Pilbara, where the rain is cyclonic and torrential, downward soakage into the soil is rapid, and ground-waters are therefore likely to be of low salinity.

Ground-water of the Perth metropolitan area

The Perth metropolitan area is underlain by deep pressure-water aquifers replenished from scattered surface intakes on the coastal plain. Shallow sands and limestone, extending downward from the surface to sometimes more than 90 m, store any available excess rainfall as unconfined, water-table ground-water, and provide domestic supplies for property holders. In a few suburbs the subsoil is silty and poorly productive, and in some the ground-water is not usable because of high salinity or other factors; but in general the coastal plain part of the metropolitan area is underlain by a sheet of water which is often at very shallow depth. It usually contains iron, and sometimes has a sulphurous odour, both of which can be removed by aeration. Overpumping of this shallow aquifer will lower the level of the water-table (i.e. the surface below which all pores in the rocks are saturated). More water may be removed in a year than can be replenished during the next winter, but it will never be entirely depleted, because of recurring replenishment and a predictable drop in usage by private landholders as the depth to water and the cost of pumping increase, and available supplies gradually diminish.

The Gnangara ground-water mound north of Perth and the Jandakot ground-water mound to the south are hydrodynamically distinct, the Gnangara area being the larger (Fig. 3.39). Both are replenished by local rainwater penetration downward through the sand. The water-level fluctuates seasonally about 0.2 to 1.5 m. The estimated outflow from Gnangara alone is about 150 million m^3 per annum, which could be extracted each year without depleting the storage.

Besides the shallow water in the superficial formations, Perth also has even larger reserves of confined, pressure water (artesian and subartesian) in the deeper Cretaceous and Jurassic aquifers. Because recharge of the deep aquifers is less direct and movement is slow, extraction will need careful management to avoid mining the water.

Ground-water in rural areas

Ground-water conditions vary widely throughout the state. The several extensive sedimentary basins which contain large supplies of pressure water include the Perth Basin, Canning Basin, and Carnarvon Basin. Much of the water in the first two is of low salinity and suitable for domestic use. In the desert regions underlain by the Officer and southern Canning Basins, ground-water is not always obtainable and may be saline, and the Eucla Basin has mainly saline pressure and non-pressure water, but isolated thin floaters of shallow fresh water may accumulate whenever heavy rains flood the depressions and penetrate the underlying cavernous limestone.

Conditions in the Pilbara vary greatly. Hard Proterozoic rocks in general are poorly productive, but calcrete and alluvium in the river and creek-beds may contain large volumes of potable water, e.g. the De Grey, Turner and Yule Rivers and the Millstream area on the Fortescue River. When the ground-water is shallow, however, evapotranspiration losses may make it highly saline.

In the agricultural districts of the Yilgarn Shield, underlain by Archaean rocks, ground-water conditions in general are a reflection of the rainfall, water supplies becoming more scarce and more saline as the rainfall decreases. North of about lat. 29 °S some ancient drainage lines contain highly-permeable calcrete, which in some areas contains quite large volumes of low-salinity ground-water used for township and industrial purposes.

Ground-water and mining activities

The non-availability of permanent surface water of low salinity has substantially affected mining activities, in both the preliminary exploration stage and the subsequent development and exploitation of orebodies. This was particularly so in the early stages of prospecting, it being quite noticeable that the first reports of mineral

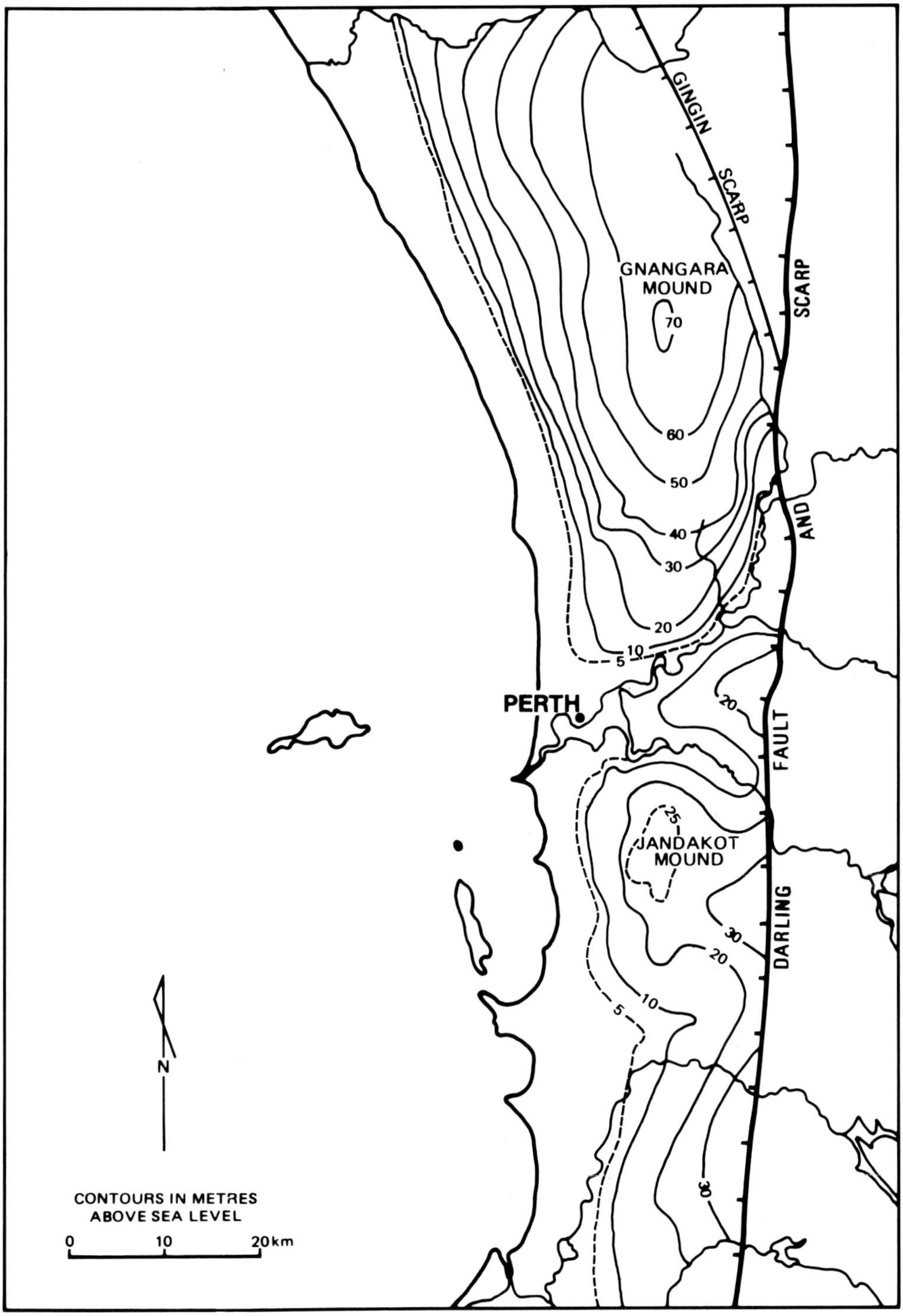

Fig. 3.39 Perth metropolitan area: contour map of the water table, showing the Gnangara and Jandakot Mounds

occurrences almost always came from the Kimberleys or Pilbara, the coastal areas such as Cossack, or districts already settled because water was available in these areas.

Between the two world wars the dependence of mining on ground-water was relatively small, with some exceptions such as at Northampton, where lead concentrates were produced, and the asbestos centre of Wittenoom, where the ground-water beneath the Fortescue River valley supplied the township. The position changed after 1945, when the rapidly expanding programme of mineral exploration and development of iron, mineral sands, alumina, nickel and even salt deposits inaugurated a widespread search for large volumes of low-salinity ground-water. The viability of more than one substantial mining venture depended on such water being available not only for beneficiating the ores, but also for township supplies. It is heartening that in nearly all cases usable water was found to be available within economic distance of where it was needed.

Because no mining company had the right to explore for and subsequently use ground-water outside its own leases, and because the Mining Act had been written long before the new requirements had been envisaged, government co-operation was needed. The Rights in Water and Irrigation Act was amended to provide that the government in effect owned all ground-water in the state; and when necessary the government then agreed to employ ground-water consultants at company expense, and to oversee and evaluate all exploratory drilling in likely areas. When a water supply was assured, the particular company was given the right to its use. Some company leases were capable of supplying all requirements without needing government help.

The lack of surface water and the generally low rainfall in so many areas of important mineral occurrence has previously been pointed out. Ground-water was therefore the only possible source of water in the large volumes needed to establish townships, to beneficiate ore (one company's stated requirement was initially 68 million litres daily of low-salinity water), and to a lesser extent to cope with dust. It is interesting to see how the problems differed because of locality, and the size and nature of mining and beneficiation operations. Speaking generally, the Pilbara was favourably supplied, but the surprising thing was that in other districts previously thought to be poorly watered, intensive search showed the existence of quite large bodies of ground-water in some localities, such as at the Windarra and Mt Keith nickel prospects.

(i) *Gold-mining*

The earlier discoveries of gold in the eastern states had stimulated interest among politicians and public alike, but inland prospecting towards the close of the nineteenth century was severely restricted because of the erratic rainfall and the scarcity of water. This was emphasized because pastoralists were slow to take up land away from the ephemeral rivers where feed was available along the banks, and where water could be obtained by digging shallow wells. Therefore, wells were not being dug by the pastoralists in areas of interest to prospectors, and penetration and exploration of the dry interior in the search for minerals was therefore possible only after rain, when the replenished gnamma-holes and seepages could be used for short periods. Some attempts were made to develop the ground-water resources, but mostly with indifferent success, because in the low-rainfall areas ground-water was often unobtainable or was too saline for drinking.

The real importance of water, and especially of ground-water, became evident with the discovery in 1887 of the Yilgarn Goldfield, when Southern Cross became a jumping-off place for prospectors venturing farther eastwards, inaugurating an era of desperate hardship, sickness and death in the almost waterless interior in conditions which today are hard to imagine.

Not all goldfields were so waterless however. By 1892 discoveries at Cue, Yalgoo, Nullagine and elsewhere in the Pilbara necessitated carting water from distant wells, but at least it was obtainable. However, by that year prospectors had pushed as far afield as Nannine, in an area where water was scarce and usually saline. Moreover, hopeful prospectors sometimes perished because of their own inexperience and lack of provision of the simplest necessities, the main one being drinking-water.

Warnings enough had been issued by the authorities. In his Annual Report for 1888-89 Government Geologist H. P. Woodward, referring to the Yilgarn, wrote: 'The great difficulty at present to successful working is the scarcity of water'; and again: 'Some rich deposits of alluvial gold have been found, but owing to the scarcity of water, little of the ground has yet been tested.'

Between 1889 and 1892 government-employed contractors unsuccessfully drilled near Southern Cross and Coolgardie, and from March 1889 to February 1895 the government spent about £6500 on drilling, but 'none of the bores served any useful purpose either for water or geologically'. The government action was needed however. Woodward had already written (1888) of Parker's Find (64 km south of Southern Cross) that, 'owing to the great scarcity of water last summer, prospectors had very little chance in this part of the field of developing their claims.'

The water-supply position was not everywhere so desperate. At places like Egina Goldfield, Ashburton Goldfield, Mallina Claim (113 km east of Roebourne), Whim Well, Yandanooka (a copper prospect), and the Nangrang Hills in the Northampton district, drinking-water was fairly readily available, and shaft-sinking was sometimes brought to a halt by an over-abundance of saline water.

In 1892 the discovery of gold at Coolgardie occasioned a frantic rush of prospectors from Southern Cross, a poorly-watered trip for which many were inadequately equipped. In the ensuing rush from as far afield as York, water sold for up to 2*s* 6*d* a gallon (4.5 litres). From the outset, and until condensers were erected, water was a crucial problem, so much so that on several occasions when an exodus from Southern Cross occurred, the men were asked to leave in small groups and to use the water along the track sparingly. The shortage of water and the almost complete lack of sanitation caused outbreaks of typhoid fever that reaped a grim harvest. In 1892 the first government drilling-rig was brought to Coolgardie, and drilled one successful bore 42 km north of the town, but in 1893 men were still paying 3*s* a gallon (4.5 litres) for water, which amounted to as much as £3 15*s* to water one riding-camel and £1 a horse. A large number of wells had been sunk in and around Coolgardie townsite, all the waters being more or less saline, varying from slightly brackish to an almost incredible 15 500 mg to the litre!

Prospecting went on, the most disastrous rush being the 1893 stampede to Siberia, about 120 km northwest from Coolgardie, across waterless country which cost an unknown number of lives when some, attacked by thirst and with no knowledge of the way, wandered off into the bush and perished. Ten bodies are known to have been found and buried, and the government had to take urgent action by sending water-carts to establish supplies along the way.

The era of the prospector, although it continued for many years, was by this time giving way to one of more organized development by mining, which brought additional water-supply problems. As mining increased, water was needed not only for men and transport but also for the more stable populations of townships, as well as for use in the crushing and extraction of gold, and even for 'civilized amenities' such as breweries.

The pressure of public opinion brought about the drilling in 1895 of a deep bore at Coolgardie, despite the opinion of Gibb Maitland that artesian water would not be obtained. At a cost of £11 692 13*s* 5*d* a diamond-drill hole in granite reached a depth of 916 m before being abandoned; its failure was almost immediately followed by C. Y. O'Connor's proposal for building the Mundaring Weir and the Goldfields Water Supply pipeline.

Though construction of the pipeline ultimately met the needs of the mines centred around Coolgardie and Kalgoorlie, this took some years, and in any case the more distant fields had to depend on water from other sources. Some were fortunate in having access to quite large volumes of ground-water from the cavernous calcretes occupying relict drainage channels, where the water was sometimes potable. Wiluna gold-mines were supplied from such a source nearby, and the Great Fingall mine at Day Dawn is reported to have carted 4500 kilolitres a day from Nallan, though the mine itself was relatively free from water, only 163.8 kilolitres being pumped daily. Few mines were greatly troubled by an excess of ground-water in the workings, one notable exception being the Emu mine at Agnew, which at a depth of about 275 m struck such a large flow that the mine was closed down for some three months while being dewatered at a rate of about 4500 kilolitres a day. The writer was told by Billy Cock, who had been employed at the mine, that the water was slightly brackish but drinkable in an emergency.

Mines in areas farther northward, extending through the Murchison and Gascoyne to the Pilbara, fared in general rather better, because of the existence of river valleys where pools occurred in the alluvium and calcrete. In 1893 Woodward commented that the Murchison Goldfield was well watered, most mines striking good water at less than 30 m, and that 'there is no immediate prospect of the water question standing in the way of the development of this field, as it has so many others in this colony.'

A government register of Mines Water Supply Wells dug between 1892 and 1902 in the Eastern Goldfields lists 144 wells, broadly of two types: very shallow holes with small soakages of fresh water; and much deeper wells than those in the central-goldfields area, and usually producing salt water. It is not uncommon to find that wells were equipped, though the water was too saline even for stock use; indeed, at Kalgoorlie the council took over a 61-m-deep well, the water from which had a salinity of 5.6 oz/gallon (over 34 000 mg/l); and Boulder Council assumed control of another well 52 m deep, with water with a salinity of 4 oz/gallon (25 000 mg/l). The records note quite commonly that a salt-water bore was equipped (with or without a caretaker), and it is evident that the water was being 'condensed' (i.e. distilled) for domestic use.

In 1902, 22 towns were using well water, and the government was actively engaged sinking wells to provide for travellers on the principal access routes. As more mines were opened up, wells were dug to supply small townships and batteries rather than for opening up the country. By 1911 another seven settlements had their own wells, but six batteries and a brewery are also listed, the total number of government wells having risen to 232, the deepest being the Credo water shaft, which was 92 m deep.

In the Central Goldfield 300 wells were sunk between 1894 and 1908, most at government expense, the usual equipment being a whip or a windlass, a 45-litre bucket, and a trough. Some were obviously for travellers, like the ones regularly spaced along the Cue-Lawlers and Cue-Lake Way roads, but 26 townships and at least one state battery, that at Black Range, were also supplied.

In time, the use of condensers became quite widespread, which meant that not only could ground-water be treated but also water from the widespread salt lakes, which were really only shallow evaporating pans replenished by occasional rain. At the ANZAAS Conference in Perth in 1926 it was reported that 'the intensely salt water of the southeastern mining region was, and still is, used largely for metallurgical processes. For boiler and domestic purposes it was condensed, and one of the most conspicuous features of each mining camp used to be its water condensing plant.'

The Pilbara Goldfield was quite favourably provided with ground-water, as the low but cyclonic rainfall resulted in a high proportion of replenishment and low evapotranspiration losses. There are also fairly numerous alluvium-filled water-courses to provide subsurface storage, and run-off is much greater because the basement rocks tend to have more surface outcrop than is the case in the Yilgarn.

Between 1895 and 1907 the most noticeable feature of the 163 wells dug in the area was that in only three localities was it necessary to exceed a depth of 30 m; and that fresh water was almost everywhere available. Only 11 townships used ground-water, and from a water-supply viewpoint the district was far less handicapped than the eastern and central goldfields, despite the rainfall being of the same order. Condensers were not used. In the Gascoyne Goldfield only four wells were recorded, and the single well at Dead Finish on the Ashburton Goldfield gave salt water at 40 m.

(ii) *Coal-mining*

Early drilling for coal naturally concentrated on sedimentary basins, which at Collie were saturated almost to surface level with low-salinity ground-water, so that subsequently the mines had to be kept dewatered by pumping. Deep drilling was also done at Carnarvon, Dongara, Yardarino and at Dardanup. It is not clear whether these were drilled primarily as water bores, but they certainly had as one objective the detection of any coal-bearing beds that might exist.

The fairly recent construction of the Muja Power Station created a need for cooling water not available from the Collie River, which supplies Wellington Dam. The extended power station will require additional ground-water, and at present (1978) an investigation is being made into the volumes available and the best sites for pumping bores.

(iii) *Tin-mining*

In the tin-producing area at Greenbushes it was early found that sluicing was the most economic method of extracting the mineral cassiterite, and local ground-water has always been used for the purpose. That this was fairly plentiful is apparent from records of the early mine shafts; one dug in 1900 to a depth of 37 m in the centre of the tin-field was reported to have contained plentiful water. As mining progressed, the excavations either filled with ground-water or were used as dams, a method also used on the tin-fields in the Pilbara, where ground-water supplies were insufficient.

(iv) *Nickel-mining*

At Mt Windarra, colluvium and alluvium overlying decomposed granitic gneiss in an area only about 20 km from the mine-site have been found to have a potential daily yield of more than 6000 m^3, the ground-water salinity being less than 1400 mg/l. The reserves are sufficient to support mining for at least 16 years. Near the mine-site itself bores drilled in small alluvium-filled drainages provided a supply for initial development, but continued pumping resulted in increased salinity in some bores.

The Agnew Nickel Project will draw its initial water supplies from the 11-Mile ultrabasic aquifer situated approximately mid-way between the mine and Leinster township. It is estimated that this source will give a daily yield of 1000 kl of 800 mg/l total dissolved solids for up to seven years. Further supplies are available from other local sources which use bores drilled into alluvium and fractured basement rock (silicified serpentinized dunite), and capable of a daily yield in aggregate of at least 3000 m^3 of potable water.

Not all nickel projects were so fortunate. Kambalda had to be supplied with water from the Goldfields Water Supply. An interesting side effect of this was that, because the water for Kambalda had to be diverted from elsewhere, exploratory drilling along the Arrowsmith River was commenced in an endeavour to supply northerly towns then dependent on the existing mains supply from the south. Morawa and Perenjori are now supplied with bore water from the Arrowsmith area.

(v) *Heavy-mineral sands mining*

Mineral sands projects on the Perth coastal plain extending from Eneabba southward to Capel have been able to obtain pressure water for all purposes including beneficiation, the aquifer being the Lower Cretaceous-Upper Jurassic Yarragadee Formation. Companies in the Eneabba area are currently (1977) using about 19 000 m^3 a day for wet-plant processing, and more water is available if needed. The companies working mineral sands in the Capel area also use pressure water from the Yarragadee Formation, from which they were extracting 15 000 m^3 a day in 1977. Future demands on the aquifer may increase considerably if the industry expands and the proposed alumina projects at Wagerup and Worsley are established.

At Laporte near Australind the water salinity is between 350 and 750 mg/l. This particular processing plant which produces titanium dioxide from ilmenite, has posed a special problem in disposing of an average daily volume of 7200 m^3 a day of effluent with a pH of 1.0 and an iron content of as much as 9 g/l. If discharged in the waters of the estuary or the sea, this iron causes severe discoloration. Research indicates, however, that the effluent can probably be satisfactorily emplaced in the coastal dunes, if proper methods are used.

(vi) *Bauxite-mining*

Between 1970 and 1976 the Alcoa alumina refinery at Pinjarra pumped 8.8 million m^3 of low-salinity (less than 700 mg/l) pressure water from the Cockleshell Gully (Jurassic) aquifer, into which it has been calculated that the recharge is 2640 m^3 a day. Future pumping at a higher rate is likely to deplete the storage.

The bauxite refined at Pinjarra is a laterite overlying Archaean rocks in the nearby Darling Range, and mining necessitates the removal of the timber cover, with resultant disturbance of the hydrologic cycle, particularly the relationship between surface run-off and downward infiltration. This affects the salt balance, and research is continuing into the likely consequences. Results so far indicate that these will probably not be serious in the high-rainfall areas, but they may become so if mining is allowed to extend into the lower-rainfall areas farther east.

(vii) *Iron-ore mining*

Probably the most interesting success story in the development of ground-water had been that of the iron-ore industry. With exceptions such as Yampi and Koolyanobbing, the economic iron-ore reserves are mainly in the Pilbara, an area where it can reasonably be said that the industry owes its successful establishment and continuing operation to the availability of large volumes of low-salinity ground-water. The cyclonic rainfall comprises heavy falls of short duration, providing ideal intake conditions for the local aquifers, even though these vary in type. Goldsworthy uses shallow ground-water from the De Grey River alluvials, which supply between 4000 and 5000 m^3 a day of potable water; and 22 km north of the townsite at Shay Gap a borefield in the Jurassic Wallal Sandstone of the West Canning Basin has an average yield of 3300 m^3 a day, the salinity being less than 500 mg/l.

Port Hedland itself, once the rather minor coastal outlet for the pastoral industry and gold-mining in the Marble Bar area, is now the largest port in Australia because of the huge tonnages of iron-ore being exported from there. It could not have developed in this way without underground water supplies, first obtained from bore fields on the Turner River to the westward, and later the Yule River. The bores obtain their water from colluvium beneath the coastal plain, replenished by river floods. More recently a third field on the De Grey River has been proved and is being developed.

At Mt Newman the company has supplied all its needs from bores in alluvium, colluvium, calcrete, Canga iron-ore, and Wittenoom Dolomite, the highest sustained yields being from the calcrete in the Fortescue River system, about 10 km east of the township. Salinities range between 500 and 1300 mg/l, with an average value of 800 mg/l. The water is treated in an ion-exchange desalination plant with a resin bed, which reduces the salinity to below 500 mg/l for domestic use. Between March 1973 and June 1977 the total consumption of water was 14 120 million m^3, of which rather more than half was used in the township.

At Pannawonica the aquifers are Maddina Volcanics (a fractured Proterozoic basalt), and Robe River alluvium. In 1973 these provided 900 m^3 a day of water with a salinity of 700-900 mg/l, half of which was used for the township and half for the mine and powerhouse. By May 1975 the total usage had increased to 1955 m^3 a day, of which 750 was for the town.

The Paraburdoo water supply comes from similar aquifers, the 1973 abstraction being almost 1 180 000 m^3, at a daily average of 2560 m^3. Once again half was used for the township. By 1977 fourteen bores in two wellfields were providing some 8080 m^3 a day.

The major source of supply for Tom Price is the Southern Fortescue wellfield (30 km to the northeast), with a calculated aquifer capacity of 22 000 m^3 a day drawn from an aquifer comprising calcrete and alluvium over cavernous Wittenoom Dolomite. In 1977 the average usage was 8700 m^3 a day, the total pumped since 1972 being some 2 million m^3. Auxiliary sources at the Hardey River and Mt Lionel should be capable of supplying more than 11 000 m^3 a day for the concentrator, the aquifer being calcrete, Canga iron-ore and fractured Proterozoic rocks.

Other mining ventures

Other ventures perhaps less spectacular than the iron-ore, are equally dependent on ground-water for their economic operation. These include:

The salt production areas at Lake MacLeod north of Carnarvon, and Useless Loop near Shark Bay, both use the Lower Cretaceous aquifer (Birdrong Sandstone) beneath the Carnarvon Basin to obtain pressure water for processing. The Newmont open-pit goldmine at Telfer has proved the availability of about 3000 m^3 a day of ground-water along fault zones in Middle Proterozoic sediments; and the Golden Grove copper prospect should be able to obtain some 3500 m^3 a day of process water from calcrete in a regional drainage, and 250 m^3 a day of potable water from fractured crystalline basement rocks. No uranium has yet been mined, but at Yeelirrie the calcrete will be capable of supplying any foreseeable need.

However, possibly the most interesting ground-water development scheme in the Pilbara is only indirectly connected with mining, in that it supplies no water to any individual mine-site but serves coastal townships owing their existence to mining activity. This development at Millstream is from an arcuate mass of cavernous calcrete some 40 km long and 10-12 km wide which occupies a former course of the Fortescue River. The present annual abstraction from it is 9 million cubic metres.

CHAPTER 4

EXPLORATION TECHNIQUES

Mineral exploration has developed from the early surface prospecting methods of following heavy mineral trails by panning surface soils to the application of diamond drilling, and geophysical and geochemical methods, for the location of sub-surface deposits. Underlying all of these methods is an appreciation of the science of geology, and geological reconnaissance or mapping is of prime importance in any systematic mineral search, whether it be for resistant heavy minerals such as gold, heavy mineral sands or gemstones; for minerals such as copper, nickel, zinc and others which do not occur as heavy residues but in chemical combination with other elements; or for liquid or gaseous accumulates such as water, oil and natural gas.

Probably most of the Western Australian ore deposits that outcrop have now been discovered. Immense tracts of the state are covered by extensive sandplains, residual soils and deeply-weathered rocks, and in such areas any economic mineral deposits which would be apparent from surface indications are obscured. As mining continues to deplete reserves of known deposits it becomes necessary to develop methods of discovering ore deposits obscured in this way. Geophysical and geochemical techniques have been applied. Remote sensing by satellite imagery (LANDSAT) is the most recent means available to the exploration geologist of delineating regions of potential mineral wealth that are promising areas for exploration.

The various methods of mineral exploration that have particular reference to Western Australia are dealt with in this chapter.

4A

Geological methods

K. J. Finucane

Consultant Geologist

Many thousands of square kilometres of semi-arid country in Western Australia have provided a favourable background of rock outcrop for mineral discovery, but this has been counter-balanced by distance and problems of transport, by lack of adequate water supplies, by the peneplanation and deep oxidation of the Archaean rocks in which most of the state's known mineral wealth is located, and by vast areas of concealed ground.

Contrary to popular beliefs, the early discoveries of viable mineral deposits since the foundation of the Swan River Settlement in 1829 were not wholly dependent on chance finds made by itinerant prospectors, but were founded primarily on well-planned exploratory expeditions undertaken by surveyors in search of agricultural and pasture lands, with the possibility of mineral discoveries as a further incentive. The most noteworthy were led by A. C. Gregory (1846-48), followed by R. Austin (1854), H. M. Lefroy (1863), and C. C. Hunt (1864-66). Hunt not only explored but also cut a track eastwards from York to Hampton Plains, enlarging soaks and rock holes on the way, thus providing an access route to the dry hinterland. These basic efforts provided the reconnaissance framework on which many subsequent gold discoveries were founded. Travel was by pack horse or on foot, and the incredible hardships endured merit pride of place in any account of mineral exploration techniques in Western Australia, particularly as survival based on some knowledge of the interior and its climatic conditions were prerequisites for future discoveries. In this context it is interesting to note that W. Hargreaves was commissioned in 1860 by the government to report on the prospect of finding gold in payable quantities, but he examined only the then-settled districts and did not penetrate the mineralized belts to the north and east.

The need for greater technical expertise and a more sophisticated appraisal of the potential for mineral discoveries led to the employment of the geologist Dr F. von Sommer from 1847 to 1851, followed by H. Y. L. Brown, 1870 to 1873, E. T. Hardman, 1883 to 1885, and H. P. Woodward, with a small staff, from 1887 to 1895. H. Y. L. Brown examined a number of lead and copper discoveries in the Geraldton and Phillips River districts, and listed seven localities which he regarded as being favourable for the occurrence of gold. E. T. Hardman's examination of the West and East Kimberley districts was the most thorough mineral exploration project carried out to that period. His careful examination of the country, systematic testing of river gravels by trial pits and panning, together with some testing of quartz reefs, established the presence of gold over an area of some 5000 km^2 of the East Kimberley district. His report and accompanying map indicated places where gold was likely to be found. It is significant that the first recorded production of gold in Western Australia was 270 fine ounces (8398 g) from the Kimberley district in 1886.

There can be little doubt that Hardman's meticulous work, his sound approach to exploration, and the advice he gave about it provided the initial stimulus which led to the wave of prospecting activity during the decade following the publication of his reports. Surface prospecting for gold spread progressively southwards through the Pilbara, Fortescue and Ashburton districts, mainly following the river systems from the coast inland, and then spreading into the more southerly parts of the state, covering the Yilgarn, Murchison and Eastern Goldfields. The broad pattern of gold mineralization outlined by the early explorers was followed progressively by an intensive search which ultimately involved thousands of men who swarmed over any likely auriferous country indicated by the presence of quartz or ironstone, searching the water-courses for alluvial gold, examining and knapping outcrops, sinking trial shafts for both alluvial and reef gold, trenching and conducting a variety of shallow test works. Many of the prospectors were experienced men from the New South Wales, Victorian and Californian alluvial fields, but a far greater number were not.

Where water was abundant, the method used for sampling the alluvium, gravel or soil was panning. Briefly this amounts to placing the fragmental sample into a shallow pan of water. This pan of saturated soil is rotated and shaken, preferably under water, so that the heavy particles sink towards the bottom. The pan is then tilted gradually to spill off the lighter substances eventually leaving a concentration of the valuable heavy minerals in the bottom. The absence of water in the southern parts of the state, in contrast to conditions in the Kimberley district, necessitated changes in technique from traditional panning and sluicing (washing) methods used in localities where there was an abundant water supply. Test work on alluvial or eluvial ground (soils) was effected by two dishes about 45 centimetres in diameter in which the broken ground was poured from dish to dish, the fine dust blowing away in the wind leaving a residue of pebbles and, hopefully, heavy specks of gold. A somewhat similar process using two dishes and a tripod (or by sieving) yielded the same product. An ingenious combination of all three processes, together with a bellows, produced the dry-blowing machine in various forms (Fig. 4.1a and b). Reef gold was tested by crushing in a dolly-pot and panning the crushed material in a small prospecting dish—a flat-bottomed dish about 23 centimetres in diameter with sloping sides. Appraisal of the many discoveries was made by mining engineers whose systematic sampling and subsequent valuations determined the course of mine development.

Transport was by pack-horse, by horse-drawn drays or wagons, or by imported camels. For those unable to afford these, bicycles or wheelbarrows, loaded with a minimum of equipment, were commonly used. Initial trenching and shaft-sinking were carried out entirely by hand labour, with picks and shovels, hammers and moils, windlasses and whip poles.

Geology plays an important role in exploration and in recording and assessing the state's mineral potential. The work of organizing a systematic geological survey covering the structure, mineral resources, mining industry and underground water supplies commenced with the appointment of A. Gibb Maitland as Government Geologist in 1896. The foundation of the knowledge of the geology and mineral resources of Western Australia is due primarily to the excellence of the bulletins and reports issued by the Geological Survey Branch of the Mines Department. These covered, and continue to cover, the whole range of known mineral occurrences. Economically and statistically the emphasis was centred on gold, even until the late 1950s and early 1960s.

The decline of the gold mining industry from a peak production of 2 064 800 fine ounces (64 422 kg) of gold in 1903 to 377 176 ounces (11 731 kg) in 1928, together with the exhaustion of a number of small base metal deposits, was accompanied by a corresponding loss of interest in mineral exploration. The established techniques of surface prospecting were still extant in some districts, but there was little change in practice, and there were no new discoveries of importance. Few operating companies employed exploration staffs. Geologists were called in only when the visible ore reserves were nearing exhaustion, funds for development non-existent, and the future of a mine was uncertain. There were exceptions, the most notable being a

Fig. 4.1 Dryblowing prospecting equipment in arid areas of Western Australia. *Left:* Prospector winnowing sample by pouring from dish to dish, with wind removing fine lightweight particles. Dryblower on left consists of shaking sieves and bellows for blowing lightweight particles out of fine fraction.

Bottom: Shaker equipment used for sieving fragmental eluvial material.

(Photographs reproduced from the Battye Library collection.)

systematic study of the Kalgoorlie field by Drs Maclaren and Thompson, and of the Sons of Gwalia Mine and the Phillips River district by Dr Maclaren over the years 1910 to 1912. Diamond drilling, both surface and underground, was used extensively by private companies and in planned government programmes with some success. Its principal function was to test the depth and/or lateral extensions of known ore deposits. Only rarely was it used as a primary exploration tool. Gossan outcrops were probed in the Yilgarn Goldfield and in the Braeside lead field without success, and there was some interest in leached outcrops.

Interest in systematic geological exploration was stimulated by the engagement of Dr Arthur Wade to report on oil prospects in the Kimberley district in 1925, and a new dimension in mineral exploration was promised by an invitation in 1928 to E. Broughton Edge to visit Australia and report on the possible use of geophysical techniques. Edge's summary of the Western Australian scene was that deep oxidation and the high salinity of the ground water would always prove inhibiting factors (Edge & Laby 1931).

The increase in the price of gold in 1929 was followed immediately by a state-wide revival of gold-mining, and emphasized the need for greater knowledge of the structure and distribution of the state's gold deposits. It was realized by some of the principal mining houses that the establishment of new mines called for the adoption of new ore-finding techniques. The initial move to a modern scientific approach was the formation in 1933 of the Western Mining Corporation in London with the expressed objective of searching for gold in Western Australia with the use of air photography and structural and regional geological studies and geophysics, in conjunction with mechanized surface exploratory units and diamond drilling. Other exploration groups adopted a more conservative and traditional approach, but in general there were intensive studies of most mining centres based on production records, grade-depth analyses, studies of available assay data and of known structural and general geological features. Intensive structural geological studies on known mining fields proved to be the most efficient and successful approach on both technical and economic grounds. During this state-wide re-appraisal it became apparent that each major field was a separate entity with individual structural and rock associations—for example the multiple lode-fracture pattern of Kalgoorlie, the dragfolded reef structures of Gwalia, cross-fracture controls in jaspilite beds at Boogardie, and the deep-seated reverse fault system of the Norseman field, with other variations at Wiluna, Laverton, Big Bell, Mount Ida and throughout the Yilgarn.

A number of regional studies based on air photography and surface mapping were completed both by private companies and the Western Australian Geological Survey. These added substantially to the overall knowledge without actually locating viable ore deposits. Some new gold finds were made by surface exploration in the latter half of the 1930s, principally by knapping quartz reef and ironstone outcrops and systematic loaming. Foremost among these were Yellowdine, Evanston, Cox's Find, the Comet Mine near Marble Bar, and the Barbara and Surprise Mines at Coolgardie. However, the thoroughness with which the early prospectors had con-

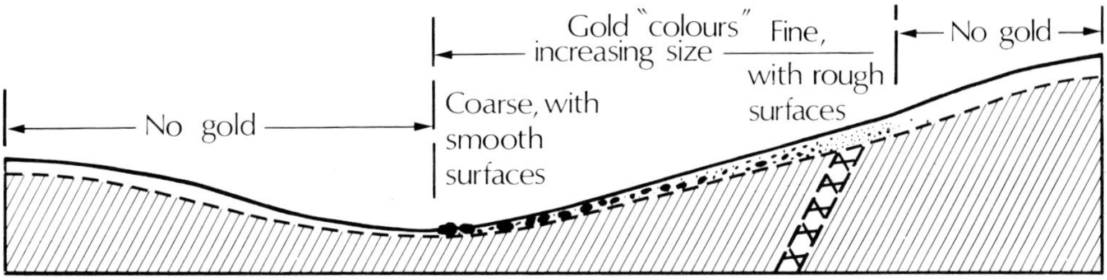

Fig. 4.2 'Loaming': diagrammatic cross-section to illustrate loaming method of prospecting for gold. The gold 'colours' decrease in size and smoothness of particle surface upslope till they cut out over, or upslope slightly from the underlying auriferous orebody.

ducted their searches, and the relatively small number of surface finds made since the early 1900s led to the adoption of loaming by those skilled and patient enough to use it.

Loaming (Cash 1934) consists of sampling the surface soil or subsoil (by shallow auger holes) and panning the samples for 'colours', the number and character of which indicate the proximity of a concealed reef or shear. Coarse 'colours' are generally distant from the source and the farther they are from the source the smoother are their surfaces. Fine gold, on the other hand, indicates proximity to the source, so the prospector works upslope until the 'colours' become very fine, or in fact disappear. The primary source of the gold is below or slightly downslope from this point (Fig. 4.2). Development of some mechanized form of loaming was tested during the 1950s but never passed beyond an experimental stage. The use of metal detectors is a more recent development in the search for alluvial gold.

The realization of an urgent need to develop and populate the northern portion of Australia through its mineral resources led to the formation of the Aerial, Geological and Geophysical Survey of Northern Australia (AGGS) in 1935. This was funded by the Commonwealth, Western Australian and Queensland Governments and covered the territory north of the Tropic of Capricorn. Emphasis was on the location of viable deposits on which centres of growth could be established. Techniques adopted included reappraisal and structural geological studies of all known mineral deposits and districts; reconnaissance examinations by aircraft, by motor-vehicle, by mule team where vehicle access was impossible, or even on foot; regional geological exploration based on air photography; and geophysical surveys and loaming of selected areas. The survey was suspended on the outbreak of the Second World War in 1939.

The post-war period from 1946 onwards was one of renewed interest in gold-mining and this stimulated a revival in scientific exploration, principally geological with the added use of geophysical and geochemical surveys in selected localities. The post-war boom in base metals and the initial search for uranium ores had no major impact on the Western Australian scene, with a resultant loss of more sophisticated techniques. However, there was considerable prospecting for mineral beach sands in the southwest of the state from 1949 onwards, with the commencement of major production at Capel in 1956. Preliminary exploration consisted of surface appraisal of potential dune deposits followed by shallow drilling, and search for former strand lines. Airborne scintillometer and aeromagnetic surveys of several areas in the state were made by the Bureau of Mineral Resources Geology and Geophysics in 1955-56. These were used extensively in later mineral exploration for nickel and uranium, but the results were difficult to interpret and evaluate at the time the survey was made.

Widespread distribution of iron-ore throughout the state had been established and recorded by the Geological Survey since its inception, but low prices, high transportation costs and a lack of markets had inhibited exploration and exploitation. Geological mapping, sampling and drilling of iron deposits was commenced in 1938, was suspended during the war years of 1939-45 and resumed in 1956. In 1959 the Western Australian Government called for tenders for mining and export of iron-ore from Mt Goldsworthy, northeast of Port Hedland, and Dowds Hill, north of Southern Cross, in defiance of an existing commonwealth ban on iron-ore export. This act, and the subsequent suspension of the commonwealth ban in 1960 immediately stimulated an intensive search for viable deposits to satisfy anticipated demands from Japanese and other overseas markets. Ground and helicopter reconnaissance, air surveys using both monochrome and colour photographs, systematic geological mapping, sampling and drilling were used throughout the state and defined many known deposits and a large number of new ones.

As with iron-ore, a ban on the export of manganese delayed exploration until 1952. Small deposits had been observed by the AGGS of Northern Australia in 1937, and many others were later found in the Gregory Ranges, mainly by surface exploration and air reconnaissance, followed by sampling and percussion drilling.

Investigation of the Darling Range bauxite deposits, which had been known for many years, was commenced by Western Mining Corporation in 1957. Initial exploration consisted of geological mapping, sampling of road and railway cuttings and laboratory analyses. Extensive bauxite deposits on the Mitchell Plateau in the Kimberley District were deduced by a study of air photos by Amax field staff in 1965, and were later confirmed by ground reconnaissance.

There was a dramatic increase in the whole field of mineral exploration in early 1966 with the announcement by Western Mining Corporation that high-grade nickel ore had been intersected in diamond drill holes at Kambalda. The Kambalda discovery, supported by the earlier developments in iron-ore, bauxite and mineral sands stimulated a wave of exploration which exceeded the revival of the gold-mining industry of the 1930s, and was comparable in many respects with that of the early gold discoveries. The conventional prospector was still active and successful, but the greatest activity was by exploration syndicates and companies. All modern techniques were utilized, including regional and structural geological studies, air photography, ground and airborne geophysics, and geochemistry supported by sampling of gossan outcrops adjacent to ultramafic contacts, mechanical trenching, percussion drilling and diamond drilling.

In areas of poor outcrop or in concealed ground, a commonly adopted exploration sequence in geologically favourable areas was a geochemical survey based on a variable grid to suit conditions, with determinations for copper, nickel, zinc and chromium by atomic absorption spectroscopy followed by a ground magnetometer survey, and supporting induced polarization. This, depending on results, would be followed by percussion drilling, backhoe or bulldozer trenching, and finally by diamond drilling.

Geological methods of exploration based essentially on geological mapping (Forrester 1946) have over the years been assisted by advances in travel (from on foot to the horse and camel to the four-wheel drive motor-vehicle, aeroplane and the helicopter); by aerial photography, both half-tone and colour (Lattman & Ray 1965); and still more recently by satellite. Landsat (satellite imagery), whose products are currently available and applicable to regional geological mapping, has the advantage that it provides a synoptic view over a wide area, although not with the same definition as the air photograph. It has been demonstrated that the use of Landsat, in conjunction with geological mapping based on air photographs, can assist in delineation of areas of hydrothermal alteration in a volcanic sequence under suitable outcrop conditions (Smith *et al.* 1978).

Geological methods of exploration have also been assisted to an immeasurable degree by the advancements in chemical and mineralogical analytic equipment such as atomic absorption and x-ray fluorescence spectrometers and x-ray diffractometers. These techniques rapidly bring abundant chemical and mineralogical data to the geologist working in the field, thus increasing his potential for the solution of problems associated with mineral exploration. Not the least important in the solution of such geological problems are the radiogenic age determination techniques which now enable actual ages to be determined for rocks and geological processes. Any such information is important in mineral exploration.

REFERENCES

Cash, S. J. 1934. *Prospecting for Gold by the Loaming System.* Perth: Frank Daniels.

Edge, A. B. Broughton & Laby, T. H. 1931. *Principles and Practice of Geophysical Prospecting.* London: Imperial Geophysical Experimental Survey.

Forrester, J. D. 1946. *Principles of Field and Mining Geology.* New York: Wiley.

Lattman, L. H. & Ray, R. G. 1965. *Aerial Photographs in Field Geology.* New York: Holt, Rinehart and Winston.

Smith, R. C. , Green, A. A., Robinson, G., & Harvey, F. R. 1978. Use of Landsat imagery in exploration for Keweenawan—type copper deposits. *J. Rem. Sens. Environ.* 7(2), 129-44.

4B

Geophysical methods

D. L. Rowston
Geological Survey of Western Australia

No account of mineral exploration in Western Australia would be complete without consideration of the contribution made by geophysical methods and techniques. Geophysics has long been accepted as a mainstay in petroleum search and its use in the exploration for other minerals is now also widely recognized.

In 1905 an Italian observer, Allessio, determined the force of gravity in Perth—probably the first authenticated geophysical measurement in Western Australia. The Dutch scientist Vening Meinesz, later established the absolute value of gravity at the University of Western Australia and at six pendulum stations on the shield in association with his submarine cruise of the Indian Ocean in 1935.

Although not related to exploration in *sensu stricto*, the Watheroo Magnetic Observatory, founded in 1919 by the Carnegie Institution of Washington, warrants mention as the earliest geophysical organization in Western Australia. Besides providing corrections essential to air and sea navigation and land surveying, the observatory recorded diurnal variations and periods of abnormal magnetic activity. Magnetic anomalies observed during these disturbances can be highly misleading. The operation of the Watheroo Observatory was transferred to the Commonwealth Government in 1947, under the direction of the Bureau of Mineral Resources (BMR), and was replaced in 1959 by the Mundaring Geophysical Observatory, where facilities now include earthquake seismology and ionospherics.

The first documented geophysical surveys in Western Australia were conducted by the Imperial Geophysical Experimental Survey (IGES) at Northampton in 1928-30 (Edge & Laby 1931). Various electrical methods, including equipotential and potential ratio, were tested over some of the lead mines near the Murchison River. Although unsuccessful in locating new mineralization, the surveys here encountered the problems of saline groundwater and deep weathering—problems which even today, with highly sophisticated instrumentation and new techniques, have not been fully overcome.

Between 1935 and 1940 the commonwealth, and the states of Western Australia and Queensland, sponsored the Aerial Geological and Geophysical Survey of Northern Australia, which, on the geophysical side, investigated a number of gold-mines and prospects near Wiluna and Kookynie and at Big Bell. Ground magnetometer measurements were made for the first time to augment the electrical and electromagnetic methods (AGGSNA Reports, Western Australia, Nos 35-39 inclusive and No 64). The Northern Australia Survey work at Wiluna was preceded by contract geophysical surveys undertaken by the Electrical Prospecting Company of Sweden for Wiluna Gold Mines Ltd in the early 1930s.

The lull in geophysical exploration during the Second World War was succeeded in the post-war years by bursts of activity associated with the various 'booms' that occurred sporadically up to the present. These periods of increased geophysical activity were superimposed on the less spectacular background of systematic data acquisitions mainly by the BMR and major oil and mining companies.

In 1947, at the request of the Government Geologist, a gravity survey was carried out over the Collie Coal Basin by the BMR (Chamberlain 1947). The Bouguer anomaly map clearly delineated the extent and structure of the basin, and although the estimated depths to basement proved to be somewhat excessive, led to the discovery of substantial coal deposits in the eastern sector. Urquhart (1956) used gravity and seismic refraction methods to define postulated deep leads in the lake system southeast of Kalgoorlie in 1948.

Two experimental gravity traverses were read by the BMR to help locate underground water supplies near Watheroo in 1949 and at Bullsbrook in 1951. The gravity profiles indicated a far greater thickness of Phanerozoic sediments than was hitherto suspected, and the observations were extended to a reconnaissance gravity survey of the entire Perth Basin during 1951-52 (Thyer & Everingham 1956). This and subsequent surveys in the Carnarvon and other basins helped to stimulate oil exploration.

Finally the commonwealth initiated the helicopter gravity survey of the continent, and these results, combined with private company and marine data, culminated in the publication of the Gravity Map of Australia in 1976. The relatively large station interval of the helicopter gravity network (about 11 km) precludes accurate definition of small gravity anomalies, but provides useful information about crustal structure and the major geological provinces and tectonic features. Although predominantly an oil-search tool, gravity surveys have been used in iron-ore and manganese exploration in the Pilbara.

As an integral part of the search for petroleum, the gravity work in the Carnarvon Basin was augmented by seismic reflection surveys by the BMR and by the first contract seismic crew introduced by West Australian Petroleum Pty Ltd in 1952. Seismic reflection techniques have since been employed in most sedimentary basins to elucidate prospective structures, although the emphasis has gradually swung from land to marine operations. As statistics for seismic activity were commonly quoted in crew-months before 1971, lineal or areal coverage by this method cannot be readily evaluated.

Another major geophysical tool, the airborne magnetometer, has been used systematically in Western Australia by the BMR since 1956. Introduced to the Eastern Goldfields in 1956 to assist gold and metals exploration and regional mapping, aeromagnetic surveys were also used in the Carnarvon and Perth Basins for petroleum exploration almost concurrently. By 1977 an estimated 900 000 line-kilometres had been flown, covering about eighty-seven 1:250 000 map sheets. (This estimate does not include the considerable amount of airborne work contracted by the mining and oil companies.) Reconnaissance surveys in the sedimentary basins are usually flown at an altitude of 450 m or thereabouts and a line spacing of 1.5 km; for regional surveys over the Precambrian shield, ground clearance is reduced to about 150 m for better resolution of the more complex rock distributions and, for detailed investigations, the ground clearance is about 70 m with a line spacing of 0.2 km.

The aeromagnetic data can provide depths to basement in the sedimentary basins and indicate major structural elements. For instance the position and extent of the Darling Fault, not readily apparent from geological mapping, is clearly indicated on the magnetic and gravity contour maps and profiles.

On the Shield the ultramafic rocks with which nickel and other base metals concentrations are sometimes associated, characteristically exhibit medium-intensity anomalies by comparison with the magnetically undisturbed granitic terrains and high-intensity anomalies of the banded iron-formations. A magnetic total field profile and related geology from the Southern Cross area is exemplified in Fig. 4.3. Although some nickel minerals are directly detectable, their indications are frequently masked by the larger anomalies arising from the ultramafic host rock. Ilmenite concentrated in beach-sand deposits can also produce magnetic perturbation of a low but significant order. The aeromagnetic contour maps otherwise provide useful guidelines in regional mapping, and for predicting rock types and their boundaries and structure, particularly where these are concealed by superficial deposits.

Ground geophysical activity in the mining sphere has fluctuated according to demand. The 'nickel boom' of 1966-71 saw the introduction of several methods relatively new to Western Australia, and an increase in the number of contract geophysical companies. Induced polarization (IP) techniques first used by McPhar Pty Ltd at Mt Ida in the early 1960s, superseded most of the earlier resistivity and electromagnetic methods. The Russian transient electromagnetic unit (MPPO-1) imported by Geotechnics in 1970 was the forerunner of SIROTEM, the transient system recently developed by CSIRO. INPUT, TURAM, TURAIR and VLF electromagnetic equipment each had its adherents and each hoped to surmount the masking and coupling difficulties engendered by deep weathering and saline groundwater. An objective evaluation of the various methods is inhibited by the reticence of many mining companies and contractors to publish their results. However, according to one of the few reports available (Liddy 1974), IP surveys successfully detected the Mt Martin, Nepean and Scotia nickel sulphide bodies despite the conductive surface layer; in fact, the discovery of the Nepean orebody is accredited directly to the use of IP and aeromagnetics.

The 'uranium boom' of 1953-59 stimulated radiometric prospecting, both ground and airborne, but these efforts failed to locate other than the sub-economic bedded uranium in Proterozoic rocks at Halls Creek, the Killi Killi Hills near the Northern Territory border, and other minor occurrences. Interest in uranium revived about 1968, and the discovery of the Yeelirrie deposit announced in 1972 was a direct result of airborne radiometrics. In addition, the development of portable spectrometers and large-volume scintillation crystals facilitated their increased use in the search for potash and heavy mineral beach-sand deposits. The latter frequently contain monazite and radioactive zircon.

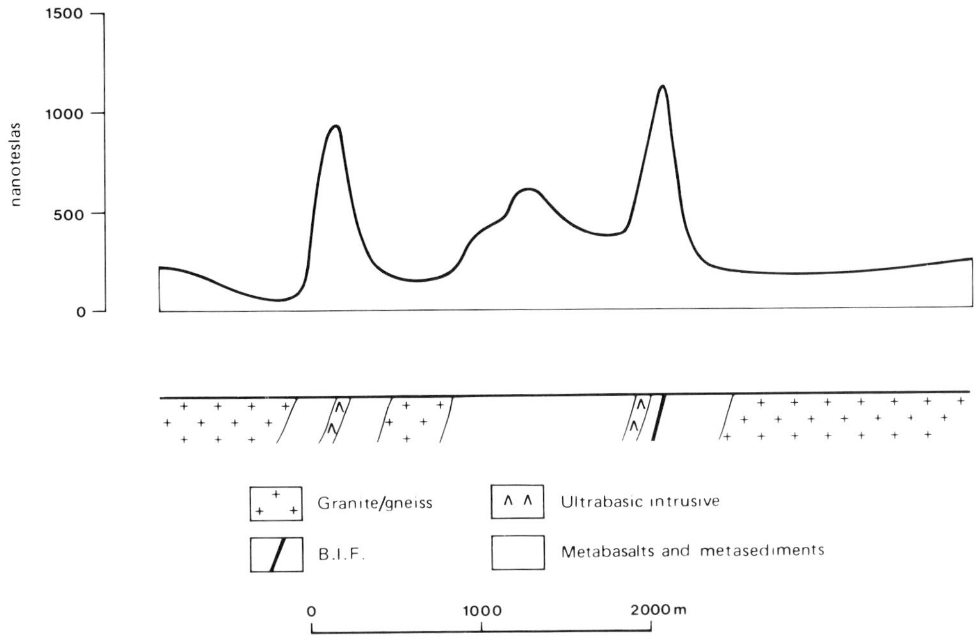

Fig. 4.3 Aeromagnetic profile, Southern Cross area, Western Australia.

In the arid conditions prevailing in Western Australia, the provision of an adequate water supply is a prerequisite to the development of any mine. Geophysical methods have been used extensively, and with success, to locate underground supplies in a variety of environments ranging from the unconsolidated sediments of the western part of the Canning Basin, and the cavernous dolomite and calcrete storages of the Fortescue Valley to the weathered Archaean terrain of the Yilgarn. Electrical resistivity and seismic refraction methods are employed most commonly for hydrogeological investigations.

Borehole logging, notably natural gamma-radiation and point-resistance methods, has been used effectively for correlating the stratigraphy of the Hamersley Basin iron-ore province, for detecting coal seams at Collie and for identifying beach-sand deposits. For instance, radioactive heavy minerals, mainly monazite, were detected during the logging of a water bore at Eneabba in 1968. A far more sophisticated suite of wireline methods has been used routinely in this state for the evaluation of oil wells since 1953.

From about 1970 onwards oil exploration has been directed mainly to the continental shelf permit areas, where the submarine geology can only be determined by geophysical means and drilling. Marine surveys usually include seismic, magnetic and gravity techniques.

Despite the many adverse environments in Western Australia, the use of geophysics as an adjunct to prospecting will continue to grow. Advances in technology will no doubt improve the response of existing methods and perhaps new and more effective ones will evolve, for example the current application of metal detectors in prospecting for gold, which at present (1979), are yielding significant results..

REFERENCES

Chamberlain, N. G. 1947. Preliminary report on the geophysical survey of the Collie coal basin. Bur. Min. Resour. Geol. Geophys. rept 1 (unpubl).

Edge, A. B. Broughton & Laby, T. H. 1931. *The principles and practice of geophysical prospecting*. Cambridge: Cambridge University Press.

Liddy, J. C. 1974, Induced polarisation in the search for buried sulphide bodies. *Australian Mining* **66**, 56-66.

Thyer, R. F. 1963. Geophysical exploration—Australia. *Geophysics* **28**, 273-305.

Thyer, R. F. & Everingham, I. B. 1956. Gravity survey of the Perth Basin, Western Australia. *Bur. Min. Resour. Geol. Geophys. Bull.* **33**.

Urquhart, D. F. 1956. The investigation of deep leads by the seismic refraction method. *Bur. Min. Resour. Geol. Geophys. Bull.* **35**.

4C

Geochemical methods

R. H. Mazzucchelli
Western Mining Corporation Limited

The earliest form of geochemical prospecting practised in Western Australia was the time-honoured technique of 'loaming' for gold, the principles of which contributed significantly to the development of modern exploration geochemistry. The discovery and exploitation of the state's gold resources prior to 1950 owed much to the skilful application of this most sensitive technique by several generations of prospectors as outlined earlier in this chapter. This section is, however, restricted to a brief summary of the application of exploration geochemistry in the modern sense, as defined by Hawkes & Webb (1962), i.e. 'any method of mineral exploration based on a systematic measurement of one or more chemical properties of a naturally occurring material . . . The purpose of the measurement is the discovery of abnormal chemical patterns, or *geochemical anomalies*, related to mineralization.' Historically this may be said to have commenced with an investigation by Sokoloff (1950) into the application of a dithizone test for gold on rocks and soils in the Eastern Goldfields area. Since that time, a variety of geochemical methods have been developed and applied in exploration in this state for gold, nickel and other base metals, and uranium.

The physiographic and past climatic conditions in Western Australia have created some unique problems for geochemical exploration, particularly in the Yilgarn Block, which includes some of the most prospective terrain for gold, nickel and copper-zinc. These conditions include oxidation depths up to 100 m in silicate rocks, vertical element zonation resulting from development and retention of lateritic soil profiles, and a variety of residual, aeolian and alluvial superficial deposits, which tend to mask mineralization occurring in bedrock. These phenomena are largely related to the low relief coupled with a semi-arid climate, which together result in incomplete erosion of the products of weathering. Other areas of active exploration for metals such as the Pilbara and Kimberley regions are characterized by much more dissected topography, and the products of past weathering cycles have in general been eroded to a far greater degree, facilitating recognition of bedrock features both visually and chemically.

Recent gold exploration in the more intensely prospected parts has been concerned with the location of 'blind', or concealed, orebodies which would not have been amenable to the methods used by prospectors. Geochemical exploration for gold has focused on the use of pathfinder elements which are spatially related to gold, but which form more easily detectable haloes in the overlying rocks, soils or sediments than does gold itself. Arsenic is of particular interest in this regard, having

transported clays containing a lithified layer (the 'Wiluna Hardpan'). No study was made of the distribution of nickel and copper in the uppermost (transported) 25 m of the overburden, but the main body of mineralization is delineated by the 4000 ppm Ni and 300 ppm Cu contours at a depth of 30 m below surface. The Cu values at this horizon represent an approximate two-fold enrichment above those in the primary zone, highlighting the usefulness of copper as an indicator for nickel/copper sulphide mineralization under lateritic conditions.

Geochemical methods have also been applied extensively in attempts to distinguish gossans (weathered surface outcrops) related to nickel sulphide mineralization from spurious ironstone outcrops. The chemical and physical attributes of some lateritic features and weathered sediment horizons make them very difficult to differentiate from nickel sulphide gossans. Clema and Stevens-Hoare (1973) and Bull and Mazzucchelli (1975) have proposed methods for discriminating significant from spurious gossans with multivariate statistical analysis based on readily determined chemical constituents. Even more sophisticated techniques have been suggested by Travis *et al.* (1976), who used metals of the platinum group as discriminators. The development of geochemical evaluation methods has proceeded in parallel with micro-textural methods and a very high level of expertise has been achieved using the two techniques in combination for the interpretation of strongly-altered outcrops.

The geomorphological and pedological problems which affect geochemical exploration for nickel in the Yilgarn Block presumably have similar adverse effects on exploration for volcanogenic massive copper/zinc sulphide deposits. However, much of the Archaean acid volcanic terrain in the Yilgarn Block (where such deposits are most likely to occur) is strongly weathered, tends to occupy topographically depressed areas, and is consequently blanketed by Recent transported overburden. This could account for the relatively low discovery rate achieved in exploration for this type of deposit and suggests that considerable scope remains for application of geochemical exploration techniques. The few examples of this type of deposit which have been discovered to date (Freddie's Well, Golden Grove and Teutonic Bore) all occur in relatively well-exposed terrain and have gossans. This applies particularly to Freddie's Well, where some 500 000 tonnes grading 10 per cent zinc and 0.25 per cent copper have been delineated. Strong geochemical indications of the main ore-shoots have been provided by zinc and copper values in the −200 mesh fraction of near-surface soils. Zinc values commonly exceed 1000 ppm, and in fact range up to 2 per cent in soils, and maximum copper values up to 2500 ppm have been recorded (Smith *et al.* 1976). Anomalous copper and zinc concentrations were also detected in the active sediments of drainage channels crossing the mineralized zones. However, such *stream sediment geochemistry* is of restricted use in the Yilgarn Block because of the generally flat terrain, wide interfluves and generally short and disconnected nature of much of the drainage.

The more dissected terrain in the northern half of the state is ideally suited to the application of stream sediment geochemistry, and large areas have been geochemically sampled in this way. Most of the work has been conducted by private companies, however, and the results have not been published. The most notable drainage reconnaissance programme was that conducted by Pickands Mather as managers for the Kimberley Ventures Corporation in 1963-72. Stream sediment samples were collected at an average density of 1 sample per 2 km^2 over a total area of 78 000 km^2 in the Halls Creek and King Leopold Mobile Belts and the −80 mesh fraction was analysed for nickel, copper, lead, zinc, arsenic and (selectively) tin. Follow-up of the many resulting anomalies located several significant base-metal occurrences such as the Bow River copper/nickel deposit described by Blain (1967).

Hydrogeochemical surveys were conducted by several companies in exploration for uranium which followed the 1972 discovery of uranium in calcreted drainage channels at Yeelirrie and elsewhere. Recognition of the fundamental role played by the transportation of uranium in groundwater led to the sampling of ground-waters and analysis for uranium on a regional basis. A geologically-based interpretation of the resulting uranium distribution patterns was found to delineate mineralized drainage catchments. Mann (1977) has suggested further refinements, taking into account the concentrations of the other ion species of the ore mineral carnotite ($K+$, H_2VO_4-) to pinpoint more precisely the likely sites for deposition and concentration of the uranium.

Geobotanical and *biogeochemical* methods (involving the chemical analysis of selected parts of plants—generally the actively growing shoots) have not been widely applied in mineral exploration in Western Australia, despite the positive results obtained in orientation studies. Elkington (1969) investigated the climatic geomorphologic and geologic factors influencing the specific distribution of vegetation in the semi-arid sclerophyllous woodlands of the Eastern Goldfields. Study areas for this investigation included gold mineralization at Coolgardie and Norseman, and nickel mineralization at Kambalda. Severne (1972) demonstrated that nickel and copper mineralization in the Murchison and Northeastern Goldfields could be detected by biogeochemical exploration methods based on the sampling and analysis of the *Acacia* species which dominate the mulga zone. The discovery that *Hybanthus floribundus*, a small shrub found growing in areas of nickel-rich rocks, accumulates nickel to an amazing degree was another result of Severne's investigations.

In summary, it may be stated that geochemical methods have found increasing application in mineral exploration in Western Australia since the early 1960s and are currently used by most mineral exploration groups seeking a wide variety of mineral deposits over the full spectrum of geological and geomorphological conditions found in this state.

REFERENCES

Blain, C. F. 1967. Petrology and mineragraphy of the Bow River copper-nickel occurrence. Hons. thesis, University of W. Aust.

Bull, A. J. & Mazzucchelli, R. H. 1975. Application of discriminant analysis to the geochemical evaluation of gossans. In *Geochemical Exploration 1974. Proc. 5th Internat. Geochem. Explor. Sympos. Vancouver* (Eds. I. L. Elliott & W. K. Fletcher), 219-28.

Butt, C. R. M. & Sheppy, N. R. 1975. Geochemical exploration problems in Western Australia, exemplified by the Mt. Keith area. In *Geochemical Exploration 1974. Proc. 5th Internat. Geochem. Explor. Sympos. Vancouver* (Eds. I. L. Elliott & W. K. Fletcher), 391-415.

Clema, J. M. & Stevens-Hoare, N. P. 1973. A method of distinguishing nickel gossans from other ironstones on the Yilgarn Shield, Western Australia. *J. Geochem. Explor.* **2**, 393-402.

Elkington, J. E. 1969. Vegetation studies in the Eastern Goldfields of Western Australia with particular reference to their role in geological reconnaissance and mineral exploration. PhD thesis, University of London.

Hall, J. S., Both, R. A. & Smith, F. A. 1973. A comparative study of rock, soil and plant chemistry in relation to nickel mineralization in the Pioneer area, Western Australia. *Proc. Australas Inst. Min. Metall.* **247**, 11-22.

Hawkes, H. E. & Webb, J. S. 1962. *Geochemistry in Mineral Exploration.* New York: Harper & Row.

Mann, A. W. 1977. Nature and origin of uranium deposits in calcrete. In Recent Geological Research Relating Low Temperature Aqueous Geochemistry to Weathering, Ore Genesis and Exploration (Eds. J. E. Glover & D. I. Groves). Univ. W. Aust. Extension Service, 14-27 (unpubl.).

Mazzucchelli, R. H. 1965. Geochemical dispersion relating to gold mineralization Western Australia. PhD thesis, University of London.

Mazzucchelli, R. H. 1972. Secondary geochemical dispersion patterns associated with the nickel sulphide deposits at Kambalda, Western Australia. *J. Geochem. Explor.* **1**, 103-16.

Mazzucchelli, R. H. & James, C. H. 1967. Arsenic as a guide to gold mineralization in laterite-covered areas of Western Australia. *Trans. Inst. Min. Metall. (Sec. B: Applied Earth Sci.)* **75**, 285-94.

Severne, B. C. 1972. Botanical methods for mineral exploration in Western Australia. PhD thesis, Massey University.

Smith, B. H. 1977. Some aspects of the use of geochemistry in the search for nickel sulphides in lateritic terrain in Western Australia. *J. Geochem. Exploration* **8**, 259-81.

Smith, R. E., O'Connell, A. M. & Edwards, R. G. 1976. Superficial Mineral Deposits and Exploration Geochemistry, Yilgarn Block, Western Australia. In *Excursion Guide No. 41C 25th Internat. Geol. Congr.* (Eds. R. E. Smith, C. R. M. Butt & E. Bettenay), 45-8.

Sokoloff, V. P. 1950. Dispersion of gold in rock and soil. *Min. Chem. Eng. Rev.* **43**, 7-10.

Travis, G. A., Keays, R. R. & Davison, R. M. 1976. Palladium and iridium in the evaluation of nickel gossans in Western Australia. *Econ. Geol.* **71**, 1229-43.

CHAPTER 5

EXPLOITATION—MINING AND METALLURY

The history of development, occurrence and exploration of Western Australian mineral deposits have been dealt with in the foregoing chapters and we come now to a consideration of their exploitation: of mining and metallurgical beneficiation to prepare the natural materials for marketable products. These differ for different materials, and this chapter is subdivided on the basis of the main products: coal, oil and gas, iron-ore, alumina, nickel, gold and other metallic minerals.

5A

Coal-mining

H. W. WILLIAMS

Manager, Western No. 2 Colliery, Collie

Coal and coal-mining have contributed strongly to the economic, social, and industrial progress of Western Australia. The only indigenous coal to be marketed in the state has come from the Collie Coal Basin. The first discovery was in 1883, but saleable production did not commence until 1898. Since then over 50 million tonnes have been supplied to consumers within the state, and 200 million dollars of value added to the Gross National Product. The market has been both hot and cold, with the importance of this fossil fuel often underrated, but at the time of writing (1978), the coalfield star is in the ascendancy, and the appreciation of the value of coal more soundly based than at any time in its turbulent history.

HISTORY

Utility

In the past, the Collie coal-mining industry and the State Railways Department were compatible opponents. Although dialogue and debate separated their spokesmen into opposing camps, it is doubtful if either would have attained or enjoyed prominence without the other's assistance. The railway demand for coal was the initial productive stimulant, and was the reason for the construction of the railway to Collie. Without this transport facility Collie coal could not have been marketed even if other markets had existed.

Unfortunately the railway required only large coal. Without a market for the slack (small coal), a large portion of the mined product was rendered waste, adding a considerable cost burden to the marketable larger sizes. To make the situation worse, the cobble coal (lump coal), being high in moisture, did not have good keeping qualities. As it dried out, it crumbled, and when the railways came to use it a further residue of fines was wasted. On the other hand, shale, an associated valueless carbonaceous rock, retained its shape better than coal and simulated lump coal, and added to the complaints about the coal supplied. The cost-conscious Railways Department was unhappy with the fuel and underlined its demand for a better product, with a threat to purchase all its requirements from Newcastle if no improvement was forthcoming. This official argument was reinforced by the railway workers union, whose members preferred the more volatile Newcastle coal with its ability to raise a 'head' of steam quickly. Farmers too added their opposition, claiming that the lighter ash of Collie coal carried sparks into adjacent crops, thus creating a fire hazard.

There was considerable merit in the Railway Department's criticisms, but on the other side of the coin there was also much merit in perservering with the local product. First, the economic stimulus to the state was indisputable and received political favour. Secondly, the supply of interstate fuel was uncertain because of industrial disputes. Thirdly, interstate suppliers tended to vary their prices according to the degree of competition from the local industry.

The debate lingered on without conclusive decisions, creating a climate of uncertainty in the developing mines, and it was only through external influences that the benefits of Collie coal were realized. The First World War, the Depression, industrial traumas on the eastern states coalfields, the Second World War, and the rapid industrial growth following the war all highlighted the need for maintaining the local industry. During and after the Second World War, the Western Australian Railways would have come to a standstill without Collie coal.

The attitude changed and the Railways Department compensated for the stated disadvantages of local coal by providing spark arrestors and eliminating the fire hazard; by improving the fire-grate system on the locomotives, thus making the Collie coal 'burn' more efficient; and by providing ponds to store coal under water, thus limiting deterioration during storage. At the same time, the State Electricity Commission's power-stations began to burn small coal, thus eliminating the waste; and, with all the problems apparently solved, the future of the industry seemed assured. However, as early as 1954 the advantages of using diesel oil to power locomotives became apparent and began to influence policy decisions, so that by the early 1960s the demand for coal by the railways began to fall. What had been a turbulent but productive marriage between the coal industry and the railways ended in divorce in 1970.

Most of the early, knowledgeable authorities had predicted that the future for Collie coal rested with stationary boilers. The SEC demand for power production began to show that that forecast was accurate, and the shift in the market made up for the declining railways requirements. Collie coal had relative advantages over its eastern states competitors: its low ash and sulphur content, and the ease with which it could be crushed and pulverized to suit modern boiler feed systems.

It was therefore a controversial policy decision when, in the mid-1960s, the SEC changed its fuel philosophy. Despite the fact that there was talk of an approaching energy crisis, the focus was shifted from coal to fuel oil, and several years of depression affected the coalfields at a time when optimism and progress should have been the order of the day. Soon, however, the energy crisis with its attendant escalation in oil prices reversed the trends, and placed Collie in a very advantageous position. The pessimism was dispelled and confidence restored, and only problems within the industry itself now could affect present enthusiasm.

Although public utilities have dominated the demand for coal, other markets have been tested. Early in the century, when steam was replacing sail on the high seas, considerable quanitities of bunker coal were loaded at Bunbury and Fremantle. The deficiencies already mentioned concerning the railways also applied, and indeed were made worse by the tendency of Collie coal to ignite spontaneously. After an initial surge the market declined except for spasmodic interest when fuel supplies were not obtainable elsewhere.

There have been many attempts to coke Collie coal (i.e. convert it, by heating, into a porous, high-fixed carbon residue suitable for metallurgical work), and there are still advocates for research into this field, despite the 'non-coking' property which characterizes all known seams in the Collie Basin. Collie coal was in fact used for some time in the gas works at East Perth, usually mixed, however, with eastern states coal.

From time to time there has been a small industrial demand for Collie coal. One of the early industries to use coal was the cement manufacturing industry, but this too later turned to fuel oil. Since the oil crisis, Cockburn Cement has swung over to coal and consumes approximately 300 000 tonnes per annum. It is expected that a much larger industrial demand will eventuate.

Administration

Coal-mining in Collie has been a twentieth-century phenomenon. Although the coal was first discovered in 1883, there was considerable delay in reporting the find, and prospecting did not begin until some seven years later. The prospecting results were sufficiently optimistic to warrant the construction of the Brunswick-Collie railway line, a spur from the Perth-Bunbury line at Brunswick. The railway was completed in 1898 and production of coal commenced with the completion of the railway.

It would be unkind to speak of the first decade as a productive decade, and consequently criticize the administration for its inability to satisfy the production expectations. The administration in those early days was primarily engaged in prospecting. A number of speculative groups formed with limited capital, hopeful of opening out a suitable mining prospect and selling it for profit to a more substantial mining company. Only one mining entrepreneur stood out: H. M. Deakin, who took over the lease of a mine developed by the government to prove the utility of the coal. Deakin called his mine the 'Wallsend', and with energetic enterprise proved that coal could be mined efficiently and economically. Unfortunately Deakin only survived long enough to see the initial progress of his work and he died before he could reinforce the foundations of his early success. Other mining ventures generally introduced a pessimism that was to burden the industry for many years.

Coal imported in Western Australia in 1898 was 132 200 tons (Lord 1952, p. 20), hardly a sufficient market to sustain more than one successful mining operation. Trying to share this market among a number of small producers was an unhealthy start. A number of problems were introduced: attempts to sell inferior coal to pay for development; inadequate haulage and pumping facilities; and higher than necessary staffing and administration costs.

A number of royal commissions set up to inquire into the problems of the industry all advocated an amalgamation of the interests. Early in the century a monopoly was handed out to Wallsend-Proprietary when the total order was allocated to that mine, but a dramatic rise in prices, owing to the monopolistic situation, soon caused a reversal of policy to re-establish the competitive market. Finally however, toward 1920 two financial entrepreneurs, Johnson and Lynn, with the aid of British capital and the backing of the Western Australian Bank (Johnson 1957, p. 33) set out to amalgamate the coal interests, and by 1920 had succeeded in establishing a monopoly under the administration of the Amalgamated Collieries Ltd, which was to remain until 1960.

A local syndicate, calling itself the Griffin Syndicate, formed by a number of local miners, prospectors and businessmen developed a mine during the 1920s, and in the face of strong opposition persevered in its attempt to claim a share of the market. Finally, the syndicate was taken over by the Griffin Coalmining Company Ltd, which, with political pressure and entrepreneural knowledge, managed to gain a small portion of the government contracts.

The Second World War, with its attendant accelerating demand for coal, found the monopolized administration wanting. The industry was unable to satisfy the demand, and the resultant pressures, tensions and publicity led to a worsening of public opinion about the coal-mining industry. The focus was unfairly directed onto the railway workers, and attitudes developed which even today have not been completely eliminated.

The urgent demand and the inefficiency of the industry combined to push up the price of coal to exorbitant heights. Attention was diverted to open-cut mining, a method which had been successfully introduced in New South Wales. Amalgamated Collieries commenced open-cutting at the Stockton Colliery in 1943, and Western Collieries Ltd entered the field in 1950, when they started an open-cut mine near Collie Burn.

High prices continued until 1960, with productivity still low and the demand still exceeding supply. The government decided to take firm action and, with the object of reducing pithead prices, issued a price ultimatum for the 1960 contracts. Amalgamated Collieries, still by far the major producer on the field, refused to accept the contract conditions and closed all its collieries. The responsiblity to supply Western Australia's coal needs was then taken up by Griffin and Western Collieries.

The reorganization resulted in a massive decline in the workforce. Employment aggregates of 1011 in 1959 fell to 582 in 1961. The obvious social impact, together with the problems of relocation, caused considerable concern over the entire Collie scene, and not unexpectedly produced acute industrial tensions. The reorganization has, however, been very successful from all points of view. The oligopolistic administration that has since prevailed has been characterized by an attitude of industrial responsibility, and there has been an absence of industrial disruptions. It is an administration that has satisfied the demand for coal.

Techniques

Following the traditional trend in coal-mining, the Collie industry began its operations with the contract system of mining. Miners were paid a tonnage rate to break down and fill the coal into skips with capacities ranging from half a ton to one ton. The skips were horse-wheeled to plats, from where they were hauled to the surface by endless rope, main-and-tail, or direct rope haulage.

Coal-cutters were the first mechanized introduction, the cutter cutting a kerf in the coal seam to provide a free face for shotfiring processes to break down the coal-face. Their early introduction was probably brought about by the restriction of the demand to large coal.

After the Second World War, intensive mechanized techniques were introduced in an attempt to improve productivity. At first, sequence-mining systems were adopted, with separate processes of cutting, boring, shotfiring, loading and transportation. High-capacity loaders filled the broken coal onto shuttle cars, which in turn delivered the coal onto conveyor belts. Later, continuous miners, which break the face mechanically and simultaneously load the coal onto the shuttle cars, replaced the sequence-mining systems. The equipment used was a prototype of the kind that was at the time gaining favour in the New South Wales coal-mines. Retrospectively, it appears that the equipment introduced was not suitable for overcoming the physical disabilities of the soft floor and tender roof. The high capitalization charges were not balanced by increases in productivity, and costs increased instead of falling, so that rather than lift the industry, mechanization placed it in considerable jeopardy.

Reorganization in 1960 produced a complete reversal in deep-mine technique. The Griffin Colliery 'Hebe' operated with scraper loaders, and Western Collieries developed a grunching (shooting off the solid) system with hand-loading onto scraper chain conveyors. Both of these systems produced better results than the intensive mechanization previously attempted.

Industrial relations

Traditionally, Collie coal miners quickly developed a strong union organization. Initially, each mine formed its own distinct institution, but by the turn of the century they had combined to form the Collie River District Miner's Association, with each mine organizing its own 'lodge' (branch). Undoubtedly the main milestone came 'with the signing of an Industrial Agreement in 1912, when a "closed shop" was introduced into the coalmining industry' (Blain 1966, p. 27). Union membership became a necessary condition of employment.

After initially developing its own welfare and accident relief funds, the union was successful in its attempts to have these institutions funded by levies on coal sales and by company contributions as additions to their own contributions. The administration of the institutions was changed to an industry function. The union retained a strong interest, however, both in the organization and distribution of funds. More recently the miner's union has become involved in a credit union activity, which appears to be a useful and appreciated innovation for the mine workers.

The miner's union has always exhibited a strong front, but industrial relations at Collie have been relatively peaceful. Specific periods can be named when abnormal unrest was experienced, but it is difficult to establish true causal influences. A number of serious disruptions occurred in the first decade in connection with co-operative movements that seemed to be motivated toward non-union labour. The depression, with its attendant reduction in wages, was not easily accepted. Politico-industrial ramifications in the early 1940s, when the local union was a member of the Federation of Coalmining Unions, was a period of considerable confrontation. The local union finally withdrew from the federation (although a close liaison between the two was maintained), a decision that was probably brought about by the problems that arose during this period of confrontation. As was to be expected, the impact of the reorganization of 1960 caused considerable chaos in the industry.

Since 1960 the industry has been peaceful. The reasons lie in the fact that top management and spokesmen from the companies have all been an integral part of the mining community, and well acquainted with the values of the industrial workforce; and that the top union executives are resident in and part of the normal community, and are influenced by community as well as industrial pressures. As a result of this, there has developed a high efficiency in communication. In the prime wages confrontation Collie is a follower, with the sword considerably blunted through following the decisions of the New South Wales industry on wages.

CURRENT TECHNIQUES

Deep mining

Only one deep mine is operational in Collie today, the Western No. 2 Colliery, which produces coal from the Wyvern seam on the Collie Burn Horizon. The seam lenses slightly, with the thickness varying from 3.5 to 4.2 m. The coal is mined on a classical bord and pillar layout, with bords (tunnels) and cut-throughs averaging from 6 to 7 m wide.

The water-bearing nature of the strata, and the presence of hydrous slurries near the surface, combined with the poorly-consolidated nature of the interbedded strata, make it prudent not to extract all the pillar coal. Extraction rates in the vicinity of 65 per cent are planned for most of the mine, but where the mine underlies the Collie River, which meanders across the area over the underground workings, extraction will be limited to 45 per cent.

The production process is by driving bords. The first activity is to drill a series of boreholes into the coal face, charge the holes with explosives with a delayed-action detonating pattern. The resultant shotfiring breaks down the coal face to achieve an advance of 1.5 m. The broken coal is loaded onto a scraper chain conveyor, which is extended to the face of the bord after each sequence. Loading is accomplished with a small diesel-powered front-end loader of the 'Bobkat' type. The face conveyor delivers the coal to a belt conveyor, which transports the coal to surface bins.

The face labour complement consists of five men referred to as 'composite' miners, who operate as a team accomplishing all the face activities, generally deciding among themselves on the distribution of the work. Each crew is required to produce 100 tonnes per shift. To effect the present target of 2500 tonnes per day, 12 crews are employed on a two-shift basis, each shift of seven hours duration. The balance of the production is won by development teams operating similar processes, but with the distinction that haul-dump vehicles replace the front-end loaders and eliminate the use of scraper-chain conveyors.

The mine environment is maintained in a satisfactory condition by four ventilating fans, two being of the axial-flow type and two centrifugal-type fans. The aggregate air-flow approximates 200 cubic metres per second. The mine is relatively wet, with a 'make' of water 13.5 million litres per day. Because of the wet conditions, the mine atmospheres are extremely humid; but as the temperatures remain steady at 20 to 21 °C, the high humidity, although uncomfortable at times, is not unhealthy.

At present the SEC is using the water from the mine at the Muja Power Station. The arrangement is a compatible one, with the SEC absorbing some of the mine's pumping costs, and simultaneously reducing their own water supply costs.

The Western 2 methodology is unique in coal-mining. The attendant physical difficulties have been compensated for by a degree of flexibility in legislative requirements, a flexibility that does not impair the safety of the mine, but which would probably be untenable if required to service a wider and more complex variety of mining operations.

The success of Western 2 operations has opened up considerable optimism for deep mining in Collie, negating the previous uncertainty. A follow-up with a new deep mine is now necessary to maintain deep-mine productivity, bearing in mind that the major portion of the reserves are classified as deep-mine prospects.

Open-cutting

Both the Griffin and Western Collieries companies are operating open-cut mines.

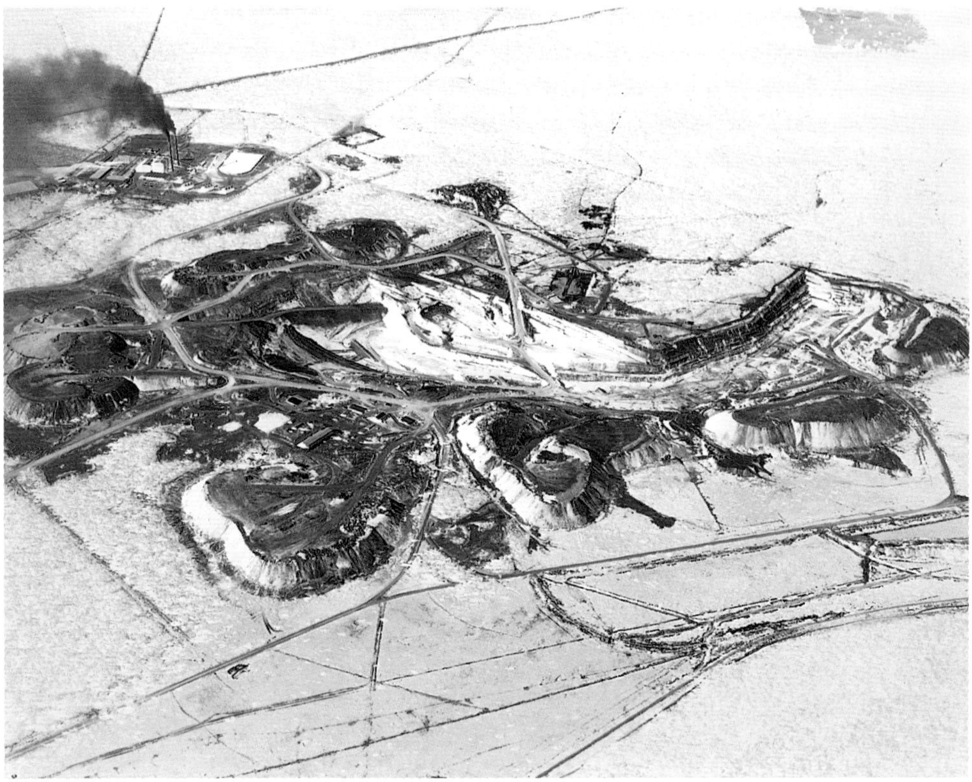

Fig. 5.1 Muja open cut, Collie Basin

On the Muja horizon the Griffin operates the Muja Open Cut (Fig. 5.1), winning coal from the many seams in that horizon down to the massive Hebe seam, which is approximately 11 m thick. On the Collie Burn and Cardiff horizons, Western Collieries operates the Western 5 Open Cut Complex, mining the Wyvern, Cardiff and Neath seams.

Open-cutting consists of physically removing the strata that overlie the coal seam, followed by extraction of the seam. Where the coal seam is sufficiently shallow, open-cut mining is generally cheaper than deep mining, and has the great advantage that it recovers almost 100 per cent of the coal.

As coal seams in the Collie Basin have 'blind' outcrops, i.e. they do not come to the surface, considerable drilling is necessary to explore the suitability of the mining area. Special attention is given to the placement of the boreholes in order to pinpoint the seams, detect the presence of faults, show the depth of the strata and thickness of the coal, and to establish coal quality. Planning layouts are then prepared, taking into consideration the haul routes for coal and overburden equipment, the gradients to be negotiated, drainage, disposal areas for overburden, and the type of equipment to be used.

Development work, after all the millable timber has been removed, is carried out in the summer months because of the soft nature of Collie soils and the heavy winter rainfall. The first exercise is to remove the top soil and separately stockpile it, so that it can be used in later reclamation processes. Overburden removal is accomplished by scrapers (large self-loading units capable of excavating 26 cubic metres per trip), and dump trucks up to an 80-tonne capacity. The trucks are loaded by face shovels, hydraulic excavators or front-end loaders.

When the coal is exposed, it is broken up by shotfiring and loaded into 35-tonne trucks for delivery to the Muja Power Station or to the crushing-station.

Finally, the rehabilitation of the mining area is considered. Backfilling of the overburden into the mined-out areas commences as soon as the working faces have advanced a safe distance. The last excavation in any particular area will fill with water once the pumps are removed, and with suitable landscaping could provide attractive lakes with recreational potential. Multiple trials, in conjunction with various government departments, are establishing the most suitable flora for replanting the waste areas, so that finally an acceptable habitat will replace the natural habitat that has been disturbed.

Rail-loading

Western Collieries deliver most of the coal mined from both the deep mine and the open cut to a crushing-station. The run of mine coal is crushed to minus 5 cm and transferred to a stockpile capable of storing 10 000 tonnes of coal. Underlying the stockpile is a reclamation tunnel, which enables the coal to be fed onto a belt conveyor, elevated into a surge bin and from there fed into rail trucks. Unit trains are loaded at a loading rate of 800 tonnes per hour, a facility that has considerably reduced the rolling stock requirements of the railway. Approximately 6000 tonnes per day are loaded onto rail.

THE FUTURE

Reserves and marketing

The future of Collie and the coal-mining industry rests heavily with the prime planning consideration: reserves. The total reserves of the Collie Basin amount to 1915 million tonnes (see page 141). The most recent drillings have modified but not substantially changed this figure. To date 405 million tonnes (286 open-cut and 119 colliery) have been defined as extractable (Table 3.19, p. 141). This figure is only 0.4 per cent of similarly stated Australian reserves at 74 340 million tonnes (Joint Coal Board 1975), demonstrating the short supply of Western Australian coal.

Extractable reserves are restricted to seams 2 m thick or more, lying at depths of less than 300 m for colliery coal and a vertical depth of 60 m (excluding the Muja Open Cut), and an overburden ration of 10:1 for open-cut coal. Not all mining engineers would agree with this premise, and would suggest that deeper and thinner seams will be extractable. Undoubtedly price increases will change the concept of economically-extractable reserves, and this is another consideration in the reserve statement. Others may be prepared to argue that the reserves are overstated and that the mining difficulties of the Collie Basin are greater than presently thought. Nevertheless, it is safe to say that the reserves are finite and limited and the situation demands a strong planning programme to eliminate waste, to achieve maximum recovery, to achieve overall efficiency, and to balance costs over time.

The necessity for a price balance is acutely demonstrated by the assessment of 119 million tonnes of deep-mine coal as compared with 286 million tonnes of open-cut coal. There is, and probably will continue to be a price differential in favour of open-cut coal, but more importantly, with increasing overburden ratios open-cut prices will increase, and with deep-mine conditions worsening, deep-mine prices will continue to rise. It will be opposed to the best interests of the industry, the town and the state if all the cheapest coal is produced first. A planning policy is necessary to develop higher cost coals to even out the price escalations, and to give more precise information to determine the accuracy of the reserve statements.

Demand statements, too, are fickle. One authority predicts that 132 million tonnes of Collie coal will have been consumed by the year 2000 (Anon. 1974, p. 35). The Fuel and Power Commission later estimate that 146 million tonnes will be consumed in the same period with a projected production rate of 8.5 million tonnes per annum by the year 2000 (Anon. 1975, p. 31). Price will be an important determinant of the demand reaching these proportions. There are alternative fuels, but coal today has an advantageous position, and to maintain this position the industry must see that relative prices do not change dramatically and erode the present advantage.

Changing techniques

Undoubtedly the increasing demand for coal will bring about corresponding changes in technique. Although the physical conditions of the deep mines may not allow the introduction of the highly productive techniques used elsewhere, other options are open. It is possible that some of the developmental continuous miners used in Europe may be operated successfully in specially devised layouts using extendable belt conveyors. This type of equipment would have the important advantage of improving the concentration of production with considerable cost advantages in both haulage and ventilation, and by the elimination of shotfiring costs. Such savings will offset the increased capital outlays.

In the open-cut field it is most likely that the present trend to larger equipment will continue, especially as the overburden ratios increase. One change may be in the motive power used. As the oil crisis worsens, the advantages of diesel-operated equipment with its extreme flexibility and manoeuvrability may give way to the more static possibilities of electrically-driven draglines and hydraulic mining techniques.

Changing personalities

The industrial relations scene will also change. The growth envisaged in size and complexity will require an influx of labour of all types, from top-line managers and supervisors to plant operators. The present intimate relationships may be replaced with the cold objectivity of specialization. Unless the lessons of the past two decades are learned and understood, problems leading to industrial confrontation may well be introduced into the Collie scene.

CONCLUSION

Only one conclusion is certain: the demand for coal in Western Australia is much stronger than it has been at any time in the past. Provided the responsibility that has characterized the industry over the last fifteen years is maintained, a bright, prosperous and harmonious future seems assured. Collie and the coal-mining industry will play an important role in the future progress of Western Australia.

REFERENCES

Anon. 1974. Environmental Assessment of Mining Coal at Collie, Western Australia. Fuel & Power Commission W. Aust.

Anon. 1975. Fossil Fuels. Fuel & Power Commission W. Aust. Rept. FP28.

Blain, A. N. J. 1966. Labour relations in the Western Australian Coal Mining Industry. Thesis, Univ. W. Aust.

Johnson, L. W. 1957. Collie Coal Mining Industry. *Univ. W. Aust. Studies Hist.* **3**, 7-51.

Joint Coal Board, 1975. 28th Ann. Rept. p. 225.

Lord, J. H. 1952. Collie Mineral Field. *W. Aust. geol. Surv. Bull.* **105**(1).

5 B

Iron-ore mining

B. E. Bursill
Mt Newman Mining Co. Pty Ltd

INTRODUCTION

The first men to talk about iron-ore in Western Australia were not geologists but pearlers. They used ironstone from the beaches of Yampi Sound, on the far north-west coastline, as ballast for their vessels. Deposits of iron-ore were discovered on two islands on the northern side of the sound—Koolan and Cockatoo—in 1880, but it was not until 1907 that mining of the hematite on Koolan Island began. Even then the relatively small amount extracted was used as a flux for copper and lead smelting, not as an ore of iron. Mining leases on Koolan Island were taken up by various interests over the years but the first real attempt to exploit the deposits was made by Yampi Sound Mining Company, the intending purchaser of the ore being the Nippon Mining Company of Japan. However, the imminence of the Second World War led the Australian Government to place an embargo on the export of minerals, including iron-ore.

Iron-ore was first mined in Western Australia about the turn of the century. It was for use as a flux in the Fremantle smelters. The deposits exploited at this time were iron-rich lateritic deposits east of Perth.

The Western Australian production of iron-ore recorded to 1919 is set out in Table 5.1.

While the vast deposits of iron-ore in the Pilbara region were first described in the 1890 annual report of the Government Geologist, H. P. Woodward, the glamour of the gold discoveries at Kalgoorlie three years later took the focus away from the Pilbara. Interest in the Pilbara revived for the second time in 1938 when government surveyors, making a ground survey of deposits at Mt Goldsworthy, estimated that these contained 6.3 million tonnes of ore with an iron content of 65.7 per cent. In 1960 the Commonwealth Government lifted its embargo on the export of iron-ore, and in 1963 the Mt Goldsworthy Joint Venture was granted a licence for the export of the total proved reserves at the rate of 4 million tonnes per year. This was the breakthrough in the exploitation of the vast iron-ore resources of the North-West. The Mt Goldsworthy operation was the first to export iron-ore from the Pilbara (in June 1966), following the lifting of the export embargo.

The potential of the Mt Whaleback deposit, some 400 km south of Mt Goldsworthy, was not realized until 1957, when a prospector, A. S. (Stan) Hilditch, searching for manganese, noted outcropping iron-ore on the then un-named hump-

Table 5.1

Western Australian iron-ore production to 1919

Locality	*Period*	*Production*		*Value*	
		Tons	*(Tonnes)*	£	($)
Whim Creek	Before 1899	100	(102)	300	(600)
Boulder	1900	450	(457)	247	(494)
Clackline	1899-1907	18 253	(18 546)	8 789	(17 578)
Coates Siding	1899	4 712	(4 788)	3 277	(6 554)
Avon*	1900-1901	22 223	(22 580)	16 241	(32 482)
Greenbushes	1899-1904	7 481	(14 962)	4 629	(9 258)
Werribee	1899	4 600	(4 674)	3 200	(6 400)
Koolan Island	1910	10.5	(10.7)	12	24

* Probably near Wundowie

Figures from *Geol. Surv. W. Aust. Mineral Resources Bull.* No. 7.

backed hill. Samples assayed at 68.8 per cent iron and in 1969 five companies which had formed the Mt Newman Joint Venture began shipping ore through Port Hedland. Meanwhile, in 1962, two CRA (Conzinc-Rio Tinto of Australia) geologists, surveying one of the many mountains that make up the Hamersley Ranges, discovered that it contained one of the world's richest deposits of iron-ore. That mountain was later named Mt Tom Price and the discovery led to the formation of Hamersley Iron, which commenced mining and shipping operations in 1966. To the north-west of Mt Tom Price, development of iron-ore reserves occurring as cappings on flat-topped mesas along the Robe River Valley began in 1962, when Cliffs Robe River Iron Associates obtained an option over the area.

In the south of the state the Kwinana iron-smelting works was being developed, and in seeking ore for the works BHP looked to the Koolyanobbing Ranges and nearby Mt Bungalbin. Tests proved a potential of some 50 million tonnes of iron-ore in the main deposit, Dowd's Hill, and the company moved to secure the rights to mine ore from this area. The comparatively small deposit at Koolanooka was worked jointly by the Western Mining Corporation and Hanna Mining between 1966 and 1975. This was the first operation to export iron-ore from the state (in March 1966) following the lifting of the ban.

Thus, the Western Australian iron-ore industry at present operates in three geographically remote areas: the Pilbara, in which four producers have invested more than $2200 million to mine vast inland deposits and rail ore over great distances to seaboard processing and shipping centres; Yampi Sound, where the close proximity of mines to berthing facilities requires no railroad construction; and Koolyanobbing, where ore is railed to Kwinana over a standard-gauge track completed in 1967.

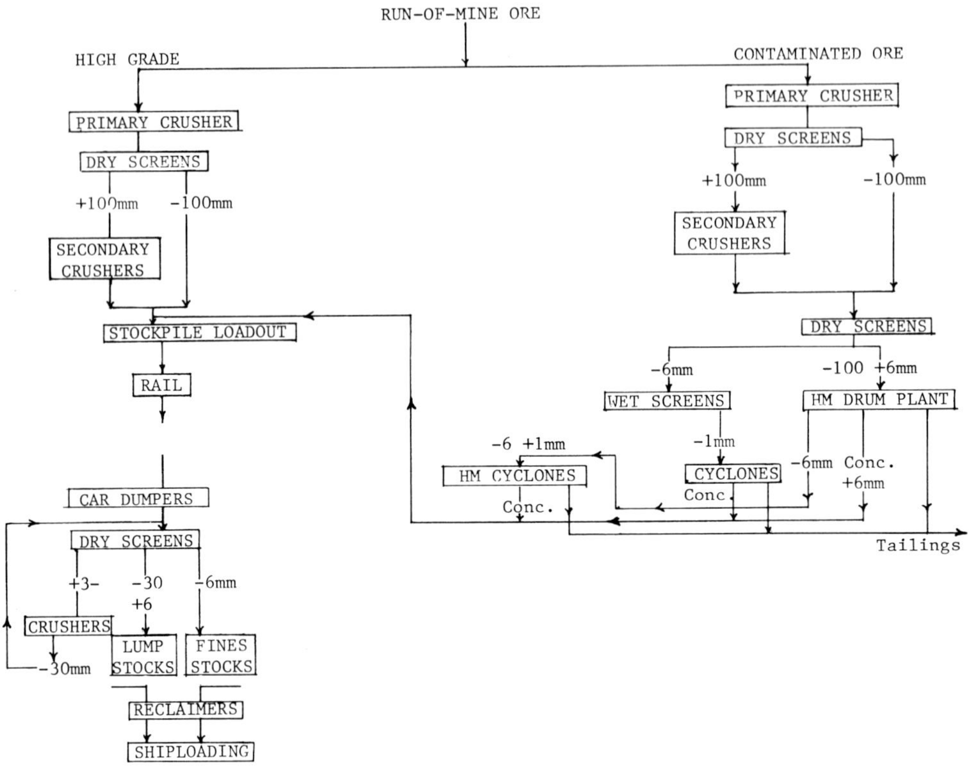

Fig. 5.2 Iron-ore-processing flowsheet

While the six producers differ in their approaches to the handling of ore mined—Robe River ore, for example, is alone in being railed to the coast uncrushed—all operate open-pit mines and employ similar extraction and processing methods (Fig. 5.2). Overlying waste material is removed, a pattern of deep holes is drilled in a layer of the ore deposit, the holes are charged with explosive and the charges detonated. The blast-shattered ore is loaded by shovels into trucks and hauled, directly in most cases, to primary crushers, which reduce it to a convenient handling and processing size. Secondary and tertiary crushing, usually carried out at the producer's port facilities, further reduces the ore to uniform shipping sizes. The crushed ore is graded, stockpiled, reclaimed in blends to meet the steel-mill requirements, and loaded into bulk carriers ranging in size to 264 000 deadweight tonnes. Ore is shipped in four forms: lump (pieces 6 mm to 30 mm), fines (smaller than 6 mm), run-of-mine (mixed) and pellets (heat-hardened balls of fines ore ranging from 9 mm to 16 mm in diameter). While lump and pelletized ore is used as direct blast furnace feed in the iron-making process, fines ore is first sintered into agglomerates at the steelworks, because as a direct feed, it would tend to smother the furnace mix. High-grade lump ore, free of fines, is used in the direct reduction steel-making process.

MINING

With mines, railroads and shipping facilities established, the Western Australian iron-ore producers entered an era of enormous production which uses giant machines able to drill, scoop, haul, crush and shipload ore on a scale undreamed of when the projects were conceived. Faced with escalating costs, the industry recognized it could hold production costs at viable levels only by scaling up operations and the size of equipment used. Accordingly the world's largest and most advanced mining equipment was commissioned—particularly in the mining of the Pilbara's vast orebodies.

Beyond the conventionality of open-pit mining (Plate 3) of scaled-up equipment, the producers' mining techniques are determined by the formations in which deposits occur (see Chapter 3C on Iron) and their nature, and more importantly, by the location of these deposits. The Yampi Sound deposits of Koolan and Cockatoo islands are mined by the Dampier Mining Company Ltd, a Western Australian subsidiary of BHP. The Cockatoo Island orebody is composed of high-grade hematite containing a minimum of 68 per cent iron. The deposit originally outcropped at the surface, some 140 m above the nearby sea-level, lending itself to the convenience of open-cut mining. After blasting, the ore is trucked to primary and secondary jaw crushers, then carried by conveyor belt to a 50 000-tonne storage bin. Koolan Island has five iron-ore deposits: Main Orebody, Acacia, Barramundi, Mullet and Eastern. In 1965 production began at the Main Orebody, which is more than 2000 m long and which has been proved to a depth of 152 m below sea-level. Here, blast-shattered ore is trucked from 12 m-high mining benches to a below-ground level crushing-station, from which it passes into a 60 000-tonne capacity storage bin.

Hamersley Iron Pty Ltd operates two iron-ore mines—Mt Tom Price and Paraburdoo—in the Pilbara Region approximately 1000 km north of Perth. At the time of writing, these mines and their associated crushing and screening facilities had an annual capacity of 40 million tonnes of ore and this was in course of being expanded to 46 million tonnes by 1979. The Mt Tom Price orebody is located at the centre of the Hamersley Iron Province, some 288 km from Dampier on the northwest seaboard, and Paraburdoo is 98 km south of Mt Tom Price. Conventional open-pit extraction methods are used at both centres, the mining operation beginning with the drilling of blastholes on 14 m-high pit benches and approximately 500 000 tonnes of material blast-shattered at a time.

At Mt Tom Price, the shovel-loaded ore is hauled away by electric-wheel, rear-dump trucks of capacities of 189 and 109 tonnes; trucks with a capacity of 150 tonnes are used at Paraburdoo. After crushing and screening, the ore is stockpiled and then fed into trains moving through a loadout tunnel. The company has two ore-loading berths at Dampier—one at Parker Point and the other on East Intercourse Island. Ore trains are unloaded by rotary car dumpers and the ore conveyed to a stacker, which blends it onto either 'lump' or 'fines' stockpiles. An iron-ore pellet plant at Parker Point agglomerates a portion of the fines generated during crushing and screening operations, producing heat-hardened balls ideal as blast fur-

nace feed. Lump ore and pellets are reclaimed from stockpiles for shiploading by a tunnel system; a rail-mounted bucket-wheel reclaimer is used for the reclamation of fines.

The Mt Newman Joint Venture, with an annual production capacity of more than 40 million tonnes (increasing to 45 million tonnes per year with the commissioning of a new beneficiation plant in 1979), produces approximately 120 000 tonnes of Mt Whaleback ore daily, which means mining crews must move about 280 000 tonnes of material each day. This is done systematically by terracing the mine into 15 m-high benches (Plate 2). To mine the ore, drilling crews first sink a series of holes on each bench, working to a pattern planned by the mining engineers. The holes are charged with ANFO (a mixture of ammonium nitrate and fuel oil) and detonated, each blast shattering up to 300 000 tonnes from the bench face. Electrically-powered shovels scoop up to 40 tonnes of broken material at a bite and load it into rear-dump trucks with capacities of 109 and 189 tonnes. The ore is hauled to crushers and other material to dumps skirting the mine. At the foot of Mt Whaleback the ore is dumped into two primary crushers, each of whose oscillating cones can crush 4000 tonnes of rock per hour into pieces smaller than 200 mm. Conveyors take the material onto secondary crushers, which reduce it to pieces smaller than 100 mm. At this stage, the ore is transferred by conveyors to surge-piles built over two train-loading tunnels. At Nelson Point, Port Hedland (Plate 2), trains feed one of two systems: in the first a rotary two-car dumper can unload a 13 000-tonne train load of ore in two and a half hours; in the second system a three-car dumper completes the unloading operation in one and a half hours. The ore is fed by conveyor to one of two tertiary crushing buildings, where it is reduced to pieces no bigger than 30 mm (lump ore). Material smaller than 6 mm (fines) is separated out by a system of screens. Lump ore and fines are conveyed out to boom stackers, which build stockpiles in blends specified by a Quality Control Department. Bucket-wheel reclaimers draw from the stockpiles according to customers' requirements and transfer the lump ore or fines to conveyors, which carry it to two shiploaders, each capable of loading carriers at around 8000 tonnes per hour.

To increase their reserves of recoverable high-grade ore and increase the life of their projects, Hamersley and Mt Newman were installing ore beneficiation plants at the time of writing (1978). These will separate lumps and particles of high-grade ore from shaly overburden material admixed during blasting and mining operations around the contact zone or edge of the main orebody. This mixture of shale and

---

Plate 2 *Top:* Mt Whaleback iron-ore mine at Newman, Pilbara. Aerial view showing 15-m high mining benches, and in centre-near distance blast-hole drilling rig and pattern of drill holes.

Bottom: Nelson Point, Port Hedland, showing a carrier taking on a cargo of iron-ore as tugs position a second vessel at the Mt Newman ore pier. Ore stockpiling area and dumping, screening and crushing plant is seen on shore. Note also dumps of white salt on left from Leslie Salt Co. works, Port Hedland, awaiting shipment.

1

2

3

4

5

6

7

8

iron-ore is known as 'contact ore'. The process thus permits the producers to beneficiate low-grade and contact ore, currently set aside on dumps, to produce a high-grade saleable product.

The nature of the Mt Goldsworthy type of ore deposits at Goldsworthy, Shay Gap and Sunrise Hill is substantially different from that of the Hamersley Range deposits. They are steeply-dipping tabular and occur in a complex geological structure. At Shay Gap and Sunrise Hill, the reserves are divided into a number of small deposits occurring in narrow, steeply-dipping layers. All deposits are mined by open-cut methods, and substantially the same equipment and techniques are used at each mine. Blast-shattered ore is scooped from mining bench faces by electric shovels, which load it into rear-dump trucks with capacities ranging from 59 to 108 tonnes for haulage to the mine's primary crushing-plant. Specially-designed, long-distance ore trucks of 66-tonne capacity carry ore from Sunrise Hill to the Shay Gap crusher. The crushing processes at Goldsworthy and Shay Gap are basically the same. The trucks tip into gyratory crushers, which discharge the primary crushed ore onto conveyors for stockpiling. It is then reclaimed through tunnels under the stockpiles and taken by conveyor to rail-loading bins, which have a capacity of 203 tonnes at Goldsworthy and 356 tonnes at Shay Gap. The ore is then railed to Finucane Island, Port Hedland, for secondary and tertiary crushing, stockpiling and shipment.

The Robe River Project members have expanded their mesa-top mining operations from one to four locations. Mining of the thick exposed cappings of the Robe River mesas [see Fig. 3.13, page 83] involves breaking the ore, loading it into 110-tonne capacity trucks with electric shovels and hauling it to either of two nearby train loading stations. Each rail loadout has a rated capacity of 3000 tonnes. Mining progresses as sections of mesa capping are drilled and fractured by blasting, ready for the shovels. Because mining operations move from mesa to mesa, no crushing facilities are provided on the mining site; these are centralized at the Cape Lambert plant on the coast. Ore delivered from the mine contains lumps up to one metre across. At the Cape Lambert complex, it is dumped through a rotary car dumper into crushers and the ore flow then splits to provide a sinter feed product or material for pelletizing. The hard-fired pellets produced are stacker-fed onto a port storage area.

Plate 3 Pilbara iron-ore mining—stages from mine to market: (1) Drill holes in orebody are charged with ammonium nitrate and fuel oil, which is (2) fired in patterns by detonating fuse. (3) Shovels load broken material into rear-dump trucks, which haul it to the crushers. (4) After crushing, the ore is railed to port, and (5) the cars are rotated, two or three at a time, to empty the ore into bins. (6) After further crushing, the ore is stacked in a series of stockpiles. (7) Bucketwheel reclaimers transfer the ore to conveyors, which carry it to (8) rail-mounted ship loaders, which feed the ore in the holds of bulk carriers. (Sequence of photos illustrating these operations by courtesy of Mt Newman Mining Co. Pty Ltd).

At Koolyanobbing mined ore is scooped by electric shovels and loaded into 45-tonne dump trucks for haulage to crushing plant. There it passes through a primary jaw crusher and secondary and tertiary gyratory crushers. Seven silo bins, with a total capacity of 10 000 tonnes, are used in train-loading operations. Development of the Koolyanobbing iron-ore mine, 48 km from the historic gold-mining centre of Southern Cross, has been linked from inception to that of the Kwinana steelworks operated by the Broken Hill Proprietary Co. Ltd. The Kwinana blast furnace was 'blown in' on 15 May 1968, and completed its first campaign life on 30 September 1974, producing approximately 4.3 million tonnes of pig iron in that period. It resumed iron production on 6 December 1974.

To meet the increasingly-exacting specifications of iron-ore buyers, whose steel-making processes are becoming more efficient, and sophisticated, the Western Australian producers are applying computer-control systems and other new technologies to ore grade and quality control. Uniformity of raw materials is essential in steel-making today, so ore consignments must be of uniform iron content and size not only from one shipment to another but also from one month to the next and from one year to the next. New blends are being evaluated by steel-makers; Mt Newman is already shipping a blend of 85 per cent Brockman Formation ore (hard, blue hematite) and 15 per cent Marra Mamba Formation ore (a soft, yellow limonitic ore). Increasing attention is being directed, also, to moisture control to facilitate ore handling and dust suppression.

TRANSPORT

All the Pilbara ore deposits are situated inland and the ore is railed to various ports (see Fig. 5.3). As mentioned earlier, Robe River ore is alone in being railed to the coast uncrushed. Mt Goldsworthy ore is railed to Finucane Island, Port Hedland, for secondary and tertiary crushing, stockpiling and shipment. Usually a train carrying 4065 tonnes is hauled by a single locomotive, but two locomotives can be used to haul loads in excess of 8130 tonnes. At the port, air-operated doors are opened in the bottom-dump cars, releasing the primary-crushed ore into a hopper. Hamersley Iron operates a 386-km long standard-gauge railway line from Paraburdoo through Mt Tom Price to Dampier. Trains generally consist of three diesel-electric locomotives hauling 184 cars carrying 17 500 tonnes of ore. At the port the trains are unloaded through rotary car dumpers at either Parker Point or East Intercourse Island. Mt Newman operates the longest railroad of all: a 426-km track from its Mt Whaleback mine at Newman to Port Hedland; it carries a greater tonnage than the entire Western Australian Government railways system. The optimum consist of each train is 138 cars drawn by three mainline diesel-electric locomotives. A train weighs approximately 17 500 tonnes, of which 13 700 tonnes is payload. The Robe River Project operates a 168-km railroad linking its mining operations with its Cape Lambert plant on the seaboard, trains averaging 120 cars of 109-tonne capacity each. Koolyanobbing ore, on the other hand, is hauled west to Kwinana along a standard-

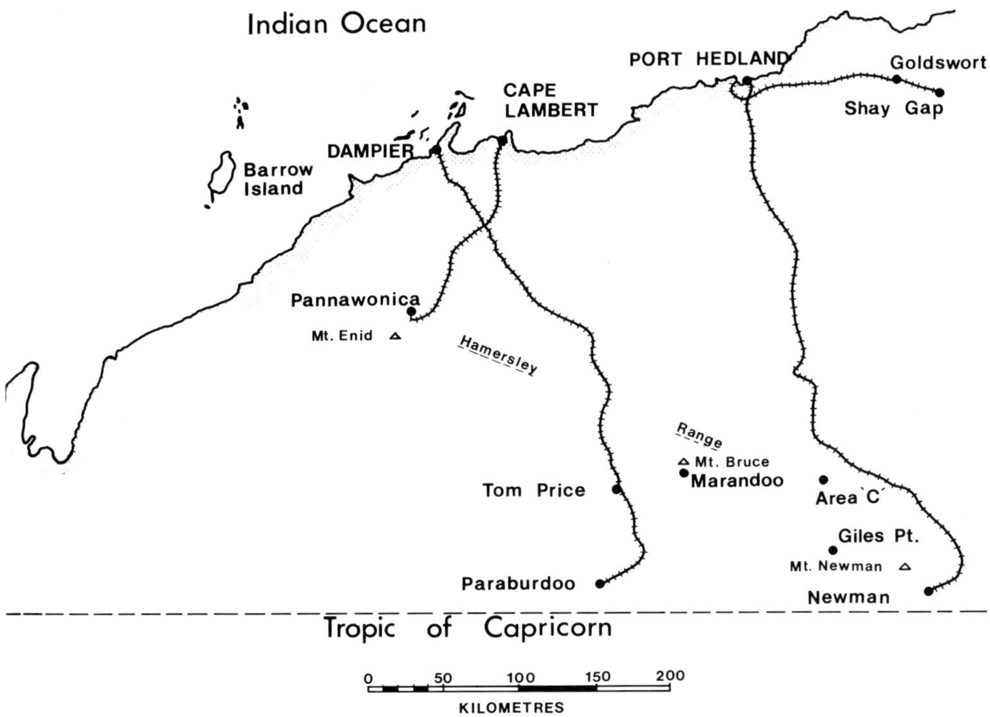

Fig. 5.3 Pilbara: mines, railways and ports

gauge railway completed in 1967. Built by the state government, this track later became part of the western half of Australia's transcontinental railway.

SHIPLOADING

The iron-ore producers' operations have had a marked impact on North-West port development, the major shipping points now being Dampier, Port Hedland (Plate 2B) and Cape Lambert. The Mt Goldsworthy and Mt Newman operations have helped to transform Port Hedland from a tidal port, restricted to ships of less than 5000 deadweight tonnes, into the biggest export port in Australia in terms of tonnage throughput. Development at Dampier and Cape Lambert has been spectacular also. The Mt Newman ore pier at Hedland is the only facility in Australia capable of berthing and loading two 160 000 deadweight bulk carriers at the same time. Here two rail-mounted shiploaders feed ore through telescopic booms into the holds of carriers. Two carriers can be loaded simultaneously or both shiploaders can operate on one carrier. In keeping with world trends, the average size of vessels calling at the Goldsworthy wharf has greatly increased. In August 1970 the first carrier in excess of 100 000 tonnes deadweight was berthed. Since that time, docking frequency of 100 000-tonne ships has increased dramatically to the extent that over 85 per cent of all vessels loaded during 1976 were of this size. The maximum-size ship that can be handled is approximately 120 000 tonnes.

Dampier's Parker Point, from which Hamersley made its first shipment in August 1966, has a berth capacity of 100 000 deadweight tonnes. However, this will increase to 120 000 tonnes, following dredging associated with new development. Maximum plant handling rate is 7200 tonnes per hour. East Intercourse Island, where shipments started in March 1972, has a berth capacity of 160 000 deadweight tonnes and the maximum plant handling rate here is 7500 tonnes per hour. The Robe River Project's harbour facilities comprise an ore wharf, which is the longest and tallest jetty in Australia and a nearby service wharf, both projecting from Cape Lambert. The wharf, supported on 400 steel piles, is 2670 m long and stands 18.6 m above low-water level. Its ore berth has two loading-faces—one designed for 100 000-tonne ships and the other for 150 000-tonne carriers—although vessels exceeding 260 000 deadweight tonnes have been safely handled. A conveyor belt travelling the length of the jetty delivers ore at 6000 tonnes per hour to a travelling shiploader, which can service either berth. During 1978 a 264 000 deadweight bulk carrier loaded Robe River ore at the jetty. It was the largest ship ever berthed anywhere in Australia.

IRON PRODUCTION IN WESTERN AUSTRALIA

Most of the iron-ore produced in Western Australia is exported, but ore from Koolyanobbing is smelted locally at Wundowie and Kwinana. Pig-iron was first produced at Wundowie (60 km east-northeast of Perth) in 1948 from a small charcoal blast furnace established by the state government. After two years the low grade Wundowie ore (assaying approximately 43% Fe) used in this furnace was supplemented by higher-grade Koolyanobbing ore (assaying approximately 62% Fe), and in 1955 was completely replaced by Koolyanobbing ore. In 1977 it produced 45 090 tonnes of pig-iron valued at $3 903 976.

The blast furnace of the Dampier Mining Co. Ltd at Kwinana was commissioned in 1968. It uses ore from Koolyanobbing, which is connected to Kwinana by a 499-km standard gauge railway. The 1977 production of pig-iron at Kwinana was 540 400 tonnes.

No steel has yet (1979) been produced in Western Australia, but steel production is envisaged before the mid-1980s.

5C

Bauxite—mining and processing

E. A. Kirke and A. M. Murray
Alcoa of Australia Limited

Historical development

A fully integrated bauxite-mining, alumina-refining and aluminium-smelting company, Alcoa of Australia Ltd, was established in 1963 to exploit the bauxite deposits in the Darling Range. The first alumina refinery was established at Kwinana, on the shores of Cockburn Sound, 35 km south of Perth, and used bauxite mined near Jarrahdale. The aluminium smelter was established at Point Henry on Port Phillip Bay in Victoria.

In 1972 a second refinery was established at Pinjarra, located 90 km south of Perth and situated at the foot of the Darling Range escarpment. Bauxite for this refinery is mined at Del Park and Huntly, 5 km and 10 km to the east, near Dwellingup.

Further exploration and evaluation have increased the bauxite reserves to 750 million tonnes, with grades in the range of 30-35.0 per cent available alumina, 18-22.0 per cent free silica and 1-2.0 per cent reactive silica.

Production

In 1963 alumina production began at Kwinana at an annual rate of 220 000 tonnes. Continuous growth has taken place, with the Kwinana refinery expanding to a production capacity of 1.4 million tonnes in 1975. The Pinjarra refinery has been progressively expanded, and during 1979 reached a production capacity of more than 2.4 million tonnes. These alumina production outputs require the mining of 5.0 million tonnes of bauxite at Jarrahdale and a further 9 million tonnes at Del Park and Huntly.

Ore definition

The bauxite of the Darling Range is an aluminous laterite containing in the order of 35 per cent available alumina, 20 per cent free silica and 1.5 per cent reactive silica.

In many areas of the Darling Range, bauxite pockets are grouped together, separated by from 30 to 300 metres of ferruginous laterite or weathered dolerite. In other areas they may be more isolated, separated by valleys, ridges, or widespread non-bauxitic laterite. Aerial photography is used to outline areas in which laterite is

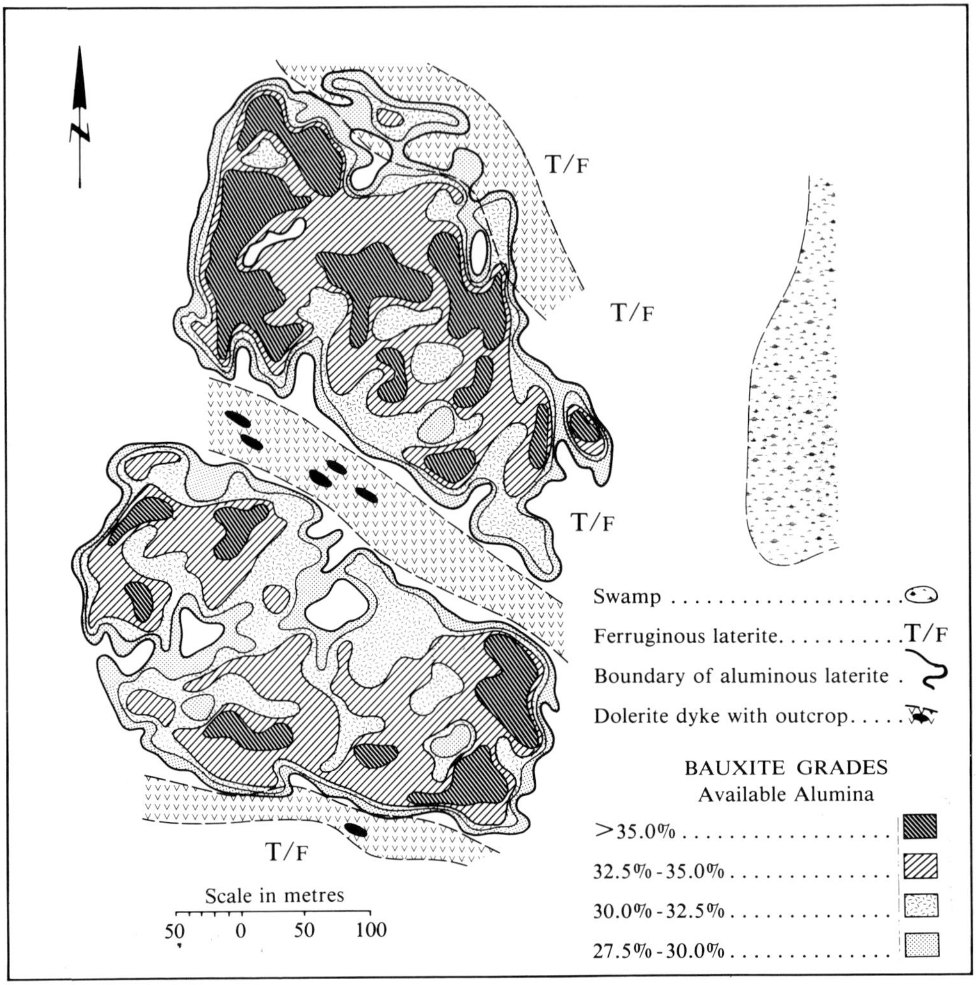

Fig. 5.4 Grade variations in a typical bauxite deposit, Darling Range, Western Australia

developed and to assess these against the background of other elements of the landscape such as soil, swamps, creeks and out-cropping basement rocks. Using the support of ground reconnaissance, the more localized areas of bauxite potential are identified and selected for drilling.

Initial testing is carried out at 120-m spacings. All holes are drilled to the base of the laterite or bauxite profile where it passes into the clay zone, usually at a depth of 5 to 10 m. Samples are taken at 0.5-m intervals and analysed for available alumina. Where three or more adjacent holes return ore-grade results over a minimum depth of 2 m, the drilling grid is closed to 60 m.

If the target areas still hold up as ore in the 60-m drilling, a final programme of holes at 15-m spacing is drilled. The depth and lateral boundaries of mineable bauxite are determined from these results, and all proven ore reserves, pit design, mine planning and grade control are based on them.

Final definition of ore boundaries and physical characteristics of the ore are obtained by geological mapping of the caprock surface. Despite the high degree of weathering and recementing effects, the caprock preserves sufficient textures and mineralogy of the parent rocks to enable principal contacts to be located (Fig. 5.4).

Drilling at all stages is carried out using rotary vacuum rigs, which are tractor-mounted and specifically designed for this work. These rigs are extremely mobile in the uneven forested terrain. A major requirement of the drilling programme is that the natural vegetation should be disturbed as little as possible and various measures are undertaken to ensure this. These include surveying the drillhole positions without clearing grid lines, washing down vehicles at suitable locations to prevent the spreading of the 'dieback' root fungus, which is common in the forest, and closing down the drilling programmes in the wet winter months to minimize wheel-rutting and the spreading of mud which might contain active 'dieback' spores.

Mining

The mining methods at the three minesites are similar with respect to overburden removal, drilling and blasting, and loading of broken bauxite into trucks. However, crushing and transportation of the bauxite to the refinery differ between sites.

Overburden removal

The overburden consists of an average 0.5 m depth of loose unconsolidated gravel. Its removal involves the use of scrapers and bulldozers with attendant front-end loaders and off-highway trucks. The soil and gravel recovered is trucked to storage piles and on completion of mining, is returned to the mined areas as part of the land-rehabilitation process.

Drilling, blasting and transport to crusher

Blasting of the caprock is carried out on a 3.5-m spacing, with hole depths averaging 3.0 m and using ammonium nitrate fuel oil explosive.

The broken caprock, together with the underlying friable bauxite, is loaded by front-end loaders into off-highway trucks of 100-tonne, 50-tonne and 35-tonne capacity for haulage to the crushing plant (see Plate 1). The 100-tonne trucks used only at Jarrahdale, are articulated with bottom-dump trailers. All others are single-unit rear-dump vehicles.

Crushing and transportation—Jarrahdale

The bauxite is delivered to a fixed crushing plant, where it undergoes three-stage crushing to a minus-25-mm product.

In the open-circuit-crushing system, under-sized material is removed by wobbler feeders, double-deck and single-deck screens. The present operating capacity of the plant is 1400 tonnes per hour.

All crushed bauxite is stacked either in the bunker above the train load-out tunnel or in a surge stockpile at one side of the train tunnel.

The ore is loaded into the rail wagons through 36 pneumatically controlled gates in the load-out tunnel below the bunker. Loading of a 4400-tonne train takes approximately one hour. It is then railed 51 km to Kwinana Alumina Refinery for blending and processing.

Two blending stockpiles are maintained at Kwinana, each holding 60 000 tonnes of bauxite. They are built up with an overhead shuttle conveyor, which lays down a series of double chevrons. Reclaiming is carried out through trapdoor bays in the stockpile floor.

Crushing and transportation—Del Park and Huntly

In 1968, during the design stage of the Del Park facility, emphasis was placed on increased mobility of mining operations. With this concept in mind, the decision was made to introduce an overland belt-conveying system as a major ore transport medium, with a mobile crusher to reduce the run-of-mine ore prior to conveying and 35-tonne trucks delivering the ore to the crusher. A similar system was designed for Huntly minesite, commissioned in 1976.

The mobile crushers consist essentially of a feed hopper, apron feeder, scalping screen, single-toggle jaw crusher and discharge conveyors mounted on a subframe. The walking mechanism is a three-point hydraulic lift and stride system. The Del Park crusher weighs 450 tonnes, with a throughput of 1500 tonnes per hour, and the Huntly crusher weighs 560 tonnes with a throughput of 1700 tonnes per hour. Both reduce the ore to minus-200 mm and feed direct onto the overland conveyors.

Each conveyor system consists of two parts, a variable upper section feeding from the mobile crusher to a transfer point and a lower fixed section from the transfer point to the Pinjarra Alumina Refinery. The upper sections are extended at will as the crushers are relocated. The lower sections are fixed and run in parallel. Where they descend the Darling Scarp to the refinery (Plate 4A) they are regenerative when loaded, providing some 600 kW of power. In 1977 the total length of each conveyor system was 11 km.

At the refinery the conveyors feed onto four blending stockpiles each holding about 170 000 tonnes. Two boom stackers build up a series of longitudinal windrow chevrons and maximum blending is ensured by relaiming transversely across each pile with a bridge-mounted bucket-wheel reclaimer.

Grade control

The bauxite deposits occur as separate pockets, each with different grade parameters, and grades within individual deposits are also variable. For this reason, planning of the development of pits is necessary to ensure that an even grade of ore

Plate 4 Bauxite mining, Darling Range. *Top:* Alumina refinery, Pinjarra, looking west along conveyor plunging down the Darling Scarp.

Bottom: Rehabilitation of mined areas: (1) Mined area with topsoil replaced being deep-ripped. Original pit face in background has been bevelled to maintain gentle slope. (2) Mined area, treated as in (1) and planted. Approximately one year after planting. (3) Mined area approximately four years after planting.

is supplied to the refineries. This is also required to reconcile trucking and service costs with considerations such as seasonal constraints caused by the wet winter season and steepness of pit floors.

On a week-to-week basis, individual stockpiles are designed to have, as nearly as possible, grades approaching the long-term ore average grade (within the limits of the yearly and seasonal mining schedules). It is frequently necessary to mine ore from several separate pits and on several faces in a single pit in order to maintain the grade balance for an individual stockpile.

Rehabilitation after mining

Alcoa accepts the responsibility of taking all possible steps to protect the environment in which the company operates, and to achieve this works in close co-operation with many state government departments. The rehabilitation of land it has mined is of major priority in the company's environmental programme.

One principal objective in rehabilitating mined areas is to restore them to maximum potential as water catchments, where this is a current or anticipated land use. Another important aim is to restore those areas affected by mining to an aesthetically acceptable state, so that they are of value to the community as a recreational resource. This is particularly important in the Jarrahdale region, which, owing to its proximity to Perth, is being used increasingly for recreation.

The process of rehabilitation involves engineering works to prevent erosion of the returned topsoil and establishment of a permanent cover of vegetation to maintain the long-term stability of the landscape. Engineering works include blasting and bulldozing of vertical pit-faces to slopes of less than one in two, and the grading of pit-floors to provide a gently undulating surface similar to the original topography. About 500 mm of topsoil is then returned to the pit-floor and deep-ripped to a depth of 2 metres with a large bulldozer. The ripping operation is to relieve subsoil compaction brought about by previous mining activity (Plate 4). It is carried out along the contours and has the effect of substantially reducing soil erosion by controlling run-off.

Since the initial experimental planting in 1966, Alcoa and the State Forests Department have developed substantial expertise in rehabilitation of mined bauxite areas. Early problems of instability and poor growth of some young trees reflected poor knowledge of techniques for revegetation of bauxite pits, for site preparation and for species selection and nutrition.

The most successful species used in the early planting programme were fast-growing eucalypts from the eastern states selected for their commercial timber potential and resistance to jarrah root rot, commonly known as 'dieback', a fungus disease which is destroying much of the jarrah forest in regions currently subject to mining. These species include red mahogany *(Eucalyptus resinifera)*, Sydney blue gum *(E. saligna)*, spotted gum *(E. maculata)* and Tasmanian blue gum *(E. globulus)*. All of these are used in current vegetation work, plus a number of local Darling Range species, which include marri *(E. calophylla)*, wandoo *(E. wandoo)*, blackbutt *(E. paterns)* and butterbark *(E. laelaei)*. The local species have little commercial potential, but do have proven tolerance of local climate and soil conditions and resistance to the fungus which causes 'dieback' disease.

Establishment of a balanced vegetation system involves more than simply planting trees as plantations or mono-cultures. Handling and stockpiling of topsoil is known to deplete its content of native seed accumulated on the forest floor over many years. For this reason Alcoa and the State Forests Department have turned their attention during the past two or three years to developing techniques for colonizing mined areas with species additional to the forest trees used to date.

Experiments have been conducted involving direct seeding of native species from jarrah forest areas and the use of agricultural clovers to provide early soil stability. These have combined with trials of mulches such as straw and cellulose fibre, plus bitumen emulsion and other binding agents to provide temporary soil stabilization during establishment of vegetation.

In addition to the question of mine rehabilitation, Alcoa and officers of the state government are applying themselves to other issues associated with the various activities in the Darling Range. Activities include production forestry, water catchment, bauxite-mining, hard-rock quarrying, recreation, beekeeping, horticulture, agriculture, urban development and fauna and flora conservation. Of particular importance are the questions of stream water quality and jarrah 'dieback' disease.

Physical pollution of water by soil can result from any operation involving the clearing of land. Such effects from bauxite-mining are adequately controlled by the engineering and revegetation works involved in land rehabilitation and mine service areas.

Stream-water salinity is a problem which can arise in some local state forest and agricultural areas. It has become serious as a result of clearing large areas of land for farming on the east of the Darling Range. Streams in those areas are showing substantial increases of salinity. One solution seems to be to re-establish vegetation with deep root systems on the cleared areas. This would involve an operation similar to that currently in progress on mined areas. No salinity problems have become apparent in areas where bauxite mining is now in progress. This has been confirmed by studies conducted by CSIRO scientists, the state government and Alcoa.

'Dieback' disease, caused by the fungus *Phytophthora cinnamomi* is of great concern to all land-users in the Darling Range. It already has accounted for the loss of 280 000 ha, or more than 10 per cent of Western Australian jarrah forests, and the infected area is increasing by 20 000 ha per year. It is the subject of substantial research activity by the State Forests Department and the CSIRO. There is no practical cure for the disease, but its spread can be reduced by eliminating vehicle activity in the forest. With this in mind, the Forests Department has quarantined a substantial area of state forest to slow the rate of spread.

Refining

Many processes have been devised for the production of pure alumina from bauxite, clay and other aluminous materials. The 100-year-old Bayer process, which utilizes bauxite, continues to be the most economic.

The basis of the process is the variation in solubility of hydrated alumina in sodium aluminate solutions with changes in concentrations of soda and temperature. This solubility variation permits efficient extraction of alumina from bauxite at elevated temperatures, together with good yields of precipitated alumina on cooling.

The bauxite is first ground to approximately minus-10 mesh and then digested with caustic liquor under suitable conditions of temperature and caustic concentration. As the hydrated alumina in the bauxite is in the form of gibbsite (alumina trihydrate), moderately mild digester conditions are sufficient, namely a caustic concentration of 150-190 g/l and a temperature of about 130 °C. In addition to the dissolution of hydrates of alumina during the digestion stage, several other important reactions affecting alumina quality and process efficiency also occur.

Following digestion, the slurry is flash cooled, stepwise, to atmospheric pressure. The flash steam from each stage is used to preheat incoming caustic liquor and bauxite slurries. Darling Range bauxite produces a residue consisting of two distinct particle size ranges: a coarse fraction rich in silica and a fine fraction, or red mud, rich in iron oxide. The coarse fraction, or sand, is separated by sand-traps which underflow to a counter-current decantation sand-washing system.

The sand-trap overflow feeds rake thickeners, where, with the aid of coagulant, the mud is separated and pumped to a series of counter-current decantation washers for recovery of entrained alumina-rich liquor.

The supersaturated sodium aluminate liquor (green liquor) overflowing the thickeners is passed through a filtration stage to remove all traces of suspended residue and then flash cooled still further. The purpose of this final flash cooling is both heat recovery and preparation of the liquor for precipitation of alumina trihydrate.

Controlled precipitation or decomposition, as the process is variously known, is achieved by mixing the green liquor with medium-sized alumina trihydrate seed in large agitated tanks. Under these conditions steady growth occurs on the seed, producing a tough crystal and without the generation of excess new nucleii. Following precipitation, the hydrate is classified hydraulically, the coarsest material being collected for calcination, and the remainder being recycled as seed.

During the calcination stage the water of hydration is driven off and a phase change occurs in the alumina. The degree of calcination is controlled to produce a non-absorbent alumina with handling characteristics satisfactory to smelters. The main physical properties affecting handling characteristics are surface area, particle size and angle of repose.

Following classification and removal of the precipitated alumina trihydrate, the spent liquor is returned for further bauxite digestion.

In view of the importance of recovery of alumina-rich liquor from the residues and the considerable amount of wash water which is required to achieve this, evaporation of a balancing amount of process liquor is an important feature of the process.

The residue after being washed essentially free of liquor is pumped to disposal areas.

Aspects of the refining process peculiar to Kwinana and Pinjarra

Darling Range bauxite has individual characteristics of grade and composition which require modifications to the Bayer process as described below.

Grade

Typical bauxite grade of 30-35 per cent is low by world standards. One obvious consequence of this is the considerable magnitude of the materials-handling operation in conveying ore to the refinery, blending, stockpiling and conveying it to the digesters.

Approximately 3 to 4 kg of bauxite are required to produce 1 kg of alumina. Consequently for a 1.4-million-tonne-per-annum alumina production at Kwinana, more than 5 million tonnes per annum of bauxite must be processed, and for 2.4 million tonnes per annum at Pinjarra up to 9 million tonnes of bauxite are required.

Residue treatment and disposal are likewise massive operations by comparison with similar refineries elsewhere. For each kilogram of alumina, approximately 2 to 3 kg of residue are produced. The usual ratio for alumina to residue is 1 to 1.

Thickening operations are complicated by the large amounts of red mud in contact with the green liquor and its tendency to initiate premature precipitation. This makes the risk of alumina loss with residue worse than generally experienced with higher grade ores. A balance has to be struck between the risk of alumina losses owing to premature precipitation arising from too long a contact time and the risk of loss arising from a short settling time, and liquor entrainment in the thickened mud.

Because large and costly losses of caustic soda and sodium aluminate can occur owing to small variations in residue-washing efficiency, considerable attention is given to maximizing underflow densities and wash-water rates.

The final stage in residue treatment is disposal. This is likewise a large and costly operation. As with mining, environmental considerations play a major part in determining the manner in which this is carried out. From an economic point of view it is desirable that the disposal area be near the refinery. However, competing potential land users, and the unavailability of suitable topographical and subsoil features, can dictate that such disposal areas be relatively remote.

As with the mining operation, the competing needs of various community interests require that such residue areas ultimately be restored for community use. Such restoration may involve ground compaction to produce bearing-load characteristics suitable for housing and industry, topsoil development and re- establishment of vegetation.

Free silica and iron in bauxite

Free silica occurs in the form of quartz grains ranging in size from 50 to 100 mesh. The amount of such free silica varies from about 18 to 22 per cent, and iron, which occurs mainly as goethite ($Fe_2O_3.H_2O$), ranges from 16 to 20 per cent.

Both free silica and iron act as dilutents of the ore and play little or no part in the chemistry of the refining process. They do, however, have a profound effect on the abrasive properties of the ore and its residue. The free silica in particular has produced severe abrasion problems. It is estimated that one-quarter of the total maintenance costs can be attributed to abrasion.

Reactive silica

Silica-containing minerals which are soluble in the process liquor yield reactive silica. In Darling Range bauxite this is mainly concentrated in the kaolinite and ranges from about 1 to 2 per cent. Reactive silica content is important for the effect it has on final alumina purity and on soda and alumina losses.

Satisfactory desilication of the process liquor, among other things, depends on an adequate reactive silica level in the bauxite to get the auto-catalytic reaction started. Usually reactive silica levels must be at least 1 per cent to ensure sufficient desilication. On the other hand, higher levels—for example 2 per cent—result in costly losses of caustic soda and alumina.

Although the reactive silica content of the Darling Range bauxite is very moderate by world standards, in view of the low alumina grade, caustic and alumina losses can be considerable for small changes in reactive silica content. It is therefore necessary that bauxite grade control with respect to reactive silica level be very precise.

Organic impurities

Darling Range bauxite, in common with most shallow deposits, has a relatively high organic content. This organic material is derived largely from decomposed vegetation and roots, and consists of many substances, such as humus, lignin, cellulose and protein. Under the alkaline oxidative conditions existing in the Bayer process, these complex organic compounds break down through a series of stages to simple compounds such as the sodium salts of succinic, acetic and oxalic acids and carbon dioxide. Predominant among these salts is sodium oxalate, which tends to reach saturation in the plant liquor and co-precipitates with alumina trihydrate, causing the generation of excessive amounts of contaminated fine hydrate.

Carbon dioxide reacts with the caustic soda in the liquor to form sodium carbonate causing increased losses of caustic soda and the need for lime addition to the liquor to regenerate caustic soda.

Conclusion

Darling Range bauxite by world standards is of low grade and possesses a number of characteristics which pose special problems in mining and refining. However, the fortunate location of the deposit and the rapid development of specialized processing knowledge have enabled this bauxite to become a source of alumina competitive in both quality and price on the world market.

5D

Underground Mining

R. A. NICHOLS
Western Mining Corporation Ltd

Although there has been great variety in the mineral products of Western Australia, gold dominated the mining industry in terms of tonnage of ore produced, persons employed and value of production until the late 1960s. Since that time, there has been a great increase in the value of mineral production, with iron-ore, bauxite, nickel and petroleum becoming significant contributors and with the contribution from gold declining. Changes in total contributions to the state's mineral production over recent years are shown in Fig. 5.5 which also highlights the role of underground mining in relation to total mining.

Clearly a description of underground-mining methods in Western Australia must be heavily biased towards gold- and nickel-mining. From the large number of mines operating under a wide range of geological environments, examples can be selected to illustrate an extensive range of underground-mining techniques.

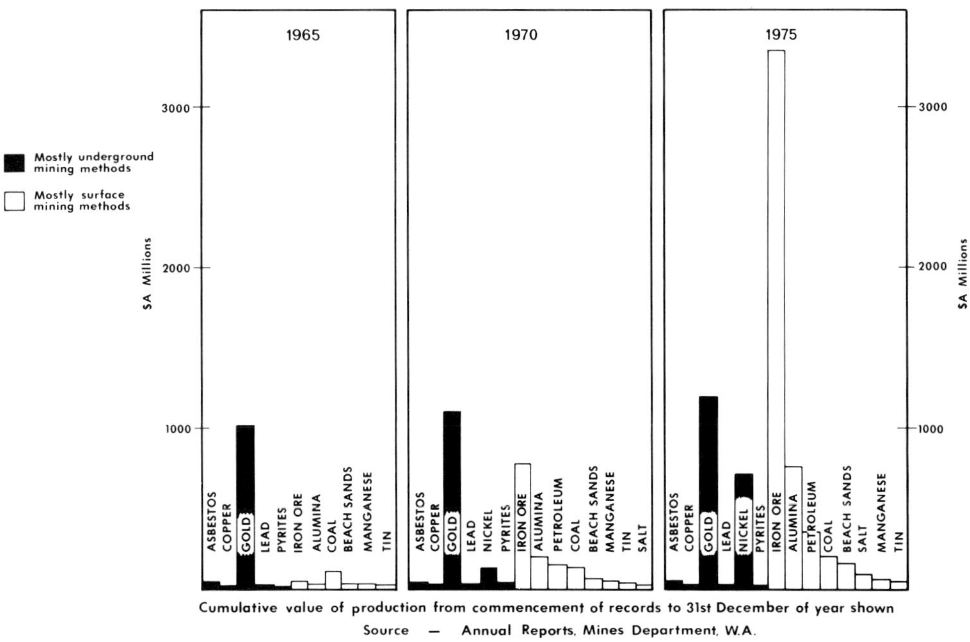

Fig. 5.5 Total value of Western Australian mineral production by underground and surface operations

The choice between surface or underground-mining methods is determined principally by the orebody to be mined. The physical nature of the orebody, its size and geometric shape, its attitude and depth below the surface, and the rocks which surround the orebody, all have an important bearing on the type of operation best-suited to its extraction. The choice of methods today is also very much influenced by the types of equipment available and adaptable to the proposed mine and the availability and skill of the workforce necessary to operate the mine.

The choice of underground-mining in the early days of gold-mining in this state was made on urgent and compelling reasons. The easily-won alluvial gold was

Table 5.2

Major Western Australian underground mining operations
(Data from W. Aust. Mines Dept)

	CURRENT OPERATING MINES (Production to 31 December 1977)	*Total production* *Ore ('000 Tonnes)*	*Gold (Kg)*
Gold	Kalgoorlie Mining Associates—		
	Mt Charlotte, Kalgoorlie	7 630	33 202
	Central Norseman Gold Corp. (NL)—		
	Norseman	5 891	82 174
Nickel	Western Mining Corp. Ltd—		
	Kambalda Nickel Ops—Kambalda	10 022	NA
	Silver Lake Mine	3 244	
	Otter Juan Mine	2 803	
	Durkin Mine	2 353	
	Fisher Mine	610	
	McMahon Mine	601	
	Hunt Mine	219	
	Jan Mine	163	
	Edwin Mine	28	
	Western Mining Corp. Ltd—		
	Mt Windarra	1 234	NA
	Metals Exploration Ltd—		
	Nepean	570	NA
	Anaconda Aust. Inc.—Redross	415	NA
	Selcast Exploration Ltd—		
	Spargoville	351	NA
	Agnew Mining Co Pty Ltd—		
	Leinster	To commence 1978	

	MAJOR MINES OF THE PAST	*Total Production* Ore ('000 Tonnes)	Gold (Kg)
Gold	Kalgoorlie Mining Associates—Fimiston		
	Closed 1976:		
	Ex Gold Mines Kalgoorlie Leases	23 909	256 985
	Ex Lake View & Star Leases	39 361	482 215
	Hill 50—Mt Magnet		
	Closed 1976	3 220	42 207
	North Kalgurli Mines Ltd—Fimiston		
	Closed 1975	14 185	162 800
	Great Boulder Mines Ltd—Fimiston		
	Closed 1969	17 075	219 529
	Sons of Gwalia—Leonora		
	Closed 1963	7 143	80 230
	Great Western Cons.—		
	Copperhead—Bullfinch		
	Closed 1963	3 373	14 243
	Big Bell Mines Ltd—Cue		
	Closed 1955	5 627	22 720
	Lancefield Mine—Laverton		
	Closed 1940	1 524	16 242
	Wiluna Gold Mines—Wiluna		
	Closed 1947	7 463	41 511
	Great Fingall Cons.—		
	Day Dawn		
	Closed 1914	1 896	36 870
Nickel	Western Mining Corp—Great Boulder Ops		
	Closed 1977:		
	Scotia	812	NA
	Carr Boyd Rocks	196	NA
Copper	Ravensthorpe Copper Mines NL—Ravensthorpe		
	Closed 1971	811 approx.	52 004 tonnes concentrate
Asbestos	Australian Blue Asbestos Ltd—Wittenoom		
	Closed 1966	2 750 approx.	153 247 tonnes fibre
Pyrites	Norseman Gold Mines—Norseman		
	Closed 1968	3 081	891 132 tonnes concentrate

rapidly being exhausted, there was neither the equipment available to work sufficiently-large open cuts to allow the deep ore to be won, and the ground holdings were not large enough to allow a large open cut. It was apparent that the gold lay at depth in the hard rock; and there was, to say the least, an adequate pool of willing labour.

Underground-mining operations entail several major fundamental activities and aspects quite different from those in surface mines:

1. All men, supplies and equipment must enter the mine through a special access way, which itself must be excavated to penetrate the earth and through which all the ore is hoisted to the surface.
2. Additional openings must be made from this major access way to the extremities of the orebody at a number of horizons to facilitate its extraction.
3. Special provision must be made for an adequate supply of ventilating air and lighting.

MINE ACCESS

The main access to an underground mine is probably the most important facility of the mine. It makes possible all the underground work necessary to mine the ore, and provides the means of hoisting the ore to the surface as well as the means of entry for the men and necessary equipment.

Table 5.3 demonstrates the variety of methods adopted in providing main access to Western Australian mines over the years.

Historically, shafts have been the accepted means of entry to Western Australia's mines, with vertical shafts greatly outnumbering inclined shafts. Rectangular shafts, supported and divided by timber sets, and having dimensions approximately 4.27 m to 4.88 m long by 1.83 m wide, have been the common or 'goldfields-type' shaft in this state.

Two notable shafts larger than the usual shafts were the North Kalgurli main shaft (8.18 m × 2.08 m) and the Copperhead Shaft of Great Western Consolidated (8.56 m × 2.07 m). Both of these shafts were divided into five compartments that provided for a large cage with a counterweight, two skips and a ladderway/pipeway, instead of the three compartments in the usual goldfields-type shaft. They were sunk to 737.6 m and 570 m respectively. The Chaffers Shaft of Lake View and Star Ltd, together with the internal shaft sunk from the 1076 m level to 1270 m, provides the deepest mine access in Western Australia.

Rectangular timbered shafts are prominent also in current operations. Western Australian hardwood timbers continue to be ideally suited for shaft work because of their strength and durability. The introduction of steel members to replace timber in shaft sets in some recent rectangular shafts has been a new development. The first of these were of typical goldfields-type dimensions and were sunk for the Anaconda/CRA partnership at Redross and Wannaway.

A new standard rectangular shaft with steel sets was sunk by Western Mining Corporation at its Jan Mine at Kambalda. This shaft has six compartments with dimen-

Table 5.3

Main access methods in current and major past underground mines in Western Australia

SHAFTS				*ADITS*	
Vertical shafts			Inclined shafts	Declined	Horizontal
Rectangular		Circular	Rectangular		
Timber sets	Steel sets	Steel sets	Timber sets		
Silver Lake (WMC-KNO) Scotia (WMC-GBO) Carr Boyd (WMC-GBO) Nepean (M. Ex.) Andrews (S.T. Spargoville) Reward and Man and Supply (K.L.V. Mt Charlotte) Chaffers (K.L.V. Fimiston) Blue Spec (Mulga Mines) Perseverance (K.L.V. Fimiston) Hill 50 Copperhead (Great Western) Big Bell Ravensthorpe Copper Mine Mt Edwards* (BHP/Inco) Norseman Gold Mines	Jan (WMC-KNO) Redross (Anaconda) Long* (WMC-KNO) Victor* (WMC-KNO) Wannaway (Anaconda)	Durkin (WMC-KNO) Agnew* (Agnew Mining)	Regent (CNGC) Princess Royal (CNGC) Edwin (WMC-KNO) Sons of Gwalia Lancefield	Otter/Juan (WMC-KNO) McMahon (WMC-KNO) Fisher (WMC-KNO) Hunt (WMC-KNO) Mt Windarra (WMC-Windarra) Agnew* (Agnew Mining) Spargoville 5B (S.T.)	Wittenoom (Aust. Blue Asbestos)
		By Raisebored method	*Inclined off vertical shaft*	*Declined from within shaft mine*	
		Otter/Juan (WMC-KNO)	Great Fingall (Vertical to 1169 ft then 59°)	Mt Charlotte (K.L.V.) Blue Spec (Mulga Mines) Nepean (M. Ex.)	

* Being sunk or not yet in production.

Key to abbreviations

WMC	Western Mining Corporation	M.Ex	Metals Exploration
KNO	Kambalda Nickel Operations	S.T.	Selection Trust
GBO	Great Boulder Nickel Operations	CNGC	Central Norseman Gold Corporation
K.L.V.	Kalgoorlie-Lake View	BHP/Inco	Broken Hill Pty Ltd/International Nickel Australia Ltd

sions, outside steel, of 6.013 m × 3.17 m, and was sunk to 619 m. Similar shafts are being sunk at Kambalda on the Long and Victor orebodies.

Circular shafts with a concrete lining have found little favour in Western Australia. Two 4.27-m-diameter shafts at the Durkin Mine, Kambalda, are the only operating circular shafts in this state. A major circular shaft is being sunk at the Agnew Nickel Project for the Selection Trust/Mt Isa Mines Ltd Joint Venture. This shaft is 7.4 m in diameter and will be sunk to 900 m.

Inclined shafts have been sunk at angles other than vertical to suit the attitude of the particular orebody to be mined. Inclinations have ranged from 22½° in the current Edwin Shaft at Kambalda and 41° and 46° at the Regent and Princess Royal Shafts at Central Norseman Gold Corporation (Fig. 5.8) to 59° in the lower part of the old Great Fingall Mine at Day Dawn.

Traditional drill-blast-muck shaft-sinking methods have been used in almost all shafts in the state. Fundamental techniques have changed very little, although equipment has been developed considerably and much of the hard manual work previously associated with shaft-sinking has been eliminated.

Mucking, or removal of the broken rock from the shaft bottom, has now been fully mechanized with the use of various types of grabs and mechanical mucking machines. The first application of the Cryderman Shaft mucker in Australia was in the Royal Shaft of Central Norseman Gold Corporation in 1958. Similar units have since had wide use in the rectangular shafts that are so common in this state, and to which they are ideally suited.

Mechanization of rock-drilling is only a recent development. A simple system of anchoring normal retractable airleg type rock-drills to a chain stretched taut across the shaft has proved particularly effective in improving drilling performance in the newest shafts at Kambalda. The shaft being sunk at the Agnew Nickel Project provides the first application in Western Australian shaft-sinking of a hydraulic jumbo mounting four large pneumatic rockdrills.

Most shafts in this state serve dual functions of ore-hoisting and mine-servicing. In most instances only one winder is used requiring ore-hoisting skips to be rapidly interchangeable with cages for transport of men and supplies. At Kalgoorlie notable exceptions to this generalization were the North Kalgurli main shaft, where separate facilities were provided, and at Mt Charlotte, where separate shafts are employed for ore haulage and for men and supply.

At Kambalda shaft design has evolved from the typical goldfields-type Silver Lake Shaft (which was subsequently provided with separate men-hoisting facilities), to the twin shafts on the Durkin orebody, and then to the six compartment shafts with separate ore-hoisting and men-hoisting facilities on the newer Jan, Long and Victor orebodies. Surface facilities for these newer Kambalda shafts are as shown for the Juan Raisebored Shaft (Fig. 5.6).

Virtually all winder engines are now powered electrically (diesel and hydraulic winders in remote locations being the only exceptions), although many of the winders at present in use began their life under steam power. As some measure of the sizes of winding installations, winder horsepower has ranged up to 1000 HP on the

Fig. 5.6 Juan shaft, Kambalda: headframe and surface facilities

Golden Mile, and 500 HP is common among the current nickel mines, with 1600 HP being installed on the Long Shaft at Kambalda. Skip capacities ranged from 2.5 tonnes to 6.3 tonnes on the Golden Mile, and 5 tonnes capacity is standard at Kambalda, with smaller capacities at the smaller nickel mines. Hoisting speeds ranged up to 12 m/s at the Perseverance Shaft on the Golden Mile, and will be equalled at the Long Shaft, Kambalda. Elsewhere among the current nickel mines, 6.35 to 7.62 m/s is commonplace.

A horizontal or near-horizontal mine entry is called an adit. The declined adit, allowing the use of diesel trucks underground, has become a significant and increasingly-important mode of entry in recent years. This method is particularly suitable to the development of orebodies close to the surface, but it has proved to be so valuable that declines have been extended to considerable depths. At the Otter/ Juan Mine at Kambalda the main decline extends from the surface to a depth of 472.5 m, and is used, in conjunction with an ore-hoisting shaft, to provide access into the mine for men, machines and equipment. Declines have also been used at the Mt Charlotte, Nepean and Blue Spec Mines to mine ore below the bottom levels of existing shafts.

Declines provide the quickest means to achieve production from close-to-surface orebodies for the minimum capital investment. Only fifteen weeks after cutting the portal, production was derived from the Otter orebody at Kambalda in 1969.

Most declines used for ore haulage have a gradient of 1 m fall for each 9 m of advance, although some sections of decline at Mt Charlotte were developed at 1:7 and access to the 5B orebody at Spargoville was provided by Selcast Exploration Ltd at 1:5.

MINE DEVELOPMENT

Mine development is a function providing access to the orebody in order to mine it, and thus embraces the shafts and declines already described. More commonly, however, the term is used to refer to the horizontal and vertical openings within the mine. Essentially, development is a repetitive drill-blast-muck operation of advancing an opening into solid rock. As development provides the first opening or entry into a particular area, it makes especial use of explosives in sufficiently large concentrations to create this opening. Depending on the type of development, parallel or inclined blast-holes are drilled into the rock-face ahead of the opening. The explosives detonated within those holes cause the rock between the holes to be shattered and ejected so that it can be mucked out to expose a new face advanced some distance from the first.

Mines in Western Australia have been traditionally developed by a series of horizontal openings at more or less regular intervals called 'levels'. Levels, in effect, subdivide the orebody into portions to be mined by a particular method, and the interval between them is determined by the mining method. Levels also provide the access way through which the ore mined will be transported out to the hoisting shaft. The process of advancing these levels is called 'driving' and the openings so created are called 'drives' (Fig. 5.8), or 'cross-cuts' if the drive cuts across the long dimension of the orebody at that level.

The drilling operation in level development has seen very significant advances in technology from essentially manual 'hammer and tap' methods to pneumatic rock-drills of increasingly-mechanized design. Hydraulic rock-drills have provided the most recent advance in drilling technology. To date only the new Agnew Nickel Project has adopted hydraulic rock-drills as normal equipment.

Until the mid-thirties mucking was carried out by hand shovels, but mechanical mucking units are now in universal application. Where railed equipment is used, pneumatically-driven rocker shovels are used to muck into Granby-type trucks. Diesel-operated front-end loaders and trucks are in use in trackless development.

A recent example of a major drive using railed equipment was the cross-cut from the Durkin Shaft at Kambalda to the Gibb ore position:

size of drive:	3.0 m × 2.7 m; length 1910 m;
rate of advance:	95 m/4-week period;
productivity:	0.6 m/man-shift;
drilling equipment:	Atlas Copco mini-rail 3-boom jumbo mounting three Cop 89 rock-drills;
mucking equipment:	Atlas Copco LM70 mechanical loader;
trucking equipment:	8-tonne Gemco loco, 3.4 m^3 Granby trucks; on 0.762 m-gauge track.

A very significant amount of vertical development is necessary for interlevel travelling ways, airways and orepasses, for exploring the orebody and as integral parts of stoping (ore-winning) operations. The direction of advance of the opening significantly affects the development technique. Movement upwards from a particular level towards the level above is known as 'raising', and the opening produced is called a 'raise', or 'rise' (Fig. 5.8); movement towards the level below is known as 'winzing' and the opening is called a 'winze' (Fig. 5.8).

In earlier years, winzing comprised the predominant method of vertical development. Mechanization of the winzing operation has not proved to be a practical reality, with the result that nowadays labour is unwilling to engage in this operation. Winzing is practised today only when the same result cannot be achieved by raising.

In raising, gravity takes care of the removal of the broken rock. The conventional method of raising is now the 'gig' method, which developed from the raises of earlier days. A platform, or gig, is supported on a hoist rope which passes through a pilot drill-hole from the level above. The raise miner is hoisted to the work-face by a hoist on the level above. The gig and hoist rope are removed before blasting takes place.

The development of vertical openings by the *raiseboring* method is probably the most dramatic move forward in mining technology in the history of hard-rock underground mining in Western Australia. Raiseboring provides the first practical application of a fully mechanized, continuous method of excavation suitable for wide use in hard-rock mines.

Since the first raise drill in Australia was commissioned at the Silver Lake Mine, Kambalda, in February 1969, the method has been extensively used in Western Australia as shown in Table 5.4.

To drill a raise by the raiseboring method, access to both top and bottom of the proposed raise is required. The raise-drill machine is sited on the upper level, and by means of conventional rotary drilling technique drills a pilot hole (approximately 30 cm in diameter) to the level below. On the completion of the pilot hole, a large reaming-bit, equipped with a number of rolling cutters, is attached to the drill rods. The capacity of the raise-drill machine is sufficient to exert a massive torque to rotate the reaming-bit and to apply the optimum loading to each of the cutters as it is pulled upwards.

Tungsten carbide inserts are necessary in the pilot hole bit and the reaming cutters

Table 5.4

Vertical development by raiseboring
(All measurements in metres)

Year	*MINE*							*Total raise-boring*	*Total vertical development in W.A.**
	Kambalda	*Mt Windarra*	*Mt Charlotte*	*Spargo-ville*	*Nepean*	*Scotia & Carr Boyd*	*Redross*		
1969	992							992	4 972
1970	1 595				625	71		2 291	11 356
1971	2 202				219	126		2 547	5 328
1972	2 080	196		77	109	111		2 573	5 803
1973	1 904	289	289	78			289	2 849	4 890
1974	1 880	840	1 346	419	150			4 635	9 775
1975	1 846	842	273	366			728	4 055	8 550
1976	1 708	933	209	322			80	3 252	7 072
1977	1 298	1 024	254	164				2 740	7 077

* Rising and winzing only; does not include shaft-sinking.

when drilling in the hardest rocks. In the softer serpentinites, disc cutters have been used with great success. The cuttings of rock produced by the reaming operation fall down to the lower level, where they are mucked out and removed by the normal mine haulage system.

The resultant raise is a smooth-walled cylindrical opening up to 2.4 m in diameter and with a maximum length to date of 389 m. The rate of advance of raises by this method is many times faster than conventional raising, especially in the case of longer raises. The method is obviously safer than conventional raising, for no miners are required to be exposed to the hazards of a normal raise and the need for competent raise miners is eliminated.

While the cost of raiseboring is being continually improved by better equipment and techniques, it is generally higher than for conventional raising. The cost is, however, more than offset by the speed and safety and by the superior result.

A particular noteworthy application of raiseboring occurred in 1974 at the Otter/Juan Mine, Kambalda, in the drilling with a Dresser 800 strata borer of a 340 m ore-hoisting shaft. Particular care in drilling the pilot hole resulted in a deviation of only 2.4 m by limiting advance to an average of 1.3 m per hour. Reaming of the pilot hole to 2.4 m diameter was achieved at an average rate of 0.43 m per hour, with the whole reaming operation being completed in six weeks. The 2.4-m-diameter raise was subsequently equipped with steel sets to accommodate two 4-tonne-capacity skips which hoist ore from the Juan 9 level crusher at approximately 140 tonnes per hour.

Future raise-drilling capacity at Kambalda will be very greatly increased by the purchase in 1978 of a Robbins Model 121R raise-drill designed to drill raises to 900 m at 3.66 m diameter in the rocks encountered at Kambalda.

STOPING

Stoping is the name given to thc operation of actually mining the orebody, i.e. the productive operation. No universal stoping method may be applied to orebodies at random. Each orebody is unique, with its own particular situation, and must be mined by a method which achieves maximum extraction efficiency with the maximum of safety and security for that situation. As shown in Table 5.5, a wide variety of methods have been adopted for use in Western Australian mines. Historically the most important stoping methods have been those adopted by the mines of the Golden Mile.

Shrink-stoping

This method has had wide application in the mines of the Golden Mile throughout its history, and still accounted for more than 30 per cent of production in the final years of operation of the major mines.

Shrink-stoping is suited to relatively steep orebodies of narrow to moderate widths, and of a generally tabular nature, enclosed between walls of strong country rock. In operation, it entails the blasting of successive horizontal slices of the

Table 5.5

Principal stoping methods in major mines in Western Australia—past and present

UNFILLED OR DELAYED FILLING METHODS				*FILLED METHODS*			
Shrink Stoping	*Up Dip Open Stoping*	*Longwall Stoping*	*Sub Level Stoping*	*Cut & Fill Stoping*	*Rill Stoping*	*Resuing*	*Sub Level Caving*
Lake View & Star Ltd. Gold Mines of Kalgoorlie North Kalgurli Ravensthorpe Copper Mine Great Western-Copperhead Scotia	Kambalda Nickel Ops. Central Norseman Gold Corp.	Aust. Blue Asbestos	Mt Charlotte Mt Windarra Carr Boyd Rocks Scotia Spargoville (S.T.) Big Bell Sons of Gwalia Great Western-Copperhead Norseman Gold Mines	Lake View & Star Ltd. Gold Mines of Kalgoorlie North Kalgurli Sons of Gwalia Kambalda Nickel Ops. Spargoville (S.T.) Redross Nepean Mt Charlotte Mt Windarra	Lake View & Star Ltd. Nepean	Nepean	Scotia Mt Charlotte Mt Windarra

orebody, commencing from the lower level, into the space made by the blasting of the previous slice. Most of the ore so broken is retained within the stope itself to provide a platform from which the miner is able to work to drill out the next slice. Because the ore occupies a larger volume after it is broken by blasting than it did when it was in the unbroken state, a proportion of the broken ore must be drawn out of the stope as mining proceeds to ensure that sufficient space remains for the miner to continue working. When all the ore has been blasted, approximately 35 per cent will have been removed as stoping progressed leaving approximately 65 per cent as broken ore within the stope. This ore, no longer required by the miner as a working platform, is then drawn off. Generally the void left after the broken ore has been removed remains unfilled, with no risk of instability in the very strong ground where this method is used.

Cut-and-fill stoping

Several variants of a cut-and-fill stoping system have been used on the Golden Mile, but common to all has been the placement of an introduced filling material after mining. The fill gives support to the weaker stope walls as well as providing the working platform for the miner, in the same way that the broken ore does in shrink-stoping.

Rill-stoping

This is a system of cut-and-fill mining in which slices of ore are blasted off in a sequence which moves generally upwards from one level to the level above. In detail, however, the stope faces are advanced downwards at an inclination which parallels the angle of repose of the dry gravel filling-material used to fill the stope. The broken ore gravitates down the rill of filling into chutes to the level below. After a face has been advanced down the full length of the rill, more filling is introduced from the level above to take up most of the void created by mining. The process is repeated until all the ore between two levels is removed and replaced by fill material.

Horizontal cut-and-fill stoping

This method became increasingly important in the mining of the Golden Mile as the depth of operations increased, with a consequent increase in rock stress and the need for more adequate wall rock support and control.

This mining operation is basically similar to shrink-stoping insofar as the ore is broken in horizontal slices, or flitches, upwards from the lower level. However, the ore broken by each firing now rests on previously placed filling material, from which it is scraped into a chute that passes through the filling to the level below (Fig. 5.7).

In Kalgoorlie, deslimed mill-tailings are used for filling, but in the current nickel mines, where this method is widely used, windblown dune sand is used for filling. In all cases the fill is placed hydraulically.

Drilling is normally carried out by light rock-drill/airleg combinations or with

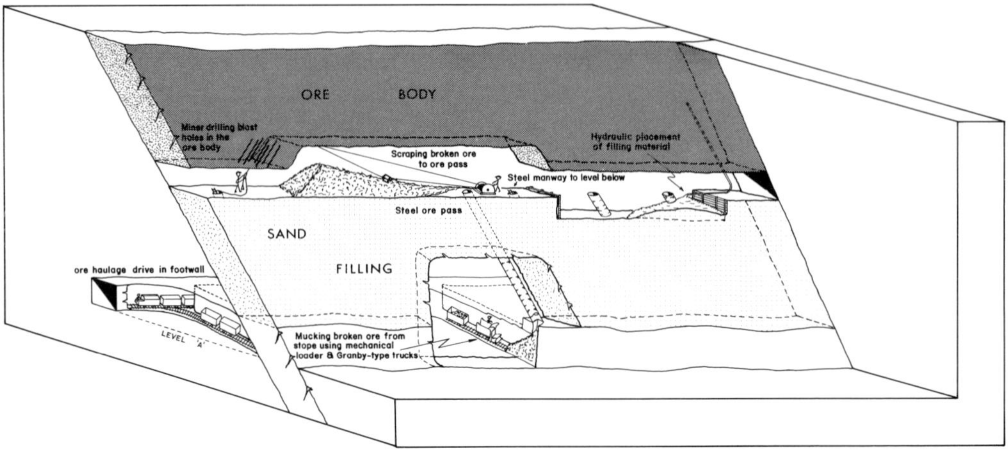

Fig. 5.7 Cut and fill stoping, Kambalda

stoper-feed rock-drills. Ammonium nitrate fuel oil (ANFO) is the normal explosive. Broken ore is normally scraped to the ore chute by pneumatic or electrically powered scraper hoists

A large-scale mechanized cut-and-fill stoping method was used to mine the ore above No. 5 level at Mt Charlotte prior to 1964. This operation was the first to use diesel-powered equipment underground in Western Australia.

Updip open stoping

This method is currently in use at Central Norseman Gold Corporation (Fig. 5.8) and is the source of approximately 35 per cent of mine production from Kambalda, where the method is called 'slot stoping'. It is suited to the mining of relatively-narrow tabular orebodies of dip less than approximately 40°. Mining proceeds up the dip of the orebody, leaving random pillars of ore where necessary to support the hanging wall.

Sublevel stoping

Sublevel stoping is the largest-scale, lowest-cost underground mining method used in Western Australia. It was introduced into the state by Big Bell Mines Ltd in the late 1930s. Since then it has been used wherever orebodies are large enough and conditions are suitable.

The method is designed to achieve production at the highest rate commensurate with the size of the orebody by permitting large tonnages of ore to be broken with each blast. Sublevels were developed between the main levels of the mine at intervals determined by the limits of accuracy of the drilling method in use. Initially diamond drills were in use, allowing a maximum sublevel interval at Big Bell of 18 m. Little variation in sublevel intervals resulted from the introduction of lower-cost, long-

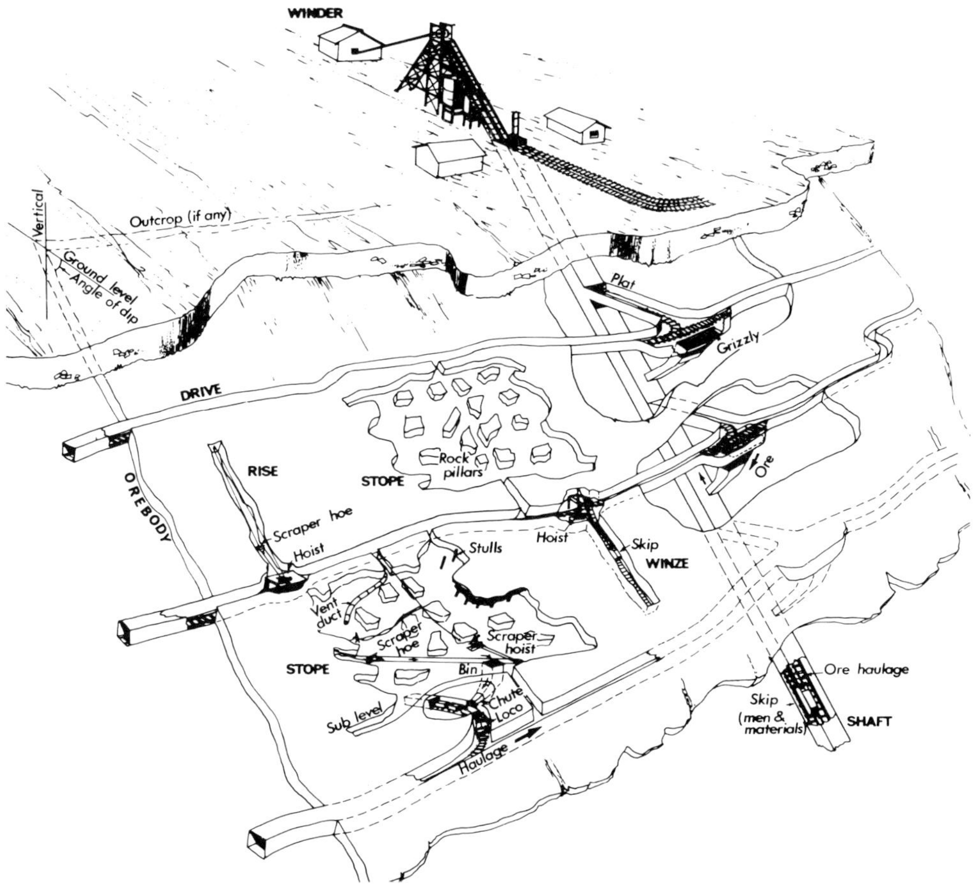

Fig. 5.8 Updip stoping and workings of Central Norseman Gold Corporation

hole percussion drilling; but the more recent development of large in-the-hole drilling techniques has virtually eliminated the need for sublevels by making practical an interval equivalent to the normal mine-level interval.

In its original concept, the method was intended to be an open stope technique, which depends upon strong wall conditions remaining open without any artificial support. Recent innovations have adapted the technique to less favourable ground conditions by incorporating filling into the sequence of mining. To achieve this, a stope is mined out in two stages—a true open stoping of relatively small primary stopes followed by the simultaneous blasting of the intervening rib-and-crown pillars. This fills the entire stope void with broken ore, which provides a support to the stope walls. As the ore is drawn off, this support is maintained by filling material which is added above the stope and which follows the ore down, ultimately filling the stope as all the ore is removed.

ORE-HANDLING

The handling of broken ore within an underground mine is one of the most significant components of the mine-operating cost. While there has been a trend towards larger and larger equipment to reduce unit ore handling costs, the opportunity to use very large equipment does not exist in underground mines.

Diesel-operated loaders and trucks are now in common usage in decline mines and have effected significant improvements in the underground ore transport system: it is now possible to handle large lumps of ore, thus eliminating the major cause of bottlenecks in conventional systems, and to reduce the number of stages in the transport system. Commonly the loading of ore into trucks at the stope is the first time that the ore is handled, and once loaded it is transferred directly to the surface ore storage. Where diesel ore-haulage systems are used in conjunction with conventional shaft-hoisting systems, as at Mt Charlotte and Otter/Juan, underground crushing ensures trouble-free ore flow.

Diesel loaders and trucks specifically designed for underground use are common and are characterized by a low profile and high manoeuvrability (Fig. 5.9). Some standard front-end loaders have been modified to lower their profile, but the tendency now has been to use standard equipment.

Fig. 5.9 Diesel front-end loader, Fisher mine, Kambalda

In conventional rail-mounted mining operations, the trend to larger ore-haulage equipment has resulted in trains of Granby trucks, each of capacity up to 3.4 m^3, hauled by battery-powered locomotives up to 8 tonnes on 76.2-cm-gauge track.

HUMAN RESOURCES

Western Australia has been blessed not only with extremely valuable mineral resources but also with highly-competent, hard-working people who have made their mining possible. The miners of this state have been, and continue to be, rugged individualists, highly competent and vigorously independent and self-sufficient. Without such men as these, the course of mining in this state could not have progressed as it has. In addition, the technical and management people in the Western Australian mining industry over the years have been remarkable for their competence, vision and dogged devotion to the task while bringing innovative as well as improvising thought to the widely-varying prevailing conditions. All have worked and co-operated together in remote and harsh conditions to forge what has really become a mining community in this country.

5E

Concentrating, smelting, refining

A. E. O'Meara
Western Mining Corporation Ltd

(a) GOLD AND SILVER

There has been a steady decline in gold production in Australia and in Western Australia since 1910, as the richer and more accessible deposits were worked out, until today there are only three operating companies in this state and a small number of independent workers who use state batteries to process tonnage parcels of their ore.

BRIEF HISTORY OF WESTERN AUSTRALIAN GOLD METALLURGY

There are very few true alluvial gold deposits in the state, most of those that are classed as such being eluvial, i.e. are superficial concentrations by wind and water close to the source rock. The primary deposits in the state vary widely in their type and mineral content, ranging from jaspilites at Hill 50, altered quartz dolerites with gold and tellurides at Kalgoorlie, to deposits containing arsenic, lead and antimony sulphides as at Wiluna and Blue Spec (Nullagine). These diverse modes of occurrence created diverse treatment problems, some of which took years to solve, and, like the silver-lead-zinc metallurgical problems in the early days at Broken Hill, led to the invention and adaptation in Western Australia, and later throughout the world, of new processes for treating gold and other ores.

When payable gold was discovered in Western Australia in 1885, stamp batteries were universally used for its recovery. Treatment plants consisted of primary crushers, which reduced the ore to about 75 mm or less. The ore was then fed underneath the reciprocating stamps. Crushed ore had to pass through a 30-mesh screen to leave the stamp-box, and then passed over amalgam plates and through amalgam boxes, where the coarse gold was caught. The pulp, still containing the finer gold, was then crudely classified into sands and slimes.

The slimes treatment problem

Sands and slimes were generally leached separately with cyanide solutions for gold recovery without much trouble, except for Kalgoorlie ores, where the colloidal nature of the slimes prevented them from being treated. The problem was solved by Lake View Consuls metallurgist, the late J. W. Sutherland, who succeeded in separating leached slimes from the cyanide solution by using a cloth filter press of the type then being used in the sugar industry. This was the first time anywhere in the world that cloth filtration had been applied to mineral pulps, and the process was universally adopted.

This major development, simple though it may sound today, meant that the way was then open to treat all Western Australian free-milling ores and get acceptable gold recovery—that is, all except the Kalgoorlie gold telluride ores, which presented yet another problem.

The gold-telluride problem

When mining reached the unoxidized sulphide zone in the Kalgoorlie area, poor recoveries were obtained, even with the addition of Wilfley tables (reciprocating riffle-surfaced sloping tables) to remove denser minerals. This was due to the presence of tellurides, which were first detected by A. G. Holroyd (1897). Holroyd showed that there were three different modes of occurrence of gold (and silver):

1. Native gold or electrum (a natural alloy of gold and silver). This was the free-milling gold.
2. Gold or electrum very finely intergrown with pyrites. This required special treatment.
3. Gold and silver present as the rare tellurides calaverite $AuTe_2$ (with some silver); sylvanite $Au(Ag,Au)Te_4$; and petzite $AuAg_3Te_2$.

The telluride problem caused a period of intense metallurgical investigation in Western Australia and Germany, and two solutions were eventually found:

1. The Diehl process, named after the inventor who found a mixture of bromocyanide plus the usual sodium or potassium cyanide would dissolve gold from tellurides. Diehl also recognized another important fact, i.e. that his process would not dissolve gold intimately intergrown with pyrites. Therefore, in his treatment process, he retained Wilfley tables to remove pyrites, which was then roasted and the calcine leached with cyanide.

Diehl's process was therefore to crush and grind the ore, pass it over Wilfley tables to remove heavy minerals and separate it into sands and slimes with normal cyanide leaching of the sands and bromocyanide-cyanide leaching of the slimes. Most of the telluride minerals, being brittle and friable, accumulated in the slimes as a result of crushing.

2. The 'dry crush and roast' process, in which the ore was crushed, dry-ground in ball mills, then roasted. The calcine was further ground in amalgam pans and the resulting slimes cyanided. The newly invented filter presses were used to separate leached slimes from pregnant solution. Eventually sand leaching was abandoned for an 'all-slime' treatment scheme.

The German Krupp mill became favoured for dry grinding and the Merton and Ropp roasters eventually gave way to the Edwards roaster. The 'dry crush and roast' option eventually proved more favourable than the battery process with bromocyanidation. In the Brown Hill mill—a battery plant—flint pebbles were used as grinding media to grind the ore in cylindrical mills, the forerunner of the ball mill.

Today, a much simpler way of treating gold-telluride ores is to use a high-lime, cyanide leach process, but this was unknown in Diehl's day.

INTRODUCTION OF FLOTATION

The steady rise in the price of firewood for the all-roast process and the increase in moisture content of the ore from underground (owing to more water being used underground to help control silicosis) both generated further metallurgical testing to help reduce treatment costs and improve efficiency. The invention and successful full-scale application of the flotation process at Broken Hill was naturally considered and applied at the Kalgoorlie School of Mines metallurgical laboratory and pilot-tested in the Oroya Links-Brown Hill mill.

By 1930 the Lake View and Star Company at Fimiston was sufficiently satisfied with flotation results to change over portion of their main dry plant to flotation, and within a few years the whole plant was converted to wet ball-milling and flotation in closed circuit, with strakes to recover the coarser gold. At the outset the omission of strakes was the main reason for the early failure of the new treatment scheme. The introduction of flotation for gold and sulphides—another first for Western Australia—significantly increased gold recovery and eliminated the costly dry-grind all-roast process.

Flotation is the basis upon which all current gold ores are treated. In general the ores are crushed and wet-ground in a ball mill, followed in some cases by straking, i.e. washing over a sloping table covered by corduroy cloth, to collect coarse, free gold. The pulp is then subjected to froth flotation to float sulphides and any free gold.

The small weight of flotation concentrate is roasted to remove sulphur, and the calcine leached in a separate circuit with cyanide for gold recovery. In some plants, the flotation tailing is also leached for maximum gold recovery.

It is not possible here to give a full coverage of gold metallurgical developments and those interested should consult excellent historical reviews by Blaskett (1951/52) and Dunkin (1953).

The other major refractory gold-ore types in Western Australia are the gold-antimony ore at Blue Spec (Nullagine) and the most complex gold-arsenic-antimony-lead ores of Wiluna. At Nullagine gold occurs in association with stibnite, arsenopyrite and pyrites. Originally the ore was ground, straked and floated, the bulk concentrate being conditioned with cyanide and re-floated. The resulting antimonial and pyrite concentrates were shipped separately for sale. More recently the flotation tailing was cyanided on site for gold recovery. Operations have recently ceased.

At Wiluna the original plant in 1931 treated a gold-pyrite-arsenopyrite ore, grinding and floating it, and roasting the concentrate. This was followed by cyanidation of the calcine. In 1936 the plant was enlarged to treat ore from Moonlight Wiluna, this ore containing significant amounts of stibnite and some galena and tetrahedrite as well as the above minerals. The antimony and lead minerals made roasting difficult and a most complex metallurgical plant was designed and operated. It involved:

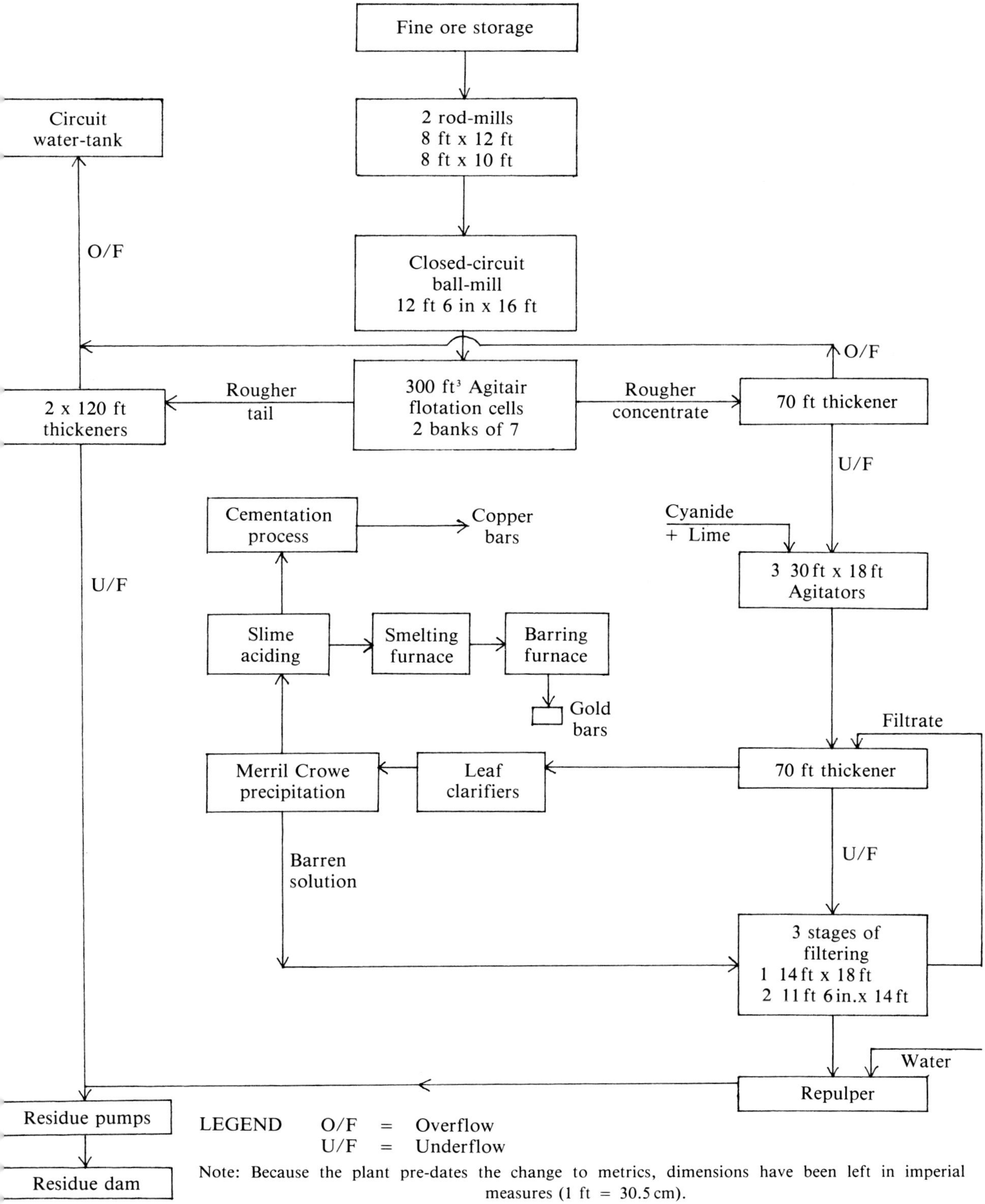

LEGEND O/F = Overflow
U/F = Underflow

Note: Because the plant pre-dates the change to metrics, dimensions have been left in imperial measures (1 ft = 30.5 cm).

Fig. 5.10 Treatment of Mt Charlotte gold ore at Kalgoorlie Oroya mill

(*a*) roasting the antimony concentrates to a clinker in a rotary kiln;
(*b*) roasting the lead-rich concentrates on a Dwight-Lloyd sinter machine;
(*c*) treating the combined calcines with fluxes in a blast furnace to produce antimonial lead, copper bullion, gold bullion and some zinc.

The three remaining Western Australian major gold operations are at Kalgoorlie, Norseman and Telfer. They are Kalgoorlie Mining Associates (Mt Charlotte), Central Norseman Gold Corporation N.L. and Newmont Pty Ltd.

Kalgoorlie Mining Associates—Mount Charlotte operations

The Mount Charlotte deposit is currently mined by an association of Kalgoorlie Lake View Limited and Homestake Australia Limited, and the ore is treated in the Oroya mill of the former Gold Mines of Kalgoorlie company. This mill is currently being modernized with new equipment and is treating 65 000 tonnes of ore per 28-day period.

The current treatment scheme is shown in Fig. 5.10 and consists of wet rod-and-ball milling followed by flotation to produce a gold/sulphide concentrate of about one-tenth the weight of the mill feed. The flotation tailings assaying less than 0.2 g of gold per tonne are now thickened and pumped to the storage dam without further treatment.

The flotation concentrate assaying 50-60 g of gold per tonne is thickened and cyanide-leached for gold recovery. A small amount of copper in solution is recovered after gold precipitation by cementation onto scrap iron.

The leached flotation concentrate, containing about 2 g of gold per tonne, is pumped to waste. Roasting is not currently practised on the Golden Mile.

Central Norseman Gold Corporation N.L.

Currently most of the ore is trucked from the Regent shaft to the mill, and a small tonnage comes from several open pits at the Princess Royal reef.

The treatment plant treats about 16 000 tonnes of ore per 28-day period by jaw crushing it to −90 mm, then to −13 mm by cone crusher before fine-grinding in rod-and-ball mills with barren cyanide solution. A noticeable feature of the plant is the mechanical endless belt strakes (invented in South Africa) which replaced the old corduroy cloth strakes, and which increased gold recovery and significantly reduced labour costs.

After straking to recover much of the free gold, the pulp is classified at 28 per cent solids into sands and slimes by hydrocyclone classifiers, the slimes going to cyanide leaching for gold recovery and the sands returning for further grinding. Occasionally a small quantity of litharge (PbO) has to be added to the mills to precipitate soluble thio salts derived from minor sulphides in the ores, otherwise gold precipitation by zinc dust becomes inefficient.

Fig. 5.11 is a flowsheet of the treatment scheme. There is no roasting at Norseman, nor flotation in the main plant, but some of the clay-rich ore from the Princess Royal reefs is floated with only a frother, during which most of the gold is concentrated into the froth.

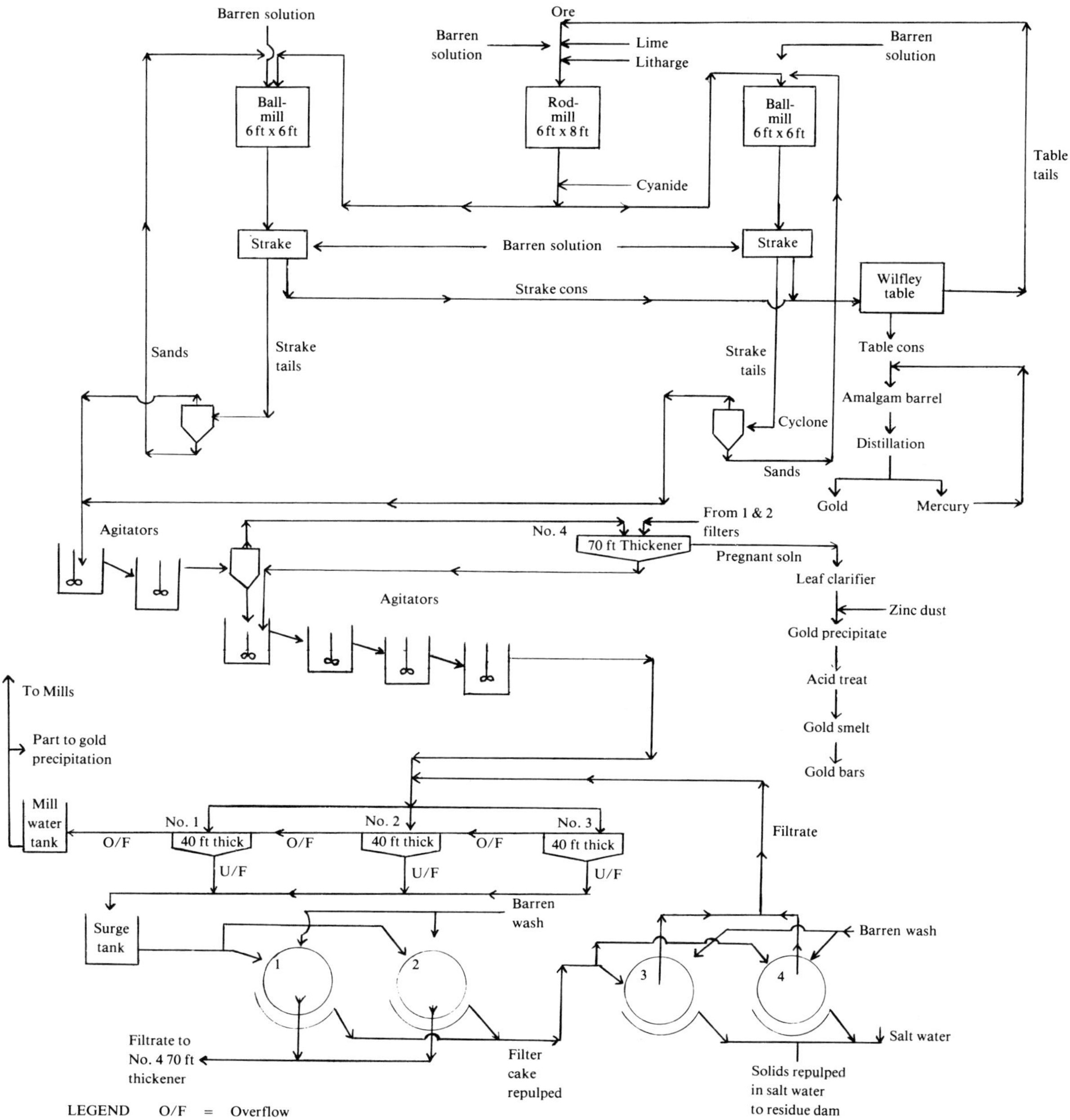

LEGEND O/F = Overflow
U/F = Underflow

Note: Because the plant pre-dates the change to metrics, dimensions have been left in imperial measures (1 ft = 30.5 cm).

Fig. 5.11 Gold-ore treatment, Central Norseman Gold Corp. N.L.

Telfer

Located in the Paterson Range on the edge of the Great Sandy Desert, about 270 km east-southeast of Marble Bar, is Newmont Proprietary Limited's gold deposit in stratabound and weathered quartz-limonite reefs containing 30 to 40 per cent of clay. The deposit was pegged on 23 May 1972. A plant has been built and is treating the 3.81 million tonnes of 8.7 g/tonne ore at the rate of 1200 tonnes per day.

Ore treatment is shown in the flowsheet of Fig. 5.12. Ore from the open-pit mine is jaw-crushed to minus 3 inch (−76 mm), cone-crushed to minus half inch (−13 mm), then wet-ball milled in leach solution and classified into sands and slimes. The sands are treated in jigs, the jig concentrate upgraded further by Wilfley table, and the table concentrate amalgam-treated for gold. About one-third of the gold is recovered by gravity concentration.

The slimes are thickened in a 100-ft (30-m) primary thickener, the overflow going to clarification and gold recovery, while the underflow is treated in flat-bottomed agitators, then by a four-stage counter-current decantation (CCD) train in 100-ft thickeners for liquid-solids separation. Overflow from the first CCD thickener is returned to the mill-head water-tank. Underflow from the fourth CCD thickener is pumped to the tailings dam.

(b) NICKEL AND COPPER

Although a number of copper orebodies have been worked in past years—for example, at Ravensthorpe, Whim Creek, Mons Cupri in the Pilbara and Whundo in the West Pilbara—all copper currently produced is as a by-product of nickel operations, 1831 tonnes having been produced for the year ended 31 December 1977.

In the future, copper may be produced again from the Ravensthorpe deposits and also from two new discoveries, namely the Teutonic Bore copper-zinc deposit near Leonora and the Golden Grove copper deposit 225 km east of Geraldton. At present, however, these deposits are still at the exploration or feasibility study stage, waiting until world copper demand improves.

NICKEL

In addition to copper as a by-product of the Western Australian nickel industry, cobalt and the precious metals gold, silver, platinum and palladium are also recovered. However, the value of nickel produced is much greater than that of the by-products, and at present nickel ranks only below iron-ore and alumina as an export revenue earner for this state.

In the 10 years between 1966 and 1976 Western Australia changed from having no nickel industry to being the world's fourth largest producer, with large ore reserves. The first discovery was at Kambalda, where shaft sinking commenced in July 1966 and the newly-constructed treatment plant produced its first nickel concentrates by June 1967. The nickel was sold as mineral concentrates overseas through the port of

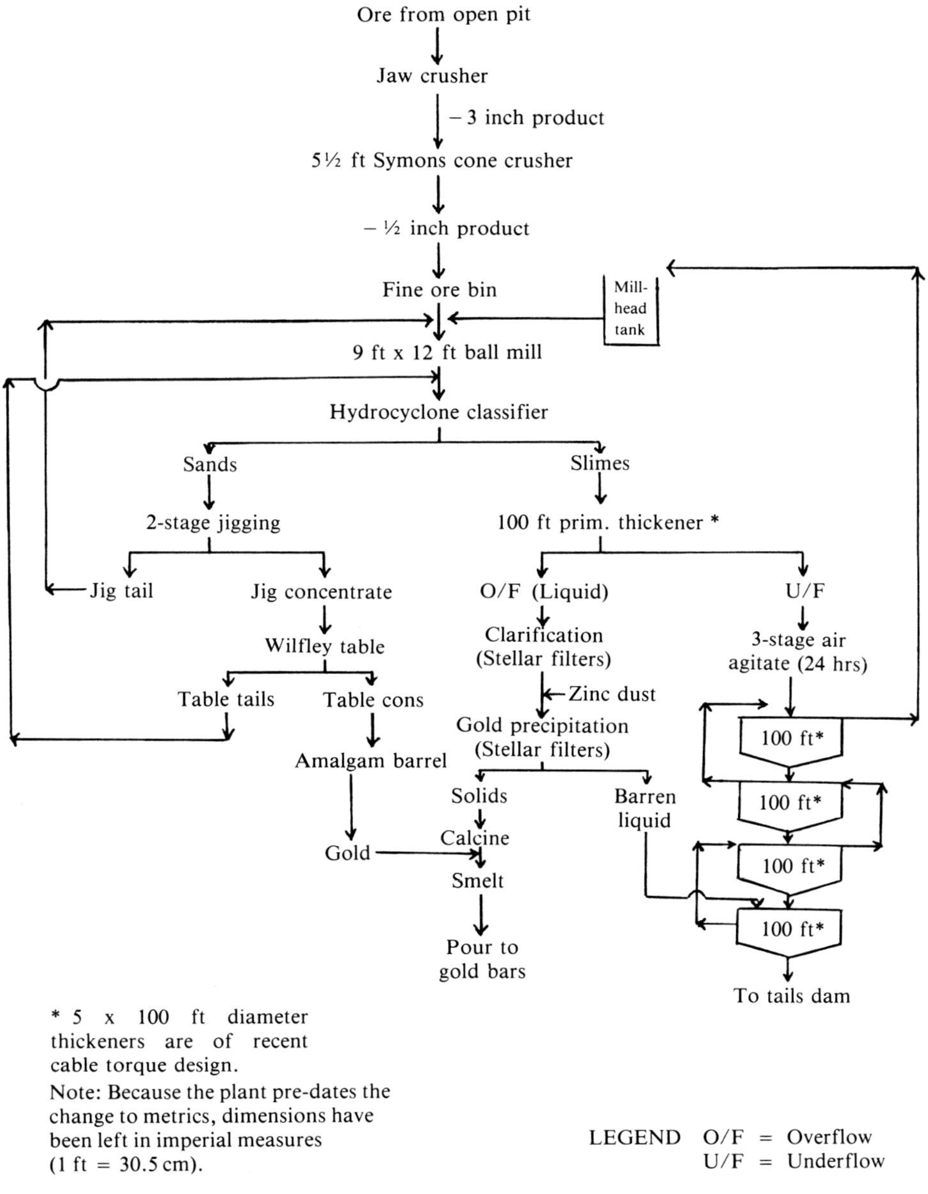

Fig. 5.12 Gold-ore treatment, Telfer

Esperance until Western Mining's ammonia leach plant was commissioned at Kwinana in 1970 to treat some of the concentrates, and thus commenced the production of refined nickel powder and briquettes.

In 1973 Western Mining commissioned an Outokumpu-type nickel flash smelter

Table 5.6

Western Australian nickel deposits

Name	*Owner*	*Ore type*	*Ore reserves tonnes × 10^6*	*Grade % Ni*	*Operation*
A. MINES					
Kambalda[1] (1966)	WMC	Sulphide	24.374	3.2	Ore treatment, flash smelter, refinery
Nepean (1967)	Metals Expl. 50% Freeport Sul. 50%	,,	0.44	4.0	Ore sold to WMC
Scotia[2] (1969)	WMC	,,	1.25	3.0	Ore milled in Great Boulder plant. Concentrates to Kambalda then refinery
Carr Boyd[2]	WMC	,,	0.56	1.52	As for Scotia but concentrates to smelter
Mount Windarra[3] (1969)	WMC 50% Shell 50%	,,	5.51	1.68	Mill at Windarra. Concentrates to WMC Smelter
South Windarra[3]	,,	,,	2.88	1.10	,, ,, ,,
Redross[4]	Anaconda 60% CRA 40%	,,	1.0	3.5	Ore toll-treated North Kalgurli Plant. Concentrates exported
Spargoville	Selcast Expl.	,,	0.51	2.61	Ore treated at Kambalda
Agnew (1971)	Western Selcast 60% MIM Holding 40%	,,	33	2.2	Concentrates toll-smelted, Kalgoorlie
Ora Banda	WMC	Laterite	93	1.0	Used as flux at Smelter
B. PROSPECTS					
Wannaway	Anaconda 60% CRA 40%		4.5	1.22	
Mount Keith	Freeport Sul. 50% Metals Ex. 33.⅓% A.C.M. 16.⅔	Sulphide	290	0.6	

Forrestania Cosmic Boy Flying Fox	Amax 51% Amoco 49%		4.0	3.0	
Forrestania Diggers Rocks	Amax 70% Endeavour Oil 30%		14	1.3	
Mount Edward	BHP 50% INCO 50%		2.8	2.15 0.2 Cu	
Mount Martin			Small	1.5 to 2.0	
Mount Clifford	WMC	Sulphide	Small	1.5 to 2.0	Millerite (NiS) main sulphide
Wingellina	Poseidon 50% Nickel Mines of Aust. N.L. 50%	Laterite	57	1.2	
East Scotia (1973)	Group Expl. 50% Sumitomo Namura Trading 50%	Sulphide	0.7	0.75	
Sherlock Bay	Texas Gulf Sulphur	,,	70	0.5 0.1 Cu	
Pykes Hill	Mid East Minerals 50% Westralia Sands 50%		10 to 15	0.1	

[1] The following shoots or deposits are mined at Kambalda: Lunnon, Durkin, Otter/Juan, Fisher, Jan, Hunt, McMahon.
[2] Originally owned by Great Boulder Mines Ltd. Company acquired by WMC. Mines and treatment plant now closed.
[3] Except for exploration and some mine development, operations suspended June 1978.
[4] Operations ceased in 1978.

Key to abbreviations

WMC	Western Mining Corporation Ltd
CRA	Conzinc Riotinto of Australia Ltd
MIM	Mt Isa Mines Ltd
ACM	Australian Consolidated Minerals Ltd
BHP	Broken Hill Pty Ltd
INCO	International Nickel Australia Ltd

at Kalgoorlie, which accepts mineral concentrates as feed and produces high-grade (72 per cent) nickel matte. Some of this is sold overseas and the remainder forms part of the Kwinana refinery feedstock.

No nickel mineral concentrates are now sold for overseas processing, the mineral concentrates being either smelted to an intermediate product nickel matte, or converted to refined nickel powder at Kwinana.

During the interval 1966-73 intensive exploration for nickel resulted in the discovery of additional significant deposits in what can be aptly termed the Western Australian nickel belt stretching from Forrestania in the south to Mount Keith in the north. These deposits are listed in Table 5.6, which shows that in addition to the Kambalda deposits, eight others have been, or are being worked to date.

Western Mining Corporation Limited has maintained its eminent position in the Australian nickel industry, having established a fully integrated operation from mining through to the production of refined nickel powder. Other West Australian nickel operators either sell their ore direct to Western Mining (Nepean), or have it toll-treated through the Kambalda mill (Spargoville), or mill their own ore and rail their dried concentrates to the smelter for conversion to matte (Windarra and Agnew).

KAMBALDA NICKEL OPERATIONS

Fig. 5.13 shows a simplified flowsheet of ore treatment, the plant having a capacity of 1.5 million tonnes of ore per year. The treatment plant has several operations unique to ore processing in Australia, and in particular to Australian nickel-ore processing. The Western Australian nickel-belt ore minerals are pentlandite, violarite, nickeliferous pyrrhotite and nickeliferous pyrites. The common gangue minerals are talc, tremolite/actinolite, sericite, chlorite and muscovite.

There is one common crushing, screening and ore-storage plant, but the grinding, flotation and drying sections were built in three stages and can be independently operated, so that if necessary, three different ore types can be separately, but simultaneously, treated.

Purchased ore, together with ore from various shoots and lenses around the Kambalda dome, is crushed, screened and stored at minus 3/8 inch (9.5 mm). During crushing and screening some 3-inch (76 mm) pebbles and 5-inch (127 mm) lumps are screened out and stored separately for use as grinding media in the No. 2 and No. 3 sections of the plant.

A unique feature of the Kambalda operation is that, in No. 2 and No. 3 concentrators, the minus 3/8-inch (9.5 mm) ore is ground to a pulp before flotation by a two-stage fully autogenous grinding system. The first stage is rock-milling in which the fine ore is partially ground using 5-inch (127 mm) lumps of ore as grinding media, and final grinding is done in a pebble mill using 3-inch (76 mm) fragments. All mills are rubber lined, and the grinding operation is almost completely automatically controlled.

The ground ore at about 75 weight per cent passing 74 microns is floated in

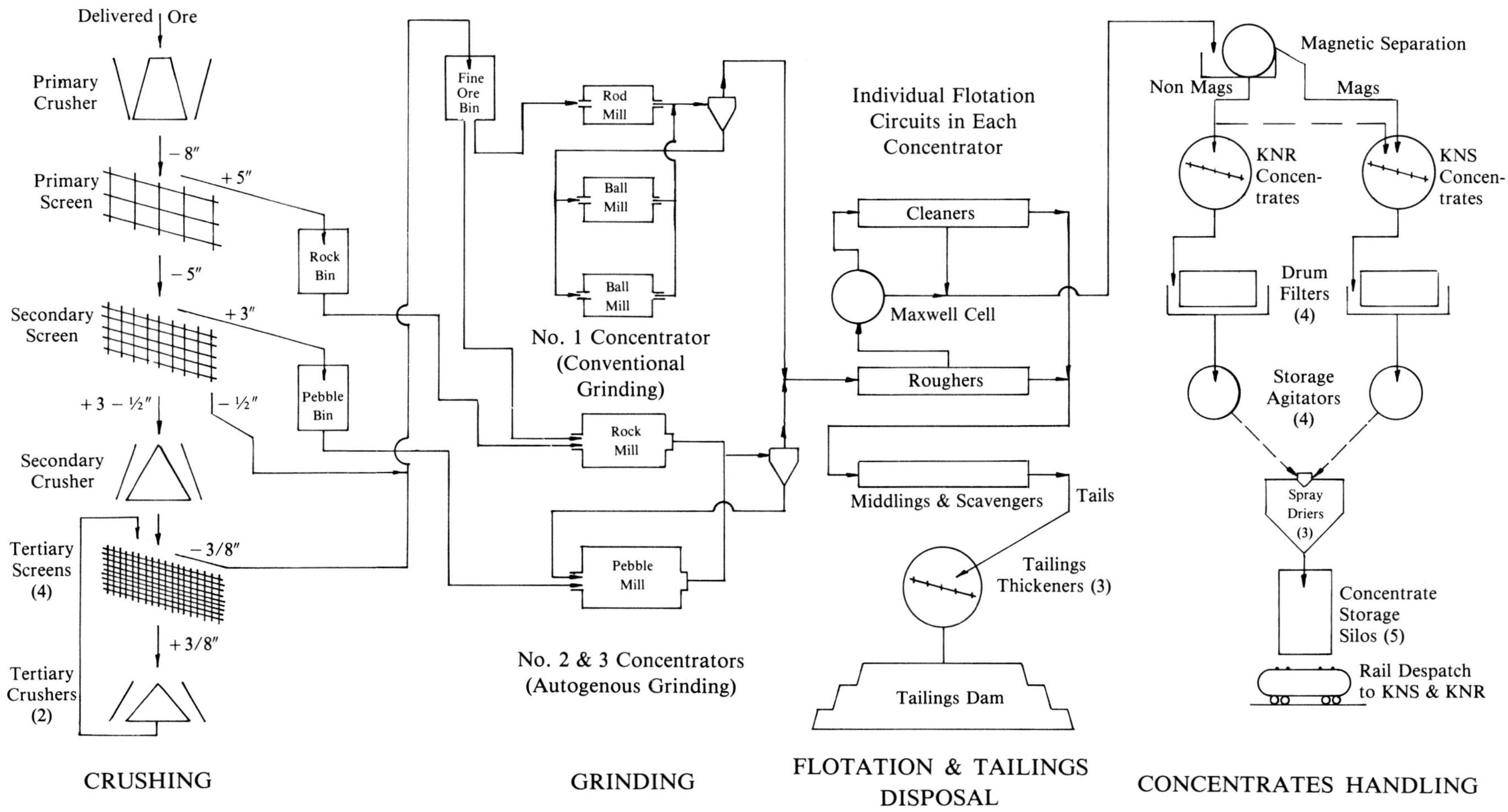

Fig. 5.13 Kambalda nickel operations: concentrator flowsheet
(Note: Because the plant pre-dates the change to metrics, dimensions have been left in imperial measures (1 ft = 30.5 cm).

Agitair 48-inch (1219 mm) flotation machines to give a nickel rougher concentrate, which is cleaned first in a Canadian-patented Maxwell flotation machine and then in additional 48-inch Agitair machines to give a final concentrate assaying 11 to 12 per cent nickel, 1 per cent copper, 0.2 per cent cobalt, 33 to 35 per cent iron and 5 per cent magnesia.

Because the Kalgoorlie nickel flash smelter requires a low magnesia, high iron and sulphur feed, the nickel concentrate slurry is passed over wet-drum magnetic separators at Kambalda, splitting the concentrate into magnetic and non-magnetic fractions.

The magnetic concentrates are higher in iron and sulphur, and lower in magnesia, containing most of the nickeliferous pyrrhotite. This concentrate is thickened, filtered and spray-dried for rail dispatch to Kalgoorlie. The non-magnetic portion of the nickel concentrates is separately dried and railed in 1500-tonne unit trains to Western Mining's Kwinana refinery.

The other unique features of Kambalda's metallurgical operations are:

1. The use in Australia for the first time of Canadian Maxwell flotation machines.
2. Use of Swedish Niro spray-driers for drying mineral concentrates. Half of the required heat for drying is derived from diesel exhaust gases from the Kambalda 31.6 MW diesel power generating station.
3. Use of Amdel/Philips radio-isotope probes in the pulp stream to give continuous nickel and iron assays for quality control.

Flotation reagents used at Kambalda are sodium ethyl xanthate for floating the sulphides, tri-ethoxy butane for frothing, locust bean gum or guar gum for magnesium-gangue mineral depression, and a small amount of copper sulphate to promote sulphide flotation. Flotation is done at the natural pH of the pulp, which is 8.0 to 8.5.

Ore-treatment operations at other plants are essentially similar to that at Kambalda, with the following exceptions:

1. Only Kambalda has autogenous grinding-mills. Other plants use ball-mills.
2. Nickel concentrates are either dispatched as a semi-dried concentrate or dried in flash-driers as at Windarra and Agnew. Flash-driers, like spray-driers, make use of waste heat from diesel engines for drying, but the flash-drier, instead of using an atomizing wheel to disperse the mineral concentrate into very fine droplets in a drying chamber, uses a long vertical drying column, and relies on the flow of hot waste gases to dry and convey the concentrates up the column and into a storage hopper.
3. Instead of natural gums for gangue depression as used at Kambalda, other treatment plants use synthetic depressants such as carboxy methyl cellulose with soda ash to maintain flotation pH at 8.5 or higher.

KALGOORLIE NICKEL SMELTER

The feasibility of building a nickel smelter was first considered in 1966, shortly after nickel was discovered. The Kwinana refinery, however, was built first because

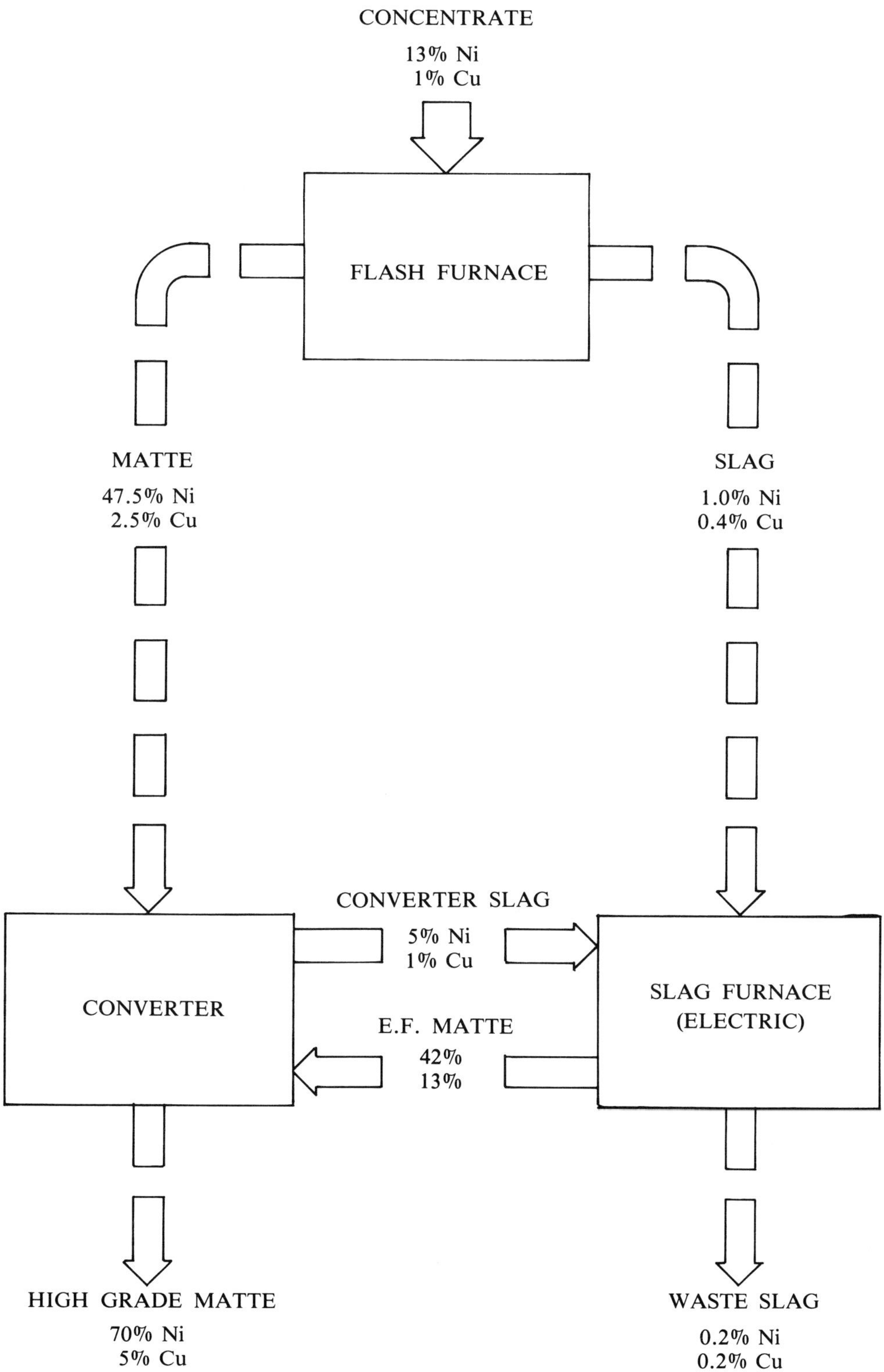

Fig. 5.14 Kalgoorlie nickel smelter: process flowchart

it was the faster route from nickel concentrates to nickel metal. However, another outlet for nickel was required and smelter matte was chosen. Several furnace types were considered, such as a blast-furnace, electric furnace, and Australia's Worcra furnace, but the Finnish Outokumpu furnace was chosen mainly because of its fuel economy. Construction commenced in early 1971, and on 5 December 1972 concentrates were fed to the furnace. This Kalgoorlie installation was the first nickel flash-furnace to be built outside Finland.

The three basic units are the flash-furnace itself, the electric slag-cleaning furnace and the converters. The flow of material to and from these units and the approximate metal values are shown in Fig. 5.14.

When it was commissioned, the plant had a design capacity to treat 200 000 tonnes of nickel concentrates per year, which at 13 per cent feed grade and 97 per cent recovery would be 25 220 tonnes per year nickel as nickel matte. In 1974 two 90-tonnes-per-day oxygen plants were purchased and installed to produce oxygen with a 95 per cent purity. This was mixed with preheated process air for combustion in the furnace, and raised the capacity of the plant to 300 000 tonnes of concentrates per year.

A new flash-furnace unit was commissioned in November 1978, initially to treat 450 000 tonnes per year of concentrates, but with the potential to increase to 700 000 tonnes per year.

Operation

Nickel concentrates, burnt lime, and high silica Ora Banda nickeliferous laterite as flux are pneumatically conveyed from storage bins and automatic weigh hoppers to furnace feed bins, from which they enter the furnace through four concentrate burners situated on top of a refractory-lined vertical circular reaction shaft. Air at 32 per cent oxygen, and preheated to 500 °C, is mixed with the solids feed, and rapid combustion occurs as the gases and solids descend the shaft. Most of the heat for combustion and smelting comes from the following exothermic reaction involving pyrrhotite:

$$\underset{\text{pyrrhotite}}{2FeS} + 3O_2 = 2FeO + 2SO_2$$

The nickel sulphides also provide heat thus:

$$3NiS + O_2 = \underset{\text{heazlewoodite}}{Ni_3S_2} + SO_2$$

Careful control is exercised of the oxygen/air/fuel oil additions to the reaction shaft, so that shaft temperature does not exceed 1350 °C-1400 °C, otherwise magnetite and accretions would form, reducing the fluidity of the slag and entrapping nickel:

$$6FeO + O_2 = 2Fe_3O_4 \text{ (magnetite)}$$

Smelting reactions continue until the charge, now molten, reaches the bottom of the shaft and enters the settler bath. Endothermic slag-forming reactions occur in the bath, and oil burners are located here to supply additional heat.

$$\underset{\text{iron oxide}}{2FeO} + \underset{\text{silica}}{SiO_2} = \underset{\text{fayalite}}{Fe_2SiO_4}$$

If too much magnetite is formed, it may be reduced by adding heat or grey cast iron or both:

$$\underset{\text{magnetite}}{Fe_3O_4} + Fe = 4FeO$$

The molten charge in the settler bath separates out by gravity into a heavy nickel copper iron sulphur matte phase and an upper calcium magnesium iron silicate slag phase.

Gases comprising residual air, suphur dioxide and burner gases leave the furnace at 1350 °C, give up much of their heat to 30-tonnes-per-hour waste heat boilers and pass through 99.8 per cent efficient electrostatic collectors, where the dust is collected and returned to the furnace feed hoppers. Cooled gases pass to the atmosphere through the 153-m high stack.

Molten slag containing about 1 per cent nickel overflows almost continuously from the settler to 6 MVA electric furnaces drawing 5 MW of peak power. Coke, crushed reverts and converter slag are added and the charge held for about three hours. The coke creates a strong reducing condition, reducing magnetite in the slag from 12 to 2 per cent weight thus:

$$Fe_3O_4 + C = 3FeO + CO$$

Nickel oxides in the slag are reduced to metallic nickel thus:

$$NiO + C = Ni + CO$$

The addition of reverts (ladle skulls, etc.) supply additional sulphur so that nickel can form matte thus:

$$2NiS + Ni = Ni_3S_2 \text{ (heazlewoodite)}$$

Any mechanically entrapped matte also has an opportunity to separate from slag in the cleaning furnace.

A low-grade matte assaying 40 to 45 per cent nickel is formed containing about 80 per cent of the nickel entering the cleaning furnace and reducing the nickel tenor of the slag to a final 0.1 to 0.2 per cent nickel. The clean slag is granulated in water and carted to waste, and the low-grade matte is periodically run off into 30-tonne ladles and poured into one of three 30-tonne nickel converters.

The converter is like a copper or steel converter in which a large volume of low-pressure air at 20 p.s.i. (138 kPa) is blown through tuyères in the bottom of the rotatable container, the air rising up through the molten matte and preferentially oxidizing the iron to oxides thus:

$$2FeS + 3O_2 = 2FeO + 2SO_2 \text{ (exothermic reaction)}$$

Because of the large amount of oxygen available, oxidizing conditions are severe and a large amount of magnetite is formed thus:

$$6FeO + O_2 = 2Fe_3O_4$$

The resulting converter slag, high in nickel, is returned to slag cleaning. Any metallic nickel present will form heazlewoodite if sulphur is available, thus:

$$2NiS + Ni = Ni_3S_2$$

Converter blowing is stopped when iron in the matte is less than 1 per cent. Further blowing would produce excess nickel oxide.

The nickel matte, assaying about 70 per cent nickel, is poured from the converter into ladles, and granulated by pouring through a stream of water. Gases leaving the converter are diluted by ingress air and any dust content settled out by gravity before the gases enter the stack.

Steam from waste heat boilers is used to drive a 10.7 MW steam turbine. The smelter power station is connected to Western Mining's power grid linking the Kambalda, Smelter and Kalgoorlie Mining Associates power stations.

KWINANA NICKEL REFINERY

Situated 32 km south of Perth, the refinery's function is to convert nickel concentrates and nickel matte into high-purity nickel powder for sale. Some of the powder is briquetted and the briquettes sintered at high temperature in a hydrogen atmosphere before sale. This process increases the apparent density of the metal and removes certain minor impurities.

As with the smelter, planning for a refinery commenced as soon as a nickel mine was established, and in late 1967 the Sherritt-Gordon patented ammonia pressure leach process was chosen because it offered the quickest route from ore to refined metal.

The metropolitan area was chosen for the site at that time because:

1. Vital raw materials were immediately available in that area.
2. Water, difficult to obtain inland, was available.
3. By-product fertilizer ammonium sulphate would be generated close to a port.
4. Capital cost and infrastructure were minimized.

The plant was commissioned in 1970 at a cost of $40 million and a designed production capacity of 15 000 tonnes of refined nickel powder per year. Since then a further $9 million has been spent to increase plant capacity to 20 000 tonnes per year and recent plant additions now allows 30 000 tonnes of nickel per year to be produced. In addition to nickel, the following by-products are also generated: up to 3000 tonnes per year of copper sulphide precipitate, assaying about 60 per cent copper, 600 tonnes per year of mixed nickel/cobalt sulphides assaying 22 per cent nickel and 14 to 15 per cent cobalt, and 120 000 tonnes of ammonium sulphate. All of these products are sold without further treatment.

SHERRITT-GORDON PROCESS

This is a continuous ammoniacal pressure leach process in which nickel mineral concentrate or ground nickel matte is pumped as a 25 per cent slurry with ammonium sulphate into stainless steel clad autoclaves at 900 kPa with air and ammonia.

The leach reactions are exothermic and temperature is maintained between 72 °C and 82 °C by cooling coils for several hours so that specific ratios of unsaturated sulphur-oxygen compounds to dissolved copper are formed in the leach solution. These compounds, which include thionates and thio-sulphates, act as precipitating agents in the subsequent copper removal operation.

Leaching is a two-stage, counter-current process with thickening, filtration, and washing of the residue at each stage. Washed residue from the second-stage leach, consisting essentially of insoluble silicates plus hydrated iron oxide, is pumped to a plastic-lined residue dam.

The pregnant solution from the first-stage leach (adjustment leach) contains copper, which must be removed before nickel precipitation. To remove copper, the solution is heated to 110 °C by direct contact with a counter-current flow of steam. This drives off excess ammonia gas, which is condensed and recycled for further leaching. As the ammonia content of the solution falls below 70 g/litre, copper in solution reacts with sulphides present and precipitates as cuprous and cupric sulphides. The copper sulphide precipitate is filtered off and sold. There is still a trace of copper in solution and this is precipitated with a small quantity of hydrogen sulphide gas.

The copper-free solution now containing only nickel and cobalt is subjected to intense oxidation at 245 °C in oxy-drolysis reactors to convert all remaining sulphides to ammonium sulphate and remove all excess ammonia. Nickel is then precipitated in reduction autoclaves with hydrogen under pressure at 200 °C and 315 kPa. Four reduction autoclaves are used in batch operation, the nickel powder being allowed to settle after each charge, the barren liquor drawn off and a fresh pregnant solution run in. Thus nickel particles are progressively built up in size by up to 40 or 50 precipitation cycles until the powder is of the required size. It is then separated, washed, dried and drummed for sale.

Some of the high-purity nickel powder is mixed with a special binder, briquetted,

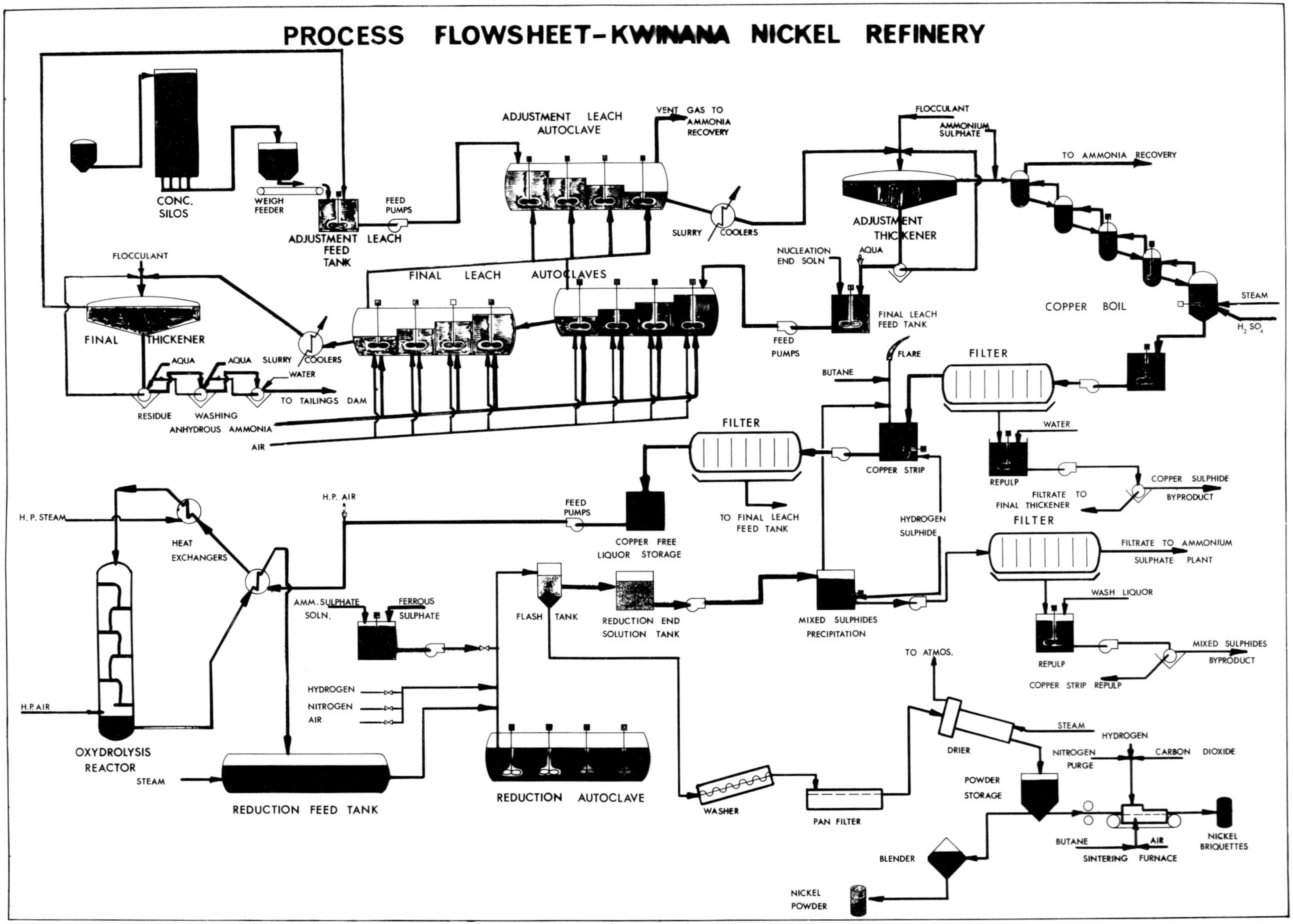

Fig. 5.15 Kwinana nickel refinery: process flowsheet

sintered in a hydrogen atmosphere in a long moving-grate sinter furnace and the nickel briquettes sold. After nickel precipitation, cobalt is precipitated with hydrogen sulphide gas as a mixed nickel-cobalt sulphide and is sold as such. Some of the ammonium sulphate is crystallized from the completely barren solution and sold as fertilizer, and the remainder of the solution returned for further leaching. Gold, silver, platinum and palladium present in the refinery feed accumulate in the copper sulphide and mixed sulphide by-products and credit is obtained for the value of these elements during sale.

A simplified flowsheet of the refinery operation is shown in Fig. 5.15.

(c) MINERAL SANDS

Mineral sands recovery operations were commenced in 1956 in the Capel area, south west of Bunbury, and in 1972 the Eneabba area south of Geraldton was opened. Products exported from the ports of Bunbury and Geraldton respectively in 1975-76 amounted to 860 000 tonnes of concentrates.

The Capel deposits provide ilmenite and zircon as major products with minor products of leucoxene and monazite. One operation also produces small quantities of rutile and xenotime, as well as upgraded ilmenite of about 92 per cent titanium dioxide content. The ilmenite is predominantly sulphation grade. The Eneabba deposits provide ilmenite, zircon and rutile. The ilmenite is mostly chlorination grade of 60-65 per cent titanium dioxide content. A small quantity of monazite concentrate is stockpiled by all three operations.

Sulphation grade ilmenite is highly soluble in sulphuric acid and this acid route is used to make titanium dioxide pigment. Chlorination grade ilmenite on the other hand is poorly soluble in sulphuric acid because it has been altered by natural oxidation towards rutile, that is, higher in titanium and lower in iron. It is, however, dissolved by hot chlorine or hydrochloric acid and thus is suitable feedstock for the chlorination route to pigment production.

In separating individual sand minerals, use is made of their properties of density, magnetic and electrostatic susceptibility (see Table 5.7) and, to a lesser extent, grain size.

Equipment employed is relatively standardized in all operations, as is the basic treatment circuit.

The mining method is described briefly in Table 5.8, and Fig. 5.16 shows a generalized flowsheet for treating beach sands. In all cases the mined sands are pulped in water and screened to remove tramp oversize. In one plant (Westralian Sands) the sands are scrubbed and fine-screened to eliminate slime. Using gravity separation through several stages of Reichert cones, a bulk heavy mineral concentrate is separated from the quartz, which makes up 85-96 weight per cent of the mined sand. This wet plant is located close to the mining operation and skid or rail mounted so that it can be periodically moved to new mining sites. The Western Mineral Sands wet plant is on a pontoon and follows the dredge.

The 90-95 per cent heavy mineral concentrate is dried in oil-fired driers before further processing. At Eneabba the gravity concentrate is dried at the wet plants and then transported to the dry plants located 150 km further north at Geraldton. At Capel, concentrates are dried at the dry plant.

Dry concentrate is passed over crossbelt electromagnetic separators to remove most of the highly magnetic ilmenite which also constitutes the greater bulk of the concentrate. This ilmenite is sulphation grade. The smaller portion of the concentrate is then split into conducting and non-conducting fractions on high tension rolls. The conducting fraction, consisting of rutile, leucoxene and less magnetic ilmenite (chlorination grade), is subjected to several stages of crossbelt separators to remove the ilmenite and leucoxene and to separate these into fractions. Induced roll electromagnetic separators clean final traces of ilmenite and leucoxene out of the final product rutile.

Residual low-density grains entrained in the heavy mineral concentrate make up a considerable portion of the non-conducting fraction after the above operations. This fraction, therefore, is repulped in water and the low-density grains removed in spiral separators. Westralian Sands supplement these with spiral and cyclone classifiers, and Allied Eneabba with wet tables.

After filtering and redrying, the gravity upgraded, non-conducting fraction is passed over induced roll electromagnets to separate the zircon and monazite fractions. The non-magnetic zircon is screened or air-tabled to remove coarse zircon, then air-tabled in several stages to remove kyanite grains from the fine zircon.

At Eneabba the magnetic monazite-rich fraction is stockpiled for future processing. At Capel it is screened to remove coarse garnet, then air-tabled to remove finer garnet from the monazite. Western Titanium also separate a small quantity of xenotime from the garnet reject.

Western Australian mineral sands are coated with iron staining and clay, which modify the magnetic and conducting properties of the grains. To obtain maximum recovery of acceptably high-grade zircon and rutile the coating must first be removed by attritioning the sand at high pulp density in acid media such as hydro-

Table 5.7

Properties of minerals of mineral sands

Ilmenite	:	S.G. approx. 4.7;	highly magnetic;	conducting
Zircon	:	S.G. approx. 4.5;	non-magnetic;	non-conducting
Rutile	:	S.G. approx. 4.2;	non-magnetic;	conducting
Leucoxene	:	S.G. approx. 4.5;	magnetic;	conducting
Monazite	:	S.G. approx. 5.2;	magnetic;	non-conducting
Xenotime	:	S.G. approx. 4.5;	magnetic;	non-conducting
Kyanite	:	S.G. approx. 3.6;	non-magnetic;	non-conducting
Garnet	:	S.G. approx. 4.0;	magnetic;	non-conducting
Quartz	:	S.G. approx. 2.6;	non-magnetic;	non-conducting

Table 5.8

Summary of Western Australian mineral sands operations

Company	Western Titanium Ltd		Westralian Sands Ltd	Cable Sands Ltd	Western Mineral Sands Pty Ltd	Allied Eneabba	Jennings Mining Ltd
Location of mines	South Capel	Eneabba	Yoganup	Wonnerup & Amber Farm	North Capel	Eneabba	Eneabba
Mining methods	Bulldozers, front-end loaders, slurry & pump	Bulldozer	Dragline, bucket exc., trucks	Front-end loaders	Dredging	Scrapers, belt conveyors	Bulldozers, front-end loaders, slurry & pump
Location of dry plant	Capel	Eneabba	Capel	Capel	None	Geraldton (Nangalloo)	Geraldton
Production capacity†							
Ilmenite	450 000	150 000	400 000	200 000	230 000	250 000	150 000
Zircon	85 000	70 000	50 000	12 000	Heavy Mineral	120 000	30 000
Rutile	33 000	30 000	—	—	concentrate	50 000	40 000
Monazite	As required	—	As required	As required	sold to	—	—
Leucoxene	As required	—	As required	As required	Westralian	—	—
Xenotime	As required	—	—	As required	Sands	—	—
Hyti 68*	—	—	20 000	—		—	—
Upgraded ilmenite	30 000						

(—) not produced * High titanium † Tonnes per annum

Note that actual plant production is a function of the composition of the heavy mineral concentrate which is fed to the plant. As this can vary widely, daily production of each mineral concentrate will not necessarily agree with plant designed capacity.

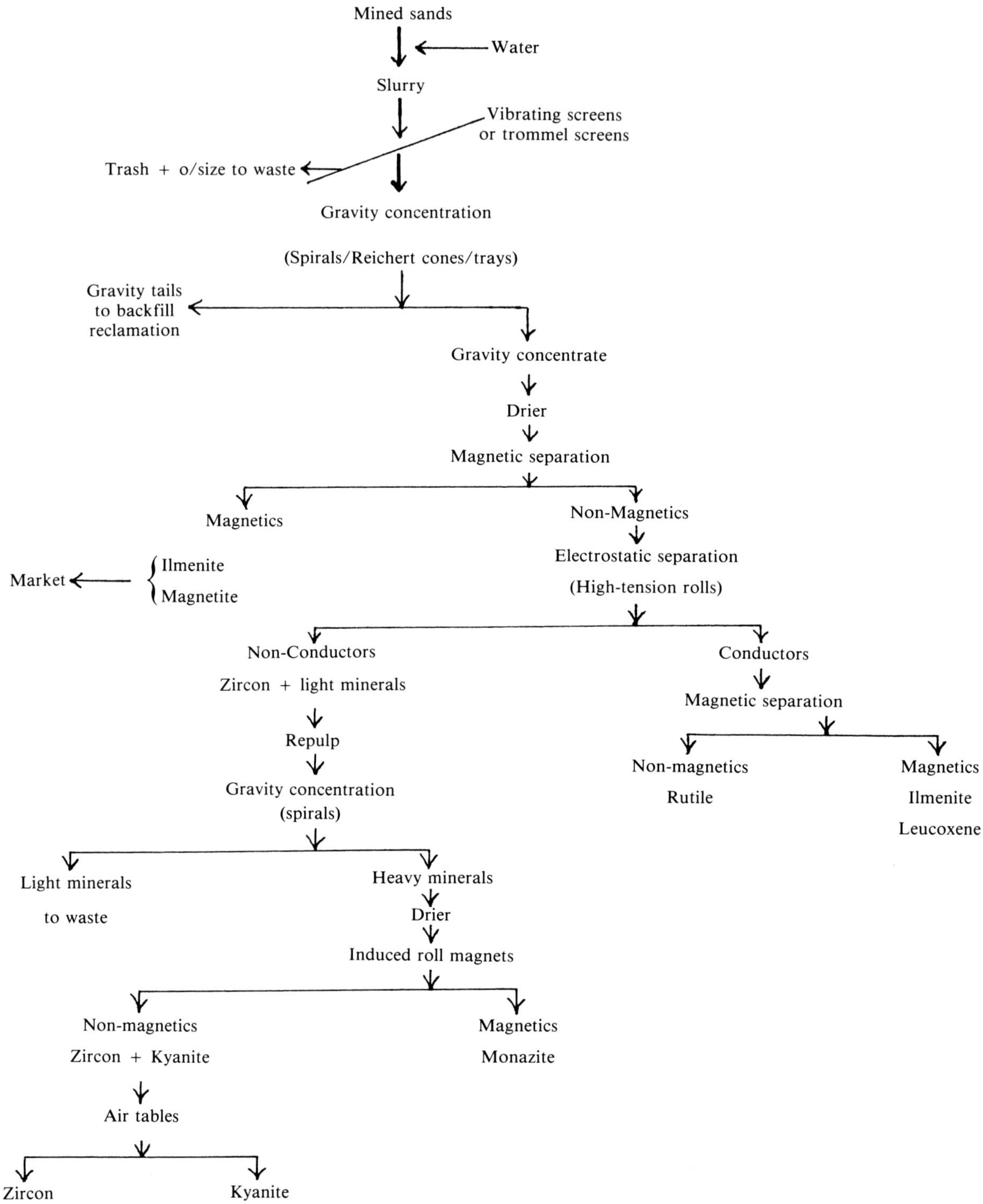

Fig. 5.16 Western Australian mineral-sand deposits: generalized treatment flowsheet

chloric and hydrofluoric acid, possibly with sodium metabisulphite added. The position in the circuit at which attritioning is done varies from operation to operation, generally being either before or after the gravity separation spirals. The above general circuit description is, of necessity, simplified. Each separation operation is generally carried out in several stages, many with a middling as well as the two product fractions. Interstage transfer patterns can be quite complex in an endeavour to maximize recovery and separation of the products. Equipment other than that stated for the duty may be employed for some of the stages. For example, Western Titanium use a Jones wet magnetic separator to remove ilmenite before drying the heavy mineral concentrate.

A significant addition to the normal mineral sands operation is the ilmenite upgrading plant of Western Titanium at Capel. Ilmenite, or altered ilmenite of chlorination grade, is oxidized in a rotary kiln at 1000-1050 °C, then cooled and mixed with coal char before transfer to a reduction kiln of special design. This rotary unit is fired by partial combustion of fine coal. Secondary air is injected at several points along the length of the kiln to maintain a temperature of 1170-1190 °C in conjunction with a controlled reducing atmosphere. During an average residence time of about 8 hours, 93-95 per cent of the iron content is reduced to metal in the form of particles segregated within the titanium-rich oxide grains.

The reduced product is cooled and unreacted char recovered by screening and magnetic separation, following which the grains are pulped in a dilute solution of ammonium chloride plus sulphuric acid, and agitated with aeration for about 13 hours at 50-60 °C. The segregated iron content rusts and separates from the grains as a slime, which is removed by cyclones, leaving an upgraded product containing about 92 per cent titania after drying.

(d) TIN, TANTALUM AND ANTIMONY

The mining and recovery of tin dates from 1889, after alluvial cassiterite was discovered at Greenbushes. Greenbushes remains as the only major producing area, mining and recovery activities now being centred on primary pegmatite and secondary alluvial deposits. The activities of Greenbushes Tin N.L. has now been expanded to recover both tantalite (tantalum/niobium oxide) and stibiotantalite (tantalum/niobium/antimony oxide).

Minor alluvial deposits of cassiterite and tantalite in the Pilbara area have been worked on a small scale from time to time but, with few exceptions, have involved little more than panning or jigging to recover a crude concentrate.

Antimony is included in this section because the only antimony being recovered in Western Australia at present is a by-product of the Greenbushes operation. In the past, Western Australia has been a substantial antimony producer as a by-product of gold ore treatment, particularly in the Wiluna, Marble Bar and Nullagine areas. Most of the antimony was made by flotation of the antimony sulphide mineral stibnite (Sb_2S_3) into a stibnite concentrate, but antimonial lead containing 7 to 9 per cent antimony was also produced at Wiluna. Stibnite concentrate production has virtually ceased since the Blue Spec mine closed in 1978.

Mining in the Greenbushes area is currently by open cut, with backhoe excavators. Ore is hauled by truck to the two wet-concentrating plants, where it is washed, screened in trommels, and crushed to − 1.4 mm by two-stage hammer mills. The heavy minerals, cassiterite and tantalite, are separated from the coarse gritty minerals, mainly quartz and tourmaline, in three-stage jigs, following which they are attritioned in caustic-soda solution to remove any iron-oxide coating. The rough concentrate is thickened by cyclone and filtered prior to transport by truck to the centralized dry plant, as a 30 per cent heavy mineral concentrate.

In the dry plant the rough concentrate is fully dried then passed over a cross-belt magnetic separator to remove highly magnetic impurities. The non-magnetics are then separated into nine size fractions on cascade screens. Stray + 1.4 mm oversize is roll-pulverized and re-treated through a small-scale auxiliary wet plant adjacent to the dry plant. The seven size fractions in the range of 1.4 to 0.12 mm are treated on air tables to separate the heavy minerals from the lighter impurities, which are now of similar size. Rejected light fractions are retreated through the auxiliary wet plant to recover any entrained heavy mineral.

The air-table concentrates are passed through 4-roll electrostatic separators to remove iron-conducting mineral, mainly stibiotantalite, and then split into tin and tantalite concentrates in rapid or induced roll electromagnetic separators. Three products are obtained from the electromagnetic separators, viz. a non-magnetic tin concentrate, a medium magnetic tantalite concentrate, and a highly magnetic reject which is recycled to the auxiliary wet plant for cleaning and retreatment.

The stibiotantalite concentrate from the electrostatic separators is further upgraded by air tabling, rapid electromagnetic separation and electrostatic separation to remove final quantities of entrained light mineral as well as tantalite, and cassiterite respectively, which are recycled to process.

Total production of tin in Western Australia as cassiterite concentrates is between 800 and 900 tonnes per year. 150 to 200 tonnes of tantalite concentrates are produced.

(e) SALT, POTASH AND MAGNESIUM SALTS

The combination of a hot and dry climate, low rainfall, sparsely-occupied land area and a number of deep-water bulk-handling ports makes Western Australia ideally suited for the recovery of salt from sea-water or evaporites. Salt production has grown over the past 15 years, with five operations now producing in excess of 5 million tonnes of salt a year. This amounts to more than three-quarters of Australia's total production.

Salt recovery operations are located at Port Hedland, Dampier, Shark Bay, Lake MacLeod, and Lake Lefroy. At the former three sites, salt is recovered from seawater, at Lake MacLeod from an extensive evaporite deposit, and at Lake Lefroy from near-saturated brine obtained by seasonal rainwater recharge of the salt lake.

Production of salt is a simple operation in which sea-water is pumped into large shallow ponds and the water removed by solar evaporation. When the water reaches saturation in salt, it is pumped to a series of smaller shallow ponds, where continued evaporation of water results in salt crystals forming and depositing on the pond floor. After salt deposition, but before the bitterns (magnesium salts) quite reach the point of saturation (at which they in turn would deposit), the bitterns are pumped back to the sea or to a soakage area. Evaporite and salt-lake deposits have the advantage of being near-saturated in salt from the start and plants processing these feedstocks only require the crystallizing ponds.

Pond construction

Like many simple operations, the details are more complex. The large shallow areas in which sea-water is evaporated must be very flat and the ground must be almost impervious so as to retain the brine. In some cases porous areas in otherwise suitable ground may be sealed by plastic sheeting. Retaining walls and dykes must be built, often a construction challenge because otherwise suitable ground for these ponds may be unable to support earth-moving equipment. These walls and dykes must be constructed to withstand the battering of waves generated in the ponds, and suitable rockfill sometimes has to be transported to the site. During evaporation of sea-water, gypsum is deposited, and this has the beneficial effect of improving the sealing of the floor of the ponds after some years.

The crystallizing ponds are constructed in such a way that, after drying out the ponds, the salt can be recovered by mechanical harvesters, portable conveyors and trucks. Thus the ponds must be constructed to support earth-moving equipment and levelled to a high degree of accuracy for harvesters. It is common to consolidate the ground by allowing a layer of salt up to 25 cm thick to form and only harvesting down to this level each year.

Details will vary from one site to the next depending upon the problems and solutions involved in pond construction, strength of feed solution, and transfer of the feed. The start of an operation may be leisurely, initial seepage losses and build-up of salt crust in the crystallizing ponds possibly delaying maximum design production for several years. Apart from these variations, the operations are similar.

Pond operation

Sea-water is pumped into the first evaporation pond at the rate required to hold its level. It then gravitates through three or more ponds in series to minimize short-circuiting or back-mixing, and thus a constant composition feed to the crystallizing ponds is ensured. Iron and aluminium impurities and most of the gypsum in the sea-water is deposited in these ponds. Brine is transferred to the crystallizing ponds at the verge of crystallization so as to obtain maximum removal of gypsum.

The saturated brine is pumped to the crystallizing ponds. In the case of Lake MacLeod and Lake Lefroy the brine feed goes straight to the crystallizers. The brine gravitates through the ponds in series, the salt crystallizing and sinking to the bottom and forming a hard layer as evaporation proceeds. Brine depth in the ponds is maintained at 30-50 cm. Discharge from the final pond is maintained at a strict specific gravity by controlling the feed rate, the gravity corresponding to incipient saturation of the brine by magnesium sulphate. This discharge is gravitated to waste, except at Lake MacLeod, where it is pumped to a second series of crystallizing ponds, allowing langbeinite (potassium magnesium sulphate) to be crystallized.

Once a year each crystallizing pond is bypassed in sequence and allowed to dry out. When dry, the salt crust is mechanically harvested down to the datum level using specially designed horizontal auger-cutters, which discharge to trucks. The salt is stockpiled, then washed as required during the full year. Washing generally involves wet screening to remove coarse crystals, followed by a classification of the finer crystals in spirals or cyclones. The oversize crystals from the screen and classifiers are further spray-washed on screen decks or conveyors before stockpiling, where they drain and dry before shipment. Incoming saturated brine to the crystallizers is used for washing, and the washings plus undersize crystals are returned to the crystallizing ponds.

The harvested crystals contain a quantity of entrapped bittern solution and this is substantially removed by washing and finally by drainage and rain-washing in the stockpiles.

A considerable amount of technical skill is required in the control, prediction, and timing of the seasonal operations.

The state produces between three and four million tonnes of salt each year.

(f) QUARRY INDUSTRIES, TALC

Within Western Australia there are a considerable number of quarry industries producing a diverse number of minerals for both local consumption and export. While many of these operations can present interesting problems in mining and beneficiation, space prevents the inclusion of details here. It must suffice to list the activities which, while small relative to other mining in the state, nevertheless *in toto* make a significant contribution to the state's economy.

For local consumption	*For export*
Limestone	Barytes
Sand	Bentonite
Marble	Diatomite
Granite and dolerite	Feldspar
Quartz	Gemstones
Clay	Gypsum
	Spodumene
	Magnesite
	Pyrolusite
	Mica
	Vermiculite

Worthy of specific mention amongst the quarry industries is talc, which is currently exported at the rate of about 90 000 tonnes per annum. Talc is mined at both Three Springs and Mt Seabrook by conventional hard-rock quarrying methods and crushed to − 125 mm. Low-grade talc and impurities are removed by hand-picking, then − 12.5 mm fines screened out. The screened rock sells overseas for premium prices. The fines are wet-screened to remove − 3 mm grit, dried and transported to Perth, where they are fine-ground in hammer mills and cleaned by screening and classification. The talc powder is sold within Australia.

REFERENCES

Blaskett, K. S. 1951, 1952. *Ore Treatment at Western Australian Gold Mines.* Melbourne: Tait Publishing Co. Pty Ltd. (Reprinted from a series of articles in *Chemical Engineering and Mining Review,* Nov. 1951, Aug. 1952.)

Dunkin, H. H. 1953. *Ore Dressing Methods in Australia and Adjacent Territories.* 5th Empire Mining and Metallurgical Congress, III. Australas. Inst. Mining and Metallurgy Melbourne.

Holroyd, A. G. 1897. The Discovery and Occurrence of Tellurides of Gold upon the Kalgoorlie Goldfields, East Coolgardie District, Western Australia *Trans. Aust. Inst. Min. Met. Eng.* **4**, 186.

5 F

Oil-drilling onshore and offshore

I. R. CAMPBELL and J. C. PARRY
West Australian Petroleum Pty Ltd

Exploration for oil in Western Australia commenced with the drilling of the Warren River bores in 1902-4. In those days the concepts of drilling for oil and the methods used to explore for it were primitive compared to those of the late 1970s. The early drilling-rigs were modifications of existing equipment: either the diamond drills used in 'hard rock' mining, or the cable-tool or percussion rig used in water-well drilling in softer rocks.

By the late 1970s drilling equipment has become far more complicated as drilling of wells to depths of 3000 to 4500 m becomes commonplace. In this advance of technology, activities in Western Australia have kept pace with world trends. We now have some of the newest and most sophisticated drilling-rigs in the world operating in this state.

The following sections summarize the changes in equipment and practice in drilling for oil in Western Australia. Although all the major activity has been confined to the last quarter of a century, it is interesting to recall that some form of exploration for oil has been going on for over half the number of years that have elapsed since Western Australia was first settled.

DRILLING EQUIPMENT AND TECHNIQUES

When the first bores were drilled for oil in Western Australia in 1902, drilling equipment was already in use in the country, exploring for water, coal and gold.

Naturally, the most common type of equipment available was water-well drilling-plant and the most popular method was the cable-tool or percussion rig. This type of rig was used for most of the oil exploration holes up to 1939, when the introduction of a standard rotary drilling-rig by Freney Kimberley Oil Company marked the end of cable-tool drilling for oil and gas in this state.

There are, however, some interesting deviations from the generalized pattern indicated above. During the disastrous oil boom of the years 1902-4 in the Warren River valley (see Fig. 5.17), three holes were drilled by a diamond drilling-rig owned by the Goldfields Diamond Drilling Company of Kalgoorlie for the Westralian Mining and Oil Corporation. The deepest well reached 524 m without any success.

In 1921 arrangements were made to rent out two calyx drilling rigs from the Public Works and Water Supply Department to Freney Kimberley Oil Company. The calyx rig was a rotary plant, capable of cutting cores, and apart from its drilling-

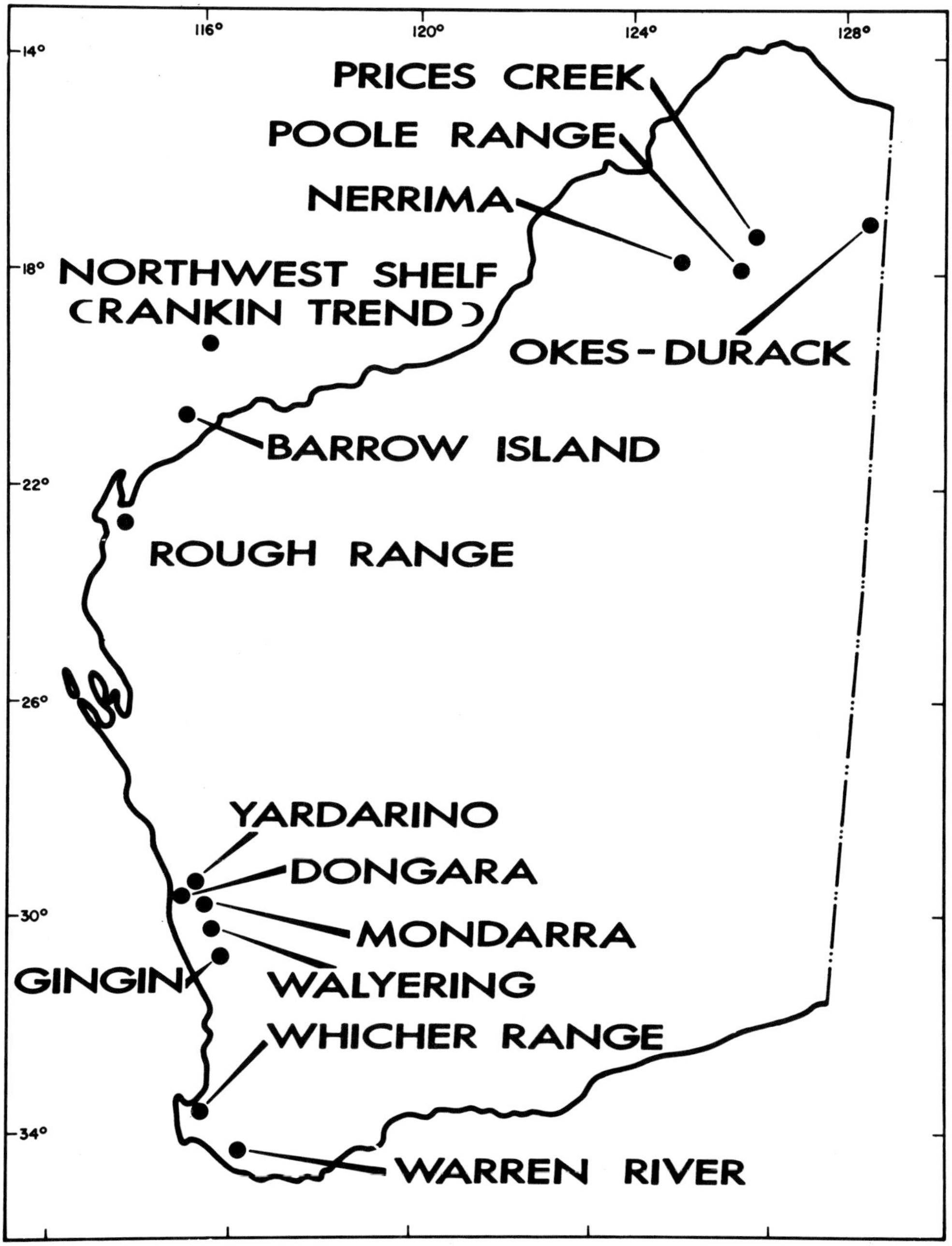

Fig. 5.17 Western Australia: oil drill-hole localities mentioned in text

head or cutting-tool, it appears to have been very similar to the forerunners of modern rotary drilling-rigs. The rigs were shipped to the Kimberleys in 1922 and one of them drilled Mt Wynne No. 1 to a depth of 273 m before being replaced by a cable-tool rig. It is not clear whether the other rig was ever used to drill any holes by Freney, or whether it was used for spare parts back-up.

All rotary rigs used for oil-drilling in Western Australia have been imported. The country of manufacture invariably was the United States of America.

Through the period to 1941 all drilling-rigs were steam-powered. The first standard rotary rig used at Nerrima No. 1 by Freney Kimberley Oil Company between September 1939 and October 1941 was driven by steam power. The rig used by West Australian Petroleum Pty Ltd (WAPET) to drill at Rough Range in the Carnarvon Basin in 1953, and all subsequent drilling-rigs, were powered by internal-combustion engines, generally by direct mechanical drive. The first diesel-electric power-driven rig was introduced in 1972 at Wonnerup No. 1 in the South-West Division.

Nearly all offshore rigs have been of the diesel-electric type. The first offshore rig (ship-shape, floating style) was the *Glomar Tasman*, which spudded Legendre No. 1 in June 1968, in water 52 m deep offshore from Port Hedland. The first offshore jack-up rig spudded Quinns Rock No. 1 in October 1968, in water 40 m deep just off Perth. This was the *Jubilee*, which was capable of drilling in waters up to 76 m deep. The first semi-submersible rig was the *Ocean Digger*, which started its first hole, North Rankin No. 1, in 1971. The deepest water (388 m) in which a rig has operated was at East Mermaid No. 1, where the Sedco 445 rig drilled to a total depth of 4069 m.

Prior to the commencement of true oil-exploration drilling some notable depth statistics for wells already existed. They included the Melville Park Estate Waterbore in the metropolitan area, drilled to 454 m in 1896, and the Cardabia Deep artesian bore drilled between Carnarvon and Exmouth in 1909-11. This latter well was drilled to 1251 m with a calyx rig. It remained this state's deepest well for the next 30 years, and was probably one of the deepest wells in the world when it was completed.

Drilling depths in early oil searches were quite modest compared to the best in water-well drilling. Of the first six holes drilled in the Kimberleys by Freney Kimberley Oil Company, the deepest was 307 m. Mt Wynne No. 3, drilled between November 1923 and September 1925, reached 657 m. Between March 1927 and November 1930 Freney Kimberley Oil Company drilled Poole Range No. 3 to 995 m, and after completing a further two shallower wells, brought to an end what can be regarded as the first round of serious drilling for oil in the state.

On 22 September 1939 the Freney Kimberley Oil Company commenced Nerrima No. 1 with a standard rotary oil-drilling rig owned by the Federal Government. This well reached a depth of 1302 m and operations ceased in October 1941, without the rig having attained its full programme depth. The reason for this early cessation was the inability of the operator to obtain spare parts, equipment and drilling-bits from the United States of America because of the Second World War.

The well did achieve one milestone, however: it was the deepest hole drilled in the state at the time.

There then followed a hiatus in drilling activity until 1953, when the discovery of oil at Rough Range (see Fig. 5.17) in WAPET's first hole sparked considerable interest in oil exploration in Western Australia.

Freney Kimberley Oil Company joined with Associated Australian Oilfields to form Associated Freney Oilfields and returned to the Kimberleys in 1955, first drilling a well on the Nerrima Structure (thus completing the original objectives of the well halted during the war). Two more wells (Myroodah No. 1 and The Sisters No. 1) were drilled in 1956-57, before the company ceased all drilling operations in the state.

A number of other companies became involved in oil-drilling over the next 20 years, but lack of success curtailed their activities until a consortium of companies commenced drilling on the North-West Shelf. This group of companies, now under the direction of Woodside Petroleum Development Pty Ltd, made a number of spectacular gas discoveries in fairly deep water (80 m to 133 m), during 1971-72 in an area which the explorationists call the 'Rankin Trend' (Fig. 5.17). Development of these large reserves is nearing reality, and Western Australia will shortly see a new era in drilling: the development of gas fields by controlled-angle drilling from fixed platforms.

Since 1953, drilling for oil and gas in this state has utilized the latest technology and equipment available in the world, and long-range planning has been an absolute necessity when one considers the unusual remoteness from supply centres of oil-drilling equipment.

Through a combination of sound practices and governmental standards, the oil-drilling industry in Western Australia has shown a keen sense of safety and a deep appreciation of the environment.

Drilling practices

A diamond drilling-rig was the first type of rig used to drill for oil in this state at Warren River, and it was by no means ideal, since this method was designed for use in hard rock and the maintenance of stable borehole conditions was not possible in unconsolidated sediments.

Cable-tool or percussion drilling held sway in the 1920s and 1930s (Fig. 5.18). The cutting-tool consisted of a chisel-shaped head at the bottom of a series of long, heavy bars. These were connected to the drilling-cable (originally manila rope, but later wire rope was—and still is—used). The drilling tools were lifted by the drilling cable off bottom for about a metre and then released, allowing the heavy cutting-assembly to drop down to chip and pulverize its way through the rocks. After about 30-50 cm had been broken up, the rock debris was recovered in a bailer lowered on another rope (referred to as the 'sand-line'). The sand-line was spooled into the hole via the 'calf-wheel', and the drilling line was spooled in and out via the 'bull-wheel'. These terms relate to the early association of drilling personnel with the cattle ranchers of Oklahoma and Texas. The up-and-down action for drilling was effected by connecting the line to a 'walking-beam' by means of the 'temper-screw'. The

Fig. 5.18 An early cable-tool rig in Western Australia. The Okes-Durack bore in the East Kimberleys, 1924

walking-beam was a huge, simple lever, its fulcrum at the central point supported by an upright called the 'samson-post'. A steam-engine drove a crank wheel, which was connected to the other end of the walking-beam by a wooden rod called the 'pitman', thus converting the circular drive of the crank wheel to the reciprocating motion of the walking-beam.

The process of 'making hole' was slow. This is indicated by the time taken to drill certain holes as described elsewhere in this article. The method did not lend itself to urgency and quite often only daylight hours were worked; in addition weekends, as days off from drilling, were sometimes observed. The early Western Australian cable-tool drilling for oil was all in the Kimberleys and supplies to the rigs were always troubled by the wet seasons. Under those conditions delays for parts and supplies, commonly brought by donkey or bullock wagons, could amount to weeks at a time.

Rotary drilling, as the name implies, involves a turning action of the cutting tool to break the rocks. Cutting tools have evolved into specialized drilling bits, generally with three revolving cones carrying specially-treated steel teeth, single-cone bits with

Fig. 5.19 A modern rotary rig in Western Australia. The jack-up *Maersk Endurer* drilling between Barrow Island and the mainland, 1977

diamonds embedded in steel, and bits designed to cut cores of rock offer a wide variety of cutting heads for all manner of rock types. Pressure is applied on top of the bit by the weight of a portion of the 'drilling string'. The drilling string, composed of tubular steel joints (each about 10 m long) screwed together, carries out two functions: one is to provide the rotary action to the bit, achieved by turning the string at the surface by a rotary 'turntable'; the other is to carry the drilling fluid ('mud') down to the bit through the hollow drilling string. The drilling fluid is of paramount importance, for it is the medium by which the broken rock is brought to the surface—on the outside of the drilling string. It also performs other duties, some of which are preventing flows of fluids from the rocks to the well-bore, cooling the bit, and stabilizing soft rock. Early drilling fluids were composed of water and clays from the rocks penetrated. These generalized principles of rotary drilling have not materially changed from the time this type of drilling was first introduced.

In 1953, with the advent of modern rotary equipment, there came a marked change in oilfield drilling practices. The high cost of rigs ($1 160 000 for the National 130 rig at Rough Range No. 1) and ancillary equipment and contracts meant that any loss of operational time was expensive. Rigs operated 24 hours a day; and roads and airstrips were built or upgraded to ensure rapid transportation of parts, supplies and personnel at any time. Of course, cost-saving was not the only reason for this new sense of urgency. The rotary drilling process, with a chemically treated mud system and an increasing knowledge of rock properties and pore fluid pressures, required that long periods of open-hole conditions be avoided because adverse reactions could occur within the rock, particularly from the effect of mud filtrate on some types of clay, causing the clay to swell. These reactions, if allowed to continue for any length of time, could result in the collapsing of the wall of the hole, or could so damage the pore spaces in the rock that no oil or gas could be produced from it.

When the oil search moved offshore, drilling costs increased again. An offshore drilling-rig has to be completely self-contained (Fig. 5.19). It has to house all the men working on it, together with repair shops and storage areas for all of the equipment and materials used in the drilling.

The only offshore drilling-rig to be constructed in Western Australia, the *Ocean Endeavour*, was completed in 1975 at a cost of $23 000 000. Drilling practices have to be tightly controlled when using such expensive equipment so that every minute of drilling time is used effectively. The hire of such a rig and its ancillary workboats and helicopters can cost up to $70 000 per day.

Drilling costs

Although contract rates for drilling in the early days are difficult to establish, some interesting figures are available. In 1902 an agreement between the Goldfields Diamond Drilling Company and Westralian Mining and Oil Corporation specified a 1500-foot (460 m) well to be drilled in the Warren River area at the following rates:

First 1000 feet (305 m) @ 18*s* ($1.80) per ft.
1000-1250 feet (305-381 m) @ 19*s* 6*d* ($1.95) per ft.
1250-1500 feet (381-458 m) @ 21*s* 6*d* ($2.15) per ft.

These rates were provisional on the section being limestone, shale or sandstone. The equipment used was a diamond drilling-rig.

In 1922 the Freney Kimberley Oil Company rented two calyx rigs from the state government. The terms of the rental agreement were £400 ($800) deposit and £5 ($10) per week per rig. These two rigs were valued at about £5000-£6000 ($10 000-$12 000) each at that time, and were capable of drilling to 1220 m.

The Okes-Durack Kimberley Oil Company used a 1220-1524 m capacity cable-tool rig (Fig. 5.18) for 14 months (1923-1924) on the only hole drilled in the Kimberleys by that company. On 24 July 1924 the company reported it had expended £12 062 ($24 124) plus £9000 ($18 000) for 'plant'. The well reached a depth of 365 m. Thus the all-up drilling cost was roughly $114 per metre. (This would have been an inflated figure, however, for it included the total value of the 'plant'. Assuming 20 per cent depreciation of the 'plant', the cost would be $76 per metre, and assuming only half the company expenses could be charged to drilling, the actual drilling cost would have been \$44 per metre.

Time, circumstances, distance and the changing value of money make cost comparison a hazardous undertaking; nevertheless, in 1977 values a modern exploratory hole to a depth of 3000 m in the Perth Basin (onshore) would cost in the region of \$1 300 000, i.e. \$433 per metre.

Offshore drilling costs are appreciably higher. A hole drilled in 1977 by a semi-submersible to 3000 m, in a water depth of 150 m, as part of a series of wells on the North-West Shelf would have cost approximately \$4 400 000, i.e. \$1470 per metre.

Detection of oil and gas

The physical detection of oil or gas in a drilling well was relatively simple in the early days of oil-drilling. If an oil accumulation was encountered, it was expected that the driller would see the oil on the drilling tools; that the oil would be bailed from the hole; or that oil would flow to the surface.

Drilling was usually initiated where hydrocarbons such as asphaltum occurred on the surface, or where a water surface had an oily appearance. Wells were also located in anticlinal areas. If these conditions were believed to have been met, there was a basic premise that discovery was certain (e.g. the Warren River bores). Any subsequent failure to find oil could be attributed to insufficient depth of drilling. As drillers and geologists began to recognize the complex nature of the in-hole conditions, they began monitoring the drilling characteristics of the strata and correlating them with the rock types which were brought to the surface either as cuttings in the mud or by cores. The important correlations provided the distinction between cap-rock and reservoir rock.

The importance of the reservoir rock as host to a possible oil accumulation led, in the late 1920s, to the measurement of physical rock properties such as porosity (percentage by volume of pore space) and permeability (ability to transmit liquid or gas when a pressure differential is applied).

At about the same time, the first wireline logging techniques were developed. They were called 'electric coring' and consisted of lowering electrodes down the borehole and measuring the electrical properties of the rock such as electromotive force and resistivity. Readings from the moving electrode were transmitted to the surface and transcribed to a paper log through an oscillating pen. The readings so obtained could help determine whether the rock contained gas, salt or fresh water, or oil.

Mudlogging techniques were developed in the 1930s and the basic instrument—the 'hotwire gas detector'—is still in use today. Gas in the rocks is absorbed in the drilling mud and carried to the surface, where it is released. A small vacuum-pump draws the released gas across a heated platinum wire which is part of a balanced Wheatstone bridge. Combustible gas raises the temperature of the wire, which throws the Wheatstone bridge out of balance and the deflection—as an indication of the presence of gas—is recorded.

Although all these tools were available to the oil industry in the late 1930s when drilling began at Nerrima No. 1, the evaluation of this well was made by the visual examination of cores. Had oil been detected in a core, one of the new sets of drill-stem-testing equipment was on hand. The testing procedure entailed sealing off the desired interval from the weight of the mud column by means of a rubber packer. This process allowed fluid or gas in the rock beneath the packer to be exposed to a pressure much less than that of the mud column and thus allowed it to enter the drill stem and be recovered at the surface.

The advent of radioactivity logging in the 1940s allowed direct downhole measurement of porosity, and tools of this type were run for the first time in Western Australia at Rough Range in 1954. From this time onwards new developments in formation evaluation have been introduced into Western Australia in step with their introduction elsewhere in the world. Of particular importance are the acoustic, density and focused resistivity tools. In varying combinations these tools can provide quite accurate determinations of rock types (such as sandstone or limestone), and their porosity, and whether the pores contain water, oil, or gas. These determinations are referred to as synergetic logging, and it is not uncommon for as many as twelve different types of surveys (electrical, radioactive, acoustic, mechanical) to be made in a well. The information obtained from these surveys is now recorded both on moving film by a photo-galvanometer and as digits on magnetic tape.

OIL AND GAS PRODUCTION

Barrow Island oilfield

Oil was discovered in August 1964 in the first well drilled on Barrow Island. It flowed at a rate of 157 m^3 per day, but immediate follow-up drilling showed that the accumulations below 1800 m in Lower Cretaceous and Upper Jurassic rocks consisted of numerous small pools. In February 1965 the Barrow No. 4 well was completed as a shallower pool discovery with an initial production rate of 20 m^3 per day from a depth of about 750 m in Lower Cretaceous rocks. This accumulation is known as the Windalia Pool and it now contributes almost the entire Barrow Island production (see Fig. 5.20). The Barrow Island oilfield was declared commercial in May 1966 and the first tankerload of crude was taken off in April 1967. In the ensuing 10 years about 21 million m^3 were shipped from the island, mainly to be refined at the BP Kwinana Refinery. The oil is a light (36° API gravity), low-sulphur, paraffinic crude.

The field extends over an area of 100 km^2 and the pay zone is about 15 m thick. Permeability of the pay zone is very low, and primary production which results from gas expansion was expected to be only 18.1 million m^3 by the year 2000. This situation prompted WAPET to make a pilot waterflood study in 1968, and the ensuing success led to a large waterflood which should allow total production of more than twice the expected primary production.

In 1977 there were 531 wells on Barrow Island. They comprised 322 producers, 11

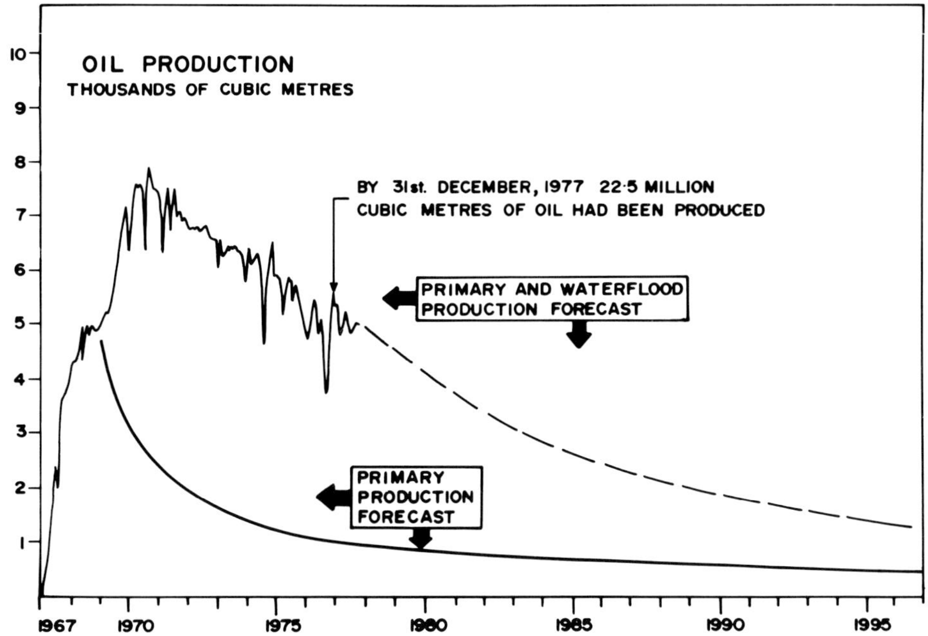

Fig. 5.20 Barrow Island oilfield: production curve

dryholes, eight water-source wells, 187 water-injection wells and three deep gas wells, which are shut in. Water from the water-source wells is pumped down the injection wells and out into the pay zone. The water flushes the oil from the reservoir, and oil and water are recovered from the nearby producers. The water, oil, and gas are separated and the oil is then pumped to storage-tanks, from which it is transferred to tankers by a 10-km submarine pipeline.

Gas fields of the Perth Basin

Drilling by WAPET in the northern Perth Basin began in 1961, and in 1964 a discovery was made at Yardarino. Yardarino No. 1 flowed 433 000 m^3 of gas per day and the No. 3 well flowed 207 m^3 of oil per day, but Nos 2 and 4 were dry. The Yardarino pool is contained in a Triassic sand about 30 m thick, where it is formed into a small anticline.

In June 1966 the Dongara Gasfield was discovered a few kilometres west of Yardarino, the two fields being separated by a large fault. Dongara No. 1 flowed 291 000 m^3 of gas, again from the Triassic sandstone. There are now 20 wells in the field, of which 10 supply the 2 266 000 m^3 consumed daily in the Perth-Pinjarra area. The gas is transported by a 355.8-mm (14-inch) pipeline, which was completed in 1971. Four wells have been plugged and abandoned as dry holes and six intersected a relatively thin oil column. Limited production of the oil began in 1976.

The Mondarra Field was discovered in 1968, south of Yardarino when Mondarra No. 1 flowed 140 000 m^3 of gas per day from Triassic sandstone. The three follow-up wells were disappointing, but Mondarra No. 1 is connected into the Dongara-Perth pipeline.

Farther south, gas was discovered at Walyering in 1971. An initial production test of the Jurassic sandstone reservoir flowed 291 000 m^3 of gas per day, but production declined rapidly. The two following wells were unsuccessful, and although Walyering No. 1 was connected into the Dongara-Perth pipeline, it declined so rapidly that it was shut in after only four months' production.

A discovery similar to Walyering had been made at Gingin in 1965. The same Jurassic reservoir yielded initially encouraging flows which declined rapidly and the follow-up well was very disappointing. In 1971 the Gingin No. 1 well was stimulated by fracturing the reservoir and it began flowing at a daily rate of 370 000 m^3 of gas with later production stabilized at around 140 000 m^3. This gas was also fed into the Dongara-Perth pipeline for a limited period.

The foregoing summary reflects the success which WAPET enjoyed in discovering gas in the Perth Basin, and it is unfortunate indeed that so frequently the impermeability of the reservoir sands prevents continuing economic flows.

CHAPTER 6

Mining law in Western Australia

B. M. Rogers
Under-Secretary for Mines, Western Australia

HISTORICAL

'Mineral Lands' in Western Australia were acknowledged in early legislation under 'Land Regulations', but the first legislation introduced specifically for mining was in 1854 when an Act (17 Vict. No. 17) was enacted as an ordinance to provide for the preservation of order in cases of any discovery of gold in Western Australia.

In the *Government Gazette* of 14 February 1860 lands under the Land Regulations were fixed into four basic categories: Town, Suburban, Country, and Mineral Lands, and in 1862 the Legislative Council offered a reward of £5000 for the discovery of a workable goldfield within 80 km of Perth.

The Land Regulations continued to be modified from time to time and contained a blend of regulations relating to the disposal of land and resources within the state and also the elements of protection of improvements on private property, the principle of which has been carried through to the present day.

The Goldfield Act of 1886 (1 October 1886)

The discovery in 1885 of payable gold at Halls Creek in the Kimberleys and the gold-rush that followed brought about a need for significant laws governing mining, the first of which was the Goldfield Act of 1886 (50 Vict. No. 18).

This was amended by the Goldfield Act 1886 Amendment Act 1888 (51 Vict. No. 24), which defined under what circumstances a reward for the discovery of a new goldfield was to be payable.

The Gold Declaration Act of 1888 (52 Vict. No. 8), which was described as an act for the purpose of ascertaining the quantity of gold found in or procured from the soil of the colony of Western Australia, followed discoveries in that year of gold at Mallina, Pilbara Creek, Cuddingwarra and the Yilgarn.

Mining and mining law were on the move as the tempo of gold discoveries increased, with further finds at Ashburton River, Oakover River, Nullagine and Roebourne in 1889, followed by the Murchison in 1891. A bill to amend the law relating to mineral lands, cited as the Mineral Land Act 1891, formed the basis for the Mineral Land Act 1892.

The Mineral Land Act of 1892 (55 Vict. No. 3)

The Mineral Land Act of 1892, assented to on 1 February 1892, repealed 'so much of the Land Regulations proclaimed on the 2nd March 1887 as are comprised in Part VI Sections 80 to 92, but such repeal shall not effect any rights, claims, or liabilities already accrued or incurred under those regulations'.

This Act was most significant in that it allowed for a lease of not less than 20 acres to be granted as private property. It had still to be read in conjunction with, as necessary, the Lands Regulations 1887, Goldfield Act 1886 and 1888 amendments, and the Gold Declaration Act of 1888.

Further gold discoveries followed at Dundas, Norseman, Coolgardie, Mt Margaret and Laverton in 1892, Kalgoorlie in 1893, The Pinnacles in 1894, and Yalgoo, Lawlers, McCaffey's and Lake Darlot in 1895.

On 1 January 1894 the Department of Mines was established to administer the laws of the growing industry.

The Goldfields Act 1895 (59 Vict. No. 40)

The Goldfields Act 1895 was assented to on 12 October 1895 as an act to make better provision for the regulation of gold-mining and management of the goldfields of the colony, and covered matters including miner's rights, business licences, gold-mining leases, and administration of justice and appeals.

The Mines Regulation Act 1895 (59 Vict. No. 37)

On the same day (12 October 1895) the Mines Regulation Act 1895 was assented to. It dealt with the practical aspects of mining and covered such matters as inspectors and mine managers and their appointment, ventilation in mines, safety precautions, use of gunpowder for blasting, etc.

Goldfields Act Amendment Act 1896 (60 Vict. No. 36)

The Goldfields Act Amendment Act 1896 was assented to on 27 October 1896. This amendment, made in the same year as gold was discovered at Mt Morgans, assisted the developing industry, with laws governing the installation, erection, maintenance, and rights of ingress and egress, of such works as electric, telephone and telegraph lines and tramways.

Mining on Private Property Act 1897 (61 Vict. No. 23)

The Mining on Private Property Act 1897 was assented to on 23 December 1897. Gold discoveries at Broad Arrow and Peak Hill extended the localities known to contain gold throughout the colony, and while these latest finds were not on private property, the need was seen for an Act to permit Mining on Private Property.

The Act covered all private land defined as 'including all land which is not Crown Land', but specifically excluding the lands granted with mineral rights to the owner of the Hampton Lands and Railways Syndicate Limited. Gold was discovered on several of that company's properties near Coolgardie, at Celebration City, Barbara and Surprise, and the company leased the land under much the same regulations as the colony's mining laws. The Act allowed for the provisional resumption of private land for mining purposes, proclamation of private land as alluvial goldfields, compensation for damage, and payment of royalty, part of which was payable to the owner. It further allowed for the issue of a Special Licence to mine, and the creation of a Compulsory Mining Lease in the event that the owner did not wish to mine the land himself, and refused to permit others to mine. At the same time it protected the owner's improvements, such as his dwelling-house, well, garden, orchard and so on.

This Act was repealed by The Mining on Private Property Act 1898 (62 Vict. No. 29), assented to on 28 October 1898. It contained many of the elements legislated in the former Act of 1897. The last amendments to the Act were assented to on 16 December 1899 (63 Vict. No. 31) as the Mining on Private Property Act 1898 Amendment Act 1899, and this Act continued until it was repealed by the Mining Act of 1904, with the exception of sections 56 to 62 and the schedule, which were carried forward into the 1904 Act.

The Goldfields Act 1895 Amendment Act 1898 (62 Vict. No. 16)

The Goldfields Act 1895 Amendment Act 1898 was assented to on 28 October 1898. It dealt with the problems of the distinctions between alluvial and deep mining, and defined alluvial as 'any earth containing or supposed to contain gold, not being a lode, dyke, reef, or vein'. It dealt with the non-employment of Asiatic or African aliens in mines, and further stipulated 'that the labour of any aboriginal native of Australia shall not be accounted *bona fide* work for the purpose of fulfilling the labour conditions upon any claim, lease, or authorised holding'. It made provision for the receipt by the Under-Secretary for Mines as Warden, of applications 'not within any proclaimed goldfield'.

This Act was to undergo amendment as the Goldfields Act Amendment Act (64 Vict. No. 23 assented to on 5 December 1900) prior to its being repealed by the Mining Act of 1904. The principal additions and amendments were the introduction of a Miner's Homestead Lease, specifying that it was not to be deemed private land within the meaning of the Mining on Private Property Act 1898, and provision for the Warden to dispose of all applications and objections in open court, after inspection by him, or on the receipt of a report by a mining surveyor or Inspector of Mines, and further that he state, in open court, his reasons for recommending or rejecting any application to which objection has been made. It further conferred powers with the Warden or Resident Magistrate to grant a Gold Dealer's Licence, and provision for the renewal of these licences.

The Mineral Land Act Amendment Act 1899 (63 Vict. No. 48)

The Mineral Land Act Amendment Act, assented to on 16 December 1899, made the last amendments to the 1892 Act prior to its being repealed by the Mining Act of 1904, and together with its companion Acts, the Goldfields Act 1895, the Mines Regulation Act 1895 and the Mining on Private Property Act 1897 was to serve the mining industry until the introduction of the Mining Act 1904.

THE PRESENT AND FUTURE ACTS

Mineral exploration and the production of minerals in Western Australia are controlled and encouraged by the mining laws of the state. The most important of the Acts now in force are:

Mining Act 1904
Petroleum Act 1967
Petroleum (Submerged Lands) Act 1967
Mines Regulation Act 1946
Coal Mines Regulation Act 1946
Mining Development Act 1902
Mine Workers' Relief Act 1932
Inspection of Machinery Act 1921
Explosives and Dangerous Goods Act 1961

In addition to the above, the Gold Buyers Act 1921 controlled the buying and selling of gold in Western Australia until it was repealed in 1977, following amendments to the Commonwealth Banking Act, which made it legal for anyone to deal in gold or gold matter, so that the provisions of the state legislation were no longer necessary.

Of the laws controlling mining in Western Australia, the Mining Act 1904, which was assented to on 16 January 1904, is the principal legislation, and was enacted to consolidate and amend the law relating to mining for gold and other minerals, which became necessary following the tremendous impetus given to the mining industry by the gold discoveries of the 1890s.

This Act has been used to administer the search for and mining of the state's mineral assets for some 75 years. During that period amendments have been made from time to time, but there have not been any major radical changes and certainly none to the original principles of the legislation.

The 1904 Act is based on the principle that all minerals are the property of the Crown, with the exception of those within some minor areas of private land alienated prior to 1 January 1899, and even in these instances gold, silver and other precious metals remain the property of the Crown.

Provision is made for mining throughout the state, and at the same time protection is given to other land usage such as private property and reserved or exempted land.

Any person, on payment of a fee of 50 cents, may be issued with a Miner's Right which entitled him to search for gold and minerals, and to take possion of, mine and occupy Crown land for mining purposes.

The holder of a Miner's Right who desires to search for gold or minerals on private land must first obtain a Permit to Enter from the Warden of the locality concerned. The permit is limited to 30 days and it entitles the holder to enter on the land to search for gold or minerals and detach samples from any vein or lode outcropping on the surface but does not entitle him to mine or otherwise disturb the surface of the land. Once having pegged and applied for a mining tenement on private land, the applicant must negotiate a compensation agreement with the owner and occupier of the land and file it with the Department of Mines before the application can be granted, or alternatively if the land is unimproved and not being used, action can be taken in the Warden's court to have compensation determined. Such compensation is not for the minerals in the ground, but is for loss of use of, and damage to the surface of the private land and any improvements thereon or thereunder for severance and for rights of way, and all consequential damages.

Notwithstanding this, however, without the consent in writing of the owner and the occupier of the land, no mining tenement can be granted comprising private land

(a) that is in *bona fide* and regular use as a yard, garden, orchard, vineyard, plant nursery, plantation or land being used for the purpose of cropping or pasturing or land whether cleared or uncleared, used for the agistment of stock in the ordinary course of management of the owners land of which the land so used forms the whole or any part;
(b) that is used as a cemetery or burial place;
(c) that is used as a reservoir;
(d) on which there is erected a substantial improvement; or
(e) of less than 2023 square metres in extent,

unless the grant or occupation is limited to a greater depth than 30 metres from the lowest part of the surface of the land.

The holder of a Miner's Right may not occupy or mine reserved or exempted land without the authorization of the Governor.

The following are the main mining tenements available for the prospecting for and mining of gold and minerals.

GOLD

1. *Prospecting Area.* Designed for the small prospector with term of 12 months only with right to renew for a further 6 months. Maximum area 48 acres outside the limits of a goldfield or mineral field and 24 acres within the limits of any such field.

2. *Gold Mining Lease.* Designed for gold production over a long period of time. Lease of 21 years with right of renewal for a further 21 years over a maximum area of 24 acres.

3. *Dredging Claim.* Designed for rivers, lakes, swamps, marshes, etc. not suitable for conventional mining methods. Term indefinite provided conditions are complied with. Maximum area: 300 acres.

4. *Temporary Reserve.* For exploration only. If gold is discovered, a Gold Mining Lease must be obtained before production can commence. Maximum area: 300 acres.

COAL

1. *Prospecting Area.* For exploration purposes. Maximum area: 3000 acres.

2. *Coal Mining Lease.* For long-term coal production. Term same as for a gold mining lease. Maximum area: 320 acres.

OTHER MINERALS

1. *Prospecting Area.* Same as for gold, designed for the small prospector.

2. *Mineral Claim.* May be used for either exploration or production of minerals. Term indefinite, as long as conditions are complied with. Maximum area: 300 acres.

3. *Mineral Lease.* Generally used where a mineral deposit has been found and production is to commence. Term 21 years with right of renewal for a further 21 years. Maximum area: 300 acres.

4. *Dredging Claim.* Same principle as a Dredging Claim for gold. Used where conventional mining operations are not appropriate. Maximum area: 300 acres.

5. *Temporary Reserve.* For exploration only. If minerals are discovered, a Mineral Lease or Claim must be obtained before production can commence. Maximum area: 77.7 square miles.

The shape of all tenements except Dredging Claims should be as near as practicable rectangular with the length not exceeding twice the width, and all except Temporary Reserves must be physically marked off on the ground and pegs or markers placed to indicate the area of ground applied for.

The Act provides generally that every mining tenement must be used for the purpose for which it is granted, unless there is a satisfactory reason for not doing so, and in such cases there is provision for the holder to be exempted from this requirement. Tenements not used and not exempted are liable to forfeiture. Warden's courts are established in Perth and in fifteen centres throughout the goldfields and mineral fields to hear applications for mining tenements and disputes which may arise from them.

During the 75 years the Mining Act 1904 has been in existence, it has stood the test of time well and it was not until the advent of the very large-scale bauxite and iron-ore mining in the early 1960s that a need for change became apparent. These projects, with their massive infrastructure requirements, were catered for under special state/company agreements ratified by Acts of parliament, such as the Alumina Refinery Agreement Act 1961 and the Iron Ore (Hamersley Range) Agreement Act 1963.

At the end of 1978 a new Mining Act was passed which updated the laws to cater for modern prospecting and mining, while at the same time retaining the sound basic fundamentals and principles which have existed under the Mining Act 1904 to encourage prospecting and development of the state's mineral resources. The new Act will not be proclaimed until regulations have been made and these are currently under consideration.

PETROLEUM

Legislation covering the search for and production of petroleum is contained in:

1. *Petroleum Act 1967*, which repealed the Petroleum Act 1936. This Act governs onshore petroleum operations, i.e. land above the mean low water (including islands) and such submerged lands as lie below waters internal to the State of Western Australia.

2. *State and Commonwealth Petroleum (Submerged Lands) Acts 1967.* These Acts are minor legislation which came into operation on 1 April 1968 and which govern the exploration and production of petroleum resources in submerged lands adjacent to the Western Australian coast, comprising the continental shelf and the seabed and subsoil beneath territorial waters. They are administered by the State Minister for Mines as the designated authority on behalf of the state and the commonwealth. The principal tenures available under the Petroleum Acts are exploration permits and production licences.

An *Exploration Permit* has a maximum area of 200 blocks under the onshore legislation and 400 blocks under the offshore legislation. (A *block* is a graticular section 5 minutes of arc of latitude by 5 minutes of arc of longitude and is approximately 65 km^2.) Onshore permits are granted for an initial period of five years and offshore permits for six years, both being renewable for further five-year periods, with 25 per cent relinquishment of onshore area (50 per cent offshore) at the first renewal, and thereafter 50 per cent relinquishments at each subsequent renewal in both cases.

Upon the discovery of petroleum within a permit, a 'location' of up to nine blocks, including the discovery block, is declared. The establishment of a location sets aside and identifies a specific number of blocks from which a *Production Licence* can be selected. The permittee has two years from the date the location is declared in which to apply for a Production Licence; however, this period may be extended by a further two years.

Production Licences are for an initial term of 21 years, with provision for renewal for a further 21 years and thereafter renewals for such periods, not exceeding 21 years, as may be determined.

HEALTH IN THE MINING INDUSTRY IN WESTERN AUSTRALIA

Safe working in mines

The Mines Regulation Act and Coal Mines Regulation Act are primarily concerned with safe working conditions for mine employees and with prevention of the contraction of occupational diseases.

The main function of these statutes is to authorize regular and thorough inspections of all mining operations to ensure that safety precautions are maintained at all times. The Regulations to the Acts provide strict guidelines which must be observed in every mine in respect of the following:

(1) management and supervision,
(2) general safety,
(3) electricity and machinery in mines,
(4) explosives and blasting agents,
(5) ventilation and control of dust and atmospheric contaminants,
(6) occupational diseases,
(7) hygiene and sanitation,

(8) transport and winding ropes and signals,
(9) safety and protection both underground and in quarries,
(10) railway operations in mines.

Occupational diseases

The occupational diseases in the mining industry in Western Australia have primarily been associated with tuberculosis, silicosis and asbestosis.

In the early days of the industry, tuberculosis took heavy toll of men working in the metalliferous mines, and so state legislation, patterned on that of South Africa in the first instance, has developed with the following objectives:

(1) to prevent men with tuberculosis from entering or remaining in the industry;
(2) to encourage men with silicosis, and more recently with asbestosis, to leave the industry;
(3) to provide compensation and/or other benefits for men who are prohibited from further work in the industry because of tuberculosis, or who leave the industry with silicosis or asbestosis.

It was not until 1925 that legislation to deal with the problem became operative, and then the sequence was as follows:

1. the Miner's Phthisis Act of 1922 was proclaimed to come into operation on 7 September 1925;
2. new regulations under the Mines Regulation Act were gazetted on 12 March 1926;
3. the 1924 amendment to the Workers' Compensation Act was proclaimed to come into operation on 15 March 1926;
4. the Mine Workers' Relief Act of 1932 was proclaimed to come into operation on 1 February 1933.

This legislation is still in force today, albeit with some changes.

However, a forerunner to the legislation was the Mine Workers' Relief Fund Incorporated, which was inaugurated in February 1915, and continued until superseded by the coming into operation of the Mine Workers' Relief Act on 1 February 1933.

CHAPTER 7

The economic and social significance of Western Australian mining developments

ALEX KERR
Murdoch University

THE MINING INDUSTRY IN AUSTRALIA'S PAST DEVELOPMENT

The coal discoveries of the late eighteenth century, the spectacular goldrushes of the mid-nineteenth century, and the oil, iron and nickel discoveries of the mid-twentieth century all demonstrate the continued importance of the role played by the mining industry in Australia's overall economic development. In this decade, indeed, many of the earlier phases of mining activity have been vividly re-created on the TV screen by Australia's burgeoning documentary film industry, and have thus made us familiar with some of the landmarks of our early and ever-so-new history.

The future of Australia as an industrial nation began with the discovery of coal near the mouth of the Hunter River, New South Wales, in 1797. By 1799 it was being exported, and by 1840 coal-mining was the nation's first established mining industry and one on which the foundations for Australia's industrial base were built. Other mining discoveries, including that of copper in South Australia in 1842, made their contributions to the development of the new colonies, but the first really great impetus to Australia's growth was given by the discovery of gold in payable quantities near Orange, New South Wales, in 1851.

Probably the most immediate and obvious effect of the mid-century gold discoveries was the upsurge in Australia's population. In the decade from 1851 to 1861 Victoria's population jumped from 77 000 to 540 000, outstripping that of New South Wales, which in the same period rose to 350 000. The problem of housing the newcomers was frantically and imperfectly solved with canvas and corrugated iron; the modern order of the day is aluminium, asbestos and air-conditioning. Gold brought avarice, crime and a crop of bushrangers who were tamed by the harsh methods of the Victorian police. The colourful diggings left their indelible mark on Victorian and, indeed, Australian life thereafter.

With the new population came a new political awareness. Among the diggers were many who had informed and vigorous political views, but who had refused to be docile in their homeland and whose independence now spawned such groups as the Ballarat Reform League, the Republic of Victoria, and those who set up the Eureka Stockade. When the gold began to run out, many of the diggers left the declining fields to try their luck elsewhere and in other pursuits. A proportion of them slipped imperceptibly into the role of transitory diggers moving on, with intervals of casual work, to New Zealand or to Queensland, and even as old men to the thirsty diggings of Kalgoorlie and Coolgardie in Western Australia, and to Alaska and the Yukon. Many, however, remained and found work equipping the newly-swollen community with hotels, schools, roads and bridges, railways and fencing for sheep-runs. Although the gold fever had passed, a permanent impression remained on the Australian economy, which was now moving steadily towards increasing industrialization and diversification.

The wave of mining discoveries hardly lost momentum during the second half of the nineteenth century, but thereafter the industry lapsed into the doldrums, where it remained for fifty years. The peak in the value of Australian mineral production reached early this century, in 1907, was not equalled again until 1937.

Neither the discovery of copper at Mt Isa in 1930, nor the more recent discovery of uranium at Rum Jungle in 1949, did much to step up the tempo of prospecting, but two other post-war events did. The first was the discovery of iron-ore deposits in the Northern Territory and in Northern Queensland, between 1950 and 1954, and the second was the discovery of enormous deposits of bauxite in Northern Queensland in 1955. These discoveries ushered in a period of mineral exploration and exploitation in the 1960s, the like of which has not been seen before in Australia nor, for that matter, in many other countries of the world. These discoveries permitted pressure to mount for the removal of an embargo which had been placed upon the export of iron-ore immediately prior to the Second World War. This embargo had remained in force in the 1940s and 1950s, but in December 1960 it was lifted because of the increased estimates of Australia's known resources of ore. Almost immediately after the lifting of the embargo, staggeringly large deposits were announced in the Pilbara, dramatically changing the state of 'known' Australian iron-ore reserves and sparking off a development movement which rapidly gathered momentum and is still continuing.

Against this brief backdrop of Australian mining developments the growth of Western Australia's mining industry may be viewed. In this growth three stages of development may be distinguished. First, in the period 1885-1900 the northern part of the state was the centre of the first gold discoveries in Western Australia, but this soon swung to the much richer Kalgoorlie-Coolgardie area to the south. Substantial quantities of the valuable mineral were won, and in the decade of the 1890s the state's population quadrupled. The increased activity brought improvements in transport and communication, and heightened activity in the Perth region as well as in the goldfields.

In the second stage, from 1900 to 1940, mining played a minor role in the state's economy, and the industry, such as it was, was still reliant on the efforts of individual prospectors. The principal significance of this period lay in the discovery of minerals which were later to become the object of company attentions and large-scale exploitation.

The third stage was the post-war period, when the large company replaced the individual miner. Australian Blue Asbestos became the sole producer of blue asbestos (crocidolite) at Wittenoom Gorge in the Hamersleys. A new township was sited and built around the industry. On Cockatoo Island, Australian Iron and Steel commenced large-scale mining of iron-ore. Once again a company township sprang up. With the introduction of large earthmoving equipment the alluvial tin fields were opened up by large companies. Manganese likewise became the object of company activity, and mining-camps mushroomed throughout the North-West. The capital resources behind those interested in oil and gas was enormous. The era of large-scale production had arrived. The exploitation of Pilbara iron-ore resources was an extension of this new type of development, though on such a scale as to overshadow completely the other developments mentioned. To these were added the exploitation and development of resources for producing salt in the Pilbara and the exploitation of bauxite deposits near Perth and nickel deposits near Kalgoorlie. Since then the discovery of massive reserves of natural gas off the northwestern coast has transformed the picture of Australia's known energy resources.

The ramifications of recent mining industry development throughout Western Australia's 2.6 million km^2 are on such a scale and of such variety as to defy adequate treatment here. For example, the impact of bauxite extraction and associated alumina-refining activities on industrial development in the Perth area has been significant. Nickel extraction and treatment in the Kalgoorlie region has boosted a flagging gold-mining town and the associated nickel refining in the Perth area has added to metropolitan development. The exploitation of mineral sands on the south-western coast to obtain rutile, zircon and ilmenite also deserves mention, as does the development of the modest-sized gas field at Dongara, from which gas is now piped to supply the Perth metropolitan area.

But the scene is a constantly changing one—witness the movements in world nickel prices and the need for Western Mining Corporation to retrench in the Kalgoorlie area—and has to be viewed in a dynamic setting. First, however, let us look briefly at the characteristics of Western Australian mineral production.

PRESENT CHARACTERISTICS OF WESTERN AUSTRALIAN MINERAL PRODUCTION

We may first deal with some of the quantitative aspects of the state's mining industries. Mineral production in Western Australia in 1976-77 had a value of $1138 million, which represents an increase of 112 per cent from five years earlier. It is one quarter (25.3 per cent) of the Australian total. The major minerals currently produced in the state, in value terms, are iron-ore, nickel matte and metal, and alumina, in that order of importance. This production has come from 134 establishments, employing 14 936 persons (mining industries, highly mechanized as they are in Australia, are not great employers of labour), and they paid $199 million in wages and salaries, an average of $13 300 per employee. They had a turnover of $1387 million, and contributed to the national income a value-added of $908 million. The contribution to Australia's gross domestic product of the mining industry has in fact doubled in ten years. They also contributed royalties in 1976-77 of $56 million to the government and $20 million to other parts of the private sector. This represents somewhat more than a quarter of the Australian total of royalties paid by Australian mining companies. Royalty payments have about doubled in the last five years.

After an unprecedented increase in exploration activity in the last half of the decade of the sixties, mineral exploration at the state level has followed the trend observable for the nation as a whole. In 1970-71 some $86 million was spent on exploration activities, and the total thereafter declined, for the most part, until it stood at $65 million in 1976-77, a fall of 24 per cent. This reflects the depressing effects of government policy in these years upon private incentives and the ability of the private companies to devote part of their cash flow to this activity. It should be noted, however, that this was a trend observable throughout the world, with the possible exception of South Africa, Brazil and the Philippines.

Some of the non-quantitative aspects of Western Australia's mining industry may also be noted. The recent mining developments that have occurred in Western Australia have been carried through largely by the companies themselves, with government intervention or guidance having been kept to a minimum, although the state government has, throughout the development period, been active in giving every kind of encouragement to the companies concerned. Secondly, the character of the operation has changed. Sophisticated exploration and development techniques have replaced the older methods. Thirdly, the finance has come mainly from outside Australia. This was due in part to the small size of the Australian capital market, which could just not cope with the capital requirements on the scale demanded for the developments in prospect, and in part to the reluctance of Australian investors to provide risk capital. Overseas ownership has been, and still is, a matter of vigorous public debate. Fourthly, with immense financial support the companies responsible for developing the great mineral deposits of the state often bore most of the huge costs of establishing the necessary social and economic infrastructures such as roads, railways, power, schools, hospitals and so on. Complete new towns and ports, constructed with company funds, sprang up with great rapidity. Goldsworthy, Shay Gap, Karratha, Wickham, Dampier, Pannawonica, Tom Price, Paraburdoo, Mt Newman and Kambalda are names that were not known twenty years ago. Now they are new, living communities, interacting with those which existed previously, such as Roebourne and Port Hedland, to the mutual benefit of all concerned. Finally, extraction and subsequent export in raw form rather than processing has been the order of the day, though some processing and up-grading is now carried out, and most of the companies, in the agreements signed with the state government, are obliged to increase their local processing activities in the future.

Other qualitative aspects, such as the change in living conditions in these previously remote areas, will be dealt with below.

THE ROLE OF GOVERNMENTS, STATE AND FEDERAL

The relaxation in 1960 of the iron-ore export ban previously imposed by the Commonwealth Government in 1938 lifted the lid on the iron-ore resources of the Pilbara and made possible the rapid expansion that followed. But the relationship between the federal government and the mining companies in the Pilbara has not been a close one; the central government did not become directly involved in the provision of developmental finance as it did, for example, with the Ord River Scheme.

Nevertheless, certain specific central government activities have recently affected the development of the mining industry. Some of these, apart from direct control over all exports, have been general monetary and fiscal measures which impinged upon the industry in common with other industries, with some oriented specifically to the mining industry, in the form of direct controls. Restrictions on overseas borrowing, which were established and changed several times in the period from December 1972 to August 1974 would be one example, and a very unpopular one with the industry. Certain taxation changes which occurred in the 1973 and 1974 federal budgets were also unpopular in the industry. These were probably prompted by the Fitzgerald Report, which had concluded that the Commonwealth Government was collecting insufficient revenue from the industry and that the industry was dominated, to its detriment, by foreign ownership. To these measures could be added the successive revaluations of the Australian dollar against the U.S. dollar in December 1973 and 1974—total changes which amounted to an effective 26 per cent revaluation over a two-year period to mid-1974. The Australian revaluations were intended to curb export earnings, which were feeding inflation, and to discourage capital inflow into Australia, which was occurring at the rate of over $1000 million annually. Other direct controls, which included limitation on the foreign ownership content of Australian mining, occurred during this period.

These measures, all taken together, dealt the mining industry a severe body blow. If the government's intention was to reduce the rate of mineral development in Australia, it certainly succeeded. Some large projects on the point of commencing were deferred, others in the pipeline were postponed indefinitely, and others currently operating were slowed down. There are good grounds for arguing that the policy of increasing Australian ownership and participation—a policy that would meet with the general, though qualified, support of all Australians—was pursued too vigorously and rapidly with a declaration of intent that frightened off many potential investors. Thus the part played by the Commonwealth Government in the development of the mineral industry in the first half of the seventies was, to say the least, unpopular within the industry.

In more recent years the central government's attitude towards the mining industry has changed. Although in some respects the present coalition government's resources-development policy appears similar to that of the previous Labor government, some significant concessions were made to mining industry development. Liberal depreciation allowances (write-off over 5 years instead of 25 years) for development expenditures were announced as well as the removal of unpopular levies on oil discoveries and coal exports, and the introduction (not restricted to mining) of trading-stock valuation adjustments.

The industry has welcomed these changes but it is too early yet to determine the extent of the impact they will have on the construction, housing and other associated industries as developments that were previously shelved get under way. The climate has changed for the better as far as the mining industry is concerned.

Details of the various arrangements made between the state government and the mining companies working in Western Australia are too numerous to repeat here, but the concept was a new one and opened up an era of government-private enterprise partnership agreements which gave a new character to the mining development and, indeed, made it possible.

In exchange for guaranteed long-term security of tenure to large, rich deposits, the companies agreed to develop them to an agreed-upon time scale, progressively increasing the local labour or capital content with processing as time went on. The agreements are formal and public, having been ratified and published by the parliament of Western Australia. They were necessary because the state government had neither the money nor the know-how to engage in the mining operation itself, nor even the money to provide all the necessary infrastructure on the scale needed.

Resources-development policy is probably one of the main areas of contention between the State and Commonwealth Governments in recent years, irrespective of the political leaning of the government of the day at either level. The nature of Australian federalism and the division of powers and functions between the central and state governments almost guarantees conflict in times of potentially rapid expansion. The Western Australian Government in particular has followed a vigorously independent policy with regard to the development of its natural resources and can be expected to do so in the future. It can thus be anticipated that the government will, within the constraints of social priorities, including examination of proposed projects by the environmental protection authority, continue to encourage expansion in the mining industry because of what it sees as the economic and social benefits it brings to the region, the state and the nation.

ECONOMIC AND SOCIAL EFFECTS OF MINING IN WESTERN AUSTRALIA

One may think of the impact of mining industry development as occurring at three levels: local, state, national. One of the most obvious immediate effects to be observed at the local level is the inflow of population associated directly with the swelling of the workforce and indirectly with the growth of local service industries. Mining and quarrying in Australia's North-West, for example, occupied a mere 1800 persons in 1966, which was, nevertheless, an increase of 66 per cent on the figure of five years earlier. Five years later the number had grown to 7200, an increase of over 300 per cent. Accompanying this was a great increase in population owing to associated service activities. This increase in the workforce was directly involved in the mining industry and indirectly involved in associated service development, and the increase in dependants accompanying this growth has transformed the region and resulted in significant changes in living styles which, for the most part, are generally acknowledged as beneficial. These changes have been exciting, convulsive and frustrating at the same time. They have created opportunities, posed problems, and offered challenges.

The main factors affecting living conditions in the region are the climate, the smallness of the region's population centres and their isolation in a country of vast distances. Climate imposes constraints on home-making, eating and dressing habits, times of rest, sporting and other activities, and house design, living costs and costs of transport and communication. The development of mining has eased many of these burdens, particularly domestic living conditions, with better generation of electricity and the widespread use of air-conditioning in company towns. All the towns, while still small by southern standards, are much bigger than they were twenty years ago and the range of amenities and services now offered in the north is much greater and has reduced the previous sense of isolation, felt particularly by the women, to tolerable levels. Decentralization, a by-product of the mineral boom, has brought a new vitality to the region and even a sense of belonging, as first generation northern town-dwellers are now appearing in growing numbers and remaining in the region.

Of course, there is still much improvement to be made in all aspects of living conditions, as the various women's action groups will hasten to point out. There are still the tensions and strains produced between company towns and open towns, within company towns, and by the high cost of living and other disabilities of life far away from the metropolis. There is also still a relatively large transient working population, lured by attractive wages, which does not identify with the region.

Apart from improvements in living conditions and general amenities, the multiplier effects—housing development, the development of ports and townships, upgrading of general infrastructure throughout the north, incomes and employment generated in service industries, the development of tourism and the opening of the north to Australians from other parts of the state, the nation, and overseas countries—are substantial and clearly to be seen by those who knew the north before these mining developments all started.

At the state level, several significant contributions of the mining developments are immediately apparent. In money terms, perhaps the most easily quantifiable are royalties, which have already been mentioned as rising to $76 million in 1976-77. The balance of state trade has, of course, also changed markedly. The state has always had a surplus on trade with overseas and a deficit on trade with the rest of Australia. Ten years ago the overseas surplus was $262 million and the interstate deficit $359 million. In 1976-77 the overseas surplus was $1767 million and the interstate deficit $1347 million.

Apart from these easily identified benefits there are others, less easy to assess in money terms but just as tangible and important. The multiplier effects, in terms of income, employment and general development, that have spilled over from the mining areas to the Perth metropolitan region are considerable. The face of Perth has been transformed as the state developed with a rapidity that was largely nurtured by the development of its natural resources. Some of its natural endowments are just now in fact being realized as the tourist industry is attracting people in larger numbers to Western Australia. To cater for these and to ensure that the many gifts that nature has bestowed on the 'Big State' are preserved, for future generations as well as those who enjoy them in the present, the Environmental Protection Authority is currently carrying out, and has almost completed, an intensive study to delineate national parks, reserves and recreation areas.

On the national level it is perhaps the balance of payments effects of Western Australian mining developments that are the most obvious. Australia's export earnings have changed dramatically in character and make-up in recent years. Ten years ago (1966-67), the major export earners were animal substances, including wool ($898 million), followed by cereals ($463 million). Now (1976-77), the major export earners are metalliferous ores and scrap ($1623 million), followed by textile fibres and their waste ($1450 million) and cereals ($1348 million). The role played by the mining industry in this metamorphosis is obvious, as is the country's reliance on the industry to contribute foreign exchange to its overseas reserves.

The contributions to the Treasury coffers by way of company income tax is equally significant though less obvious because of the changing tax provisions, with respect to the mining industry, that have occurred over the last decade. The situation has been confused, to say the least. At times of less than full employment, the contribution of personal income tax can also be regarded as a credit for the industry, which has maintained its demand for labour.

The change in the contribution of the mining industry to Australia's GDP has already been mentioned. This may be seen as significant, and the fact that there has not been an equivalent change in the industry's contribution to overall employment in the nation has already been noted. It is on this score that the industry has come under fire from various quarters and the general question of the effect of mining developments on Australia's agricultural and import-competing industries has been, and still is, the subject of public debate.

The national multiplier effects, in the terms mentioned above, are less obvious or far-reaching than those at the local or state level, but are nevertheless still tangible and important, and the same may be said of the associated development of tourism, which is now beginning to mount.

SOME ISSUES OF IMPORTANCE

Mining, particularly on the scale on which it is carried out in Australia, is an activity on which the public interest is almost always focused in one context or another. It is often praised and condemned at the same time. It is difficult to be neutral about mining, particularly in the present climate of natural-resource preservation in which the problems are compounded by the energy crisis.

In the Australian context some issues affecting mining have a current, short-term flavour, such as how and where will the natural gas won from the North-West Shelf be processed and whence and to where will it be exported. Others, however, are timeless and probably incapable of resolution to the satisfaction of all. They will provide fuel for argument for decades to come, irrespective of their individual character at any point in time. Such issues are: the question of overseas ownership and control of Australia's mineral resources; the question of the taxation of rents (excess profits) from mining, a new version of which has appeared in the guise of a proposed 'resources tax'; the matter of environmental despoliation; and the question of whether mining industry development disadvantages established rural and domestic import-competing manufacturers and, if so, what policies should the government adopt either to arrest the trend or to adjust to it.

These and other major issues related to mining will form the subject for continuing debate in public forums for decades to come in this country as much as in others. It is impossible, here, to discuss them in any detail, but an overview of the economic and social significance of the mining industry is incomplete and perhaps even misleading without mention of them. While current issues will arise from time to time to displace them temporarily, they can be expected to be just as alive in the twenty-first century as they are today.

CONCLUSIONS

In the space available it has been impossible to deal satisfactorily with many economic and sociological facets of the mining industry. But one may hope that sufficient has been said to demonstrate the economic and social significance to the nation and to the State of Western Australia of the changes which have taken place in the state's mining industry in the last two decades.

While future rates of growth may not be as startling as those of the recent past—if only because they will relate to a much larger base—there are still so many new developments to come, such as new discoveries and domestic beneficiation and other processing of local raw materials, that the 'Big State' can be expected to continue to push ahead with the vigour and sense of excitement that has characterized recent years.

As far as future growth is concerned, let us hope that it will bring with it rising standards of living and a richer life not only for Western Australians but also for their trading partners and near neighbours.

FURTHER READINGS

Crowley, F. K. 1960. *Australia's Western Third.* MacMillan: London.

Kerr, Alex, 1965. *The South-West Region of Western Australia.* Univ. W. Aust. Press: Nedlands.

——— 1975. *Australia's North-West.* Univ. W. Aust. Press: Nedlands.

CHAPTER 8

Careers in the mining and minerals industry

I. O. Jones
Western Australian Institute of Technology

The mining industry offers a challenging and rewarding career to those who are not averse to living and working in country areas. The discovery of a workable mineral deposit is a fairly rare occurrence and mines are therefore not very numerous. Indeed, Game (1977) estimates that the number of payable deposits may be as low as one in every 1000 properties examined. It is clear that, wherever such deposits are found, that is the place where the mine must of necessity be developed. People associated with the mining and minerals industry are therefore committed to a life of travel and adventure, often on an international scale. It is true to say that geologists, mining engineers and metallurgists are far less numerous than their counterparts in mechanical, civil or electrical engineering. However, the rewards, both economic and social, which come the way of the small and elite group associated with the exploration and exploitation of mineral and fossil fuel deposits are substantial.

Those of us fortunate enough to live in Western Australia enjoy a standard of living and a quality of life which are the envy of many and beyond the wildest dreams of our forefathers. The continued well-being of Australians, and the aspirations of people in the underdeveloped parts of the world for similar living conditions, underline the need for continued growth in industrial activity throughout the world. If it were not for the mining industry, there would be nothing to manufacture, since it produces the raw materials which are used to make such items as refrigerators, motor-cars and air-conditioners. The current downturn in the world economy can only be transitory, although there is no denying its debilitating effect on a country like Australia, which depends on the activity of the mining and minerals industry for almost half of its export income.

Western Australia is blessed with an abundance of minerals. Its reserves of iron-ore, nickel, bauxite and mineral sands are of world significance and there is probably as much to be discovered as has hitherto been proved. The other minerals which exist in significant quantities in the state are almost too numerous to mention. They include ore minerals of silver, zinc, tin, molybdenum, vanadium, platinum, copper, magnesium, gold and chromium. It is also possible that significant diamond deposits may be found in the near future.

Looking to the fossil fuels, the reserves of coal, while not comparable with the proved reserves in Queensland, Victoria and New South Wales, are of significant proportions and the production of deep-mined coal will undoubtedly be increased in the future. It is also inevitable that the production of natural gas from the continental shelf will provide a base for the further expansion of industry in this state. Not only will it provide an important indigenous source of fuel, but it can provide the basis for the development of a new petro-chemical industry.

An indication of the importance of the Western Australian mining and minerals industry to the economy of Australia is given in the Minerals and Mineral Development 1977 report of the Department of Industrial Development, Government of Western Australia (Anon. 1977, p. 53):

Mineral	Western Australian production as % of Australian production
Alumina	47.0
Crude oil	8.3
Gold	41.7
Iron ore and concentrates	90.2
Ilmenite	72.8
Natural gas	15.3
Nickel concentrates	71.6

Career prospects in the industry are therefore good, even though the industry is more sensitive than most to the fluctuating fortunes of national and international economics.

There are three disciplines which lead to professional careers in the mining and minerals industry. First there is the *geologist*, whose main task is to search for new mineral and/or fossil-fuel deposits and, having discovered them, define their nature and extent. In this work, the exploration geologist is closely supported by specialists in such areas as geophysics, geochemistry and geobotany. Also important in the mining industry are *hydrogeologists* (geologists specializing in the occurrence of ground-water), and *engineering geologists* (who have specialized in the application of geological principles for assuring that geological factors affecting the location, design, construction and maintenance of engineering works have been recognized and adequately provided for).

Having discovered workable deposits, it is the task of the *mining engineer* to design a suitable extraction system, whether it be a surface or an underground operation. It is the mining engineer too who is responsible for the overall supervision of the mine when it is in production. Naturally, he is supported in this work by a wide range of specialists, including electrical and mechanical engineers, as well as mine surveyors and advisors in such fields as rock mechanics and environmental engineering.

The ore, once it has been produced from the mine, becomes the responsibility of the *metallurgist*, who designs and supervises the plant for concentrating the metallic minerals and their subsequent treatment to produce refined metals. A variety of chemical and physical processes are involved in the concentration of metallic minerals, whereas the refining of crude metal is based on smelting or chemical treatment.

One of the oldest institutions of education in Western Australia is the Western Australian School of Mines at Kalgoorlie. From its inception in 1902 until it became a branch college of the Western Australian Institute of Technology in 1969, it was the responsibility of the Mines Department and catered for a wide range of entrants, some of whom had very little in the way of an academic background. In 1976 it became the major part of the School of Mining and Mineral Technology, which is one of the eight schools of study making up the Western Australian Institute of Technology. This school of study is composed of the departments of geology, metallurgy, and mining and engineering. Degree courses are available in all these fields of study, as well as a range of post-graduate and two-year diploma courses.

The University of Western Australia also has a long tradition of providing qualified geologists, as well as civil, electrical and mechanical engineers for the mining and minerals industry. Its main contribution to the minerals industry has been in the field of geology. The chair of geology, because of the importance of geology to the mining industry, was one of the seven foundation chairs in the university when it was established in 1913. Since its establishment, the department of geology has made eminent contributions to the industry in the supply of geologists to mineral exploration companies, geological survey departments (both state and federal), and academic (teaching and research) institutions engaged in training geologists for the mining and minerals industry. Post-graduate research has also increased our knowledge of Western Australian mineral deposits. Murdoch University, established in 1974, has initiated a course of study in mineral science.

Also an important part of the educational scene is the Technical Education Division of the Education Department of Western Australia, which provides a wide range of courses for the craftsmen and para-professionals required by industry. It is generally recognized that for every technologist with a degree or equivalent qualification at least three para-professionals are required to provide support services.

History has already proved the ability of the educational system in this state to produce geologists, metallurgists and mining engineers capable of establishing themselves in the forefront of their profession. The current developments taking place in the tertiary education scene will ensure that the future diplomates and graduates will be no less adaptable or successful than their predecessors.

REFERENCES

Anon. 1977. *Minerals and mineral development 1977.* Perth: W. Aust. Dept Indust. Devel.

Game, M. 1977. Key to the future. *Mining Review* (June issue), 1-3.